W9-ACR-219

COMPUTER ARCHITECTURE

Software Aspects, Coding, and Hardware

COMPUTER ARCHITECTURE

Software Aspects, Coding, and Hardware

John Y. Hsu

CRC Press
Boca Raton London New York Washington, D.C.

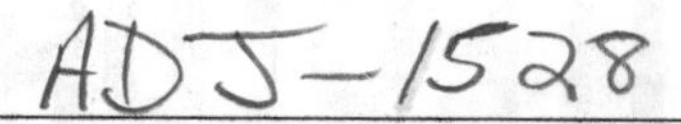

Library of Congress Cataloging-in-Publication Data

Hsu, John Y.
Computer architecture : software aspects, coding, and hardware / John Y. Hsu.
p. cm.
Includes bibliographical references and index.
ISBN 0-8493-1026-1 (alk. paper)
1. Computer architecture. 2. Computer software. I. Title.

QA76.9.A73 H758 2001
004.2′2—dc21 00-050741

Direct all inquiries to CRC Press LLC, 2000 N.W. Corporate Blvd., Boca Raton, Florida 33431.

International Standard Book Number 0-8493-1026-1
Library of Congress Card Number 00-050741
Printed in the United States of America 1 2 3 4 5 6 7 8 9 0
Printed on acid-free paper

Preface

Motive for Writing This Book

After having published my first book, *Computer Networks: Architecture, Protocols, and Software,* friends asked me how long it took to write the book. My reply was that on the surface it took about three years from beginning to end; below the surface it took more like thirty years. Yet, my job is not done unless I write a book on computer architecture and discuss some of the background materials. Most first generation computer architects are physicists who learned everything about computers on the job. Second generation computer architects studied the basics in school and later practiced in industry. My academic training enabled me to read the design documents of IBM 360 Operating Systems in the 1970s. This painstaking effort broadened my horizons about real issues, and to this day I feel very much obliged. In the 1990s, while I studied the blue book on the telecommunication network design by ITU-T (International Telecommunications Union — Telecommunication Standardization Sector), I was able to make suggestions for improving the design.

As you may not know, in 1962 I came to this great country without a penny. My life has changed ever since my late friend Bob Chen convinced me to study computers. Back then we knew so little about computers and it took us three months to find out that a compiler is software, not hardware. Today, a compiler can be embedded in hardware. Technologies come and go, but theories remain. May this book bring you confidence and success.

Who Should Read This Book

This book discusses computer architectural topics from a beginner's level to an advanced level and explains the reasons behind certain computer design. Preferably, readers should be familiar with at least one programming language and Boolean algebra. The intended audience mainly consists of:

- Undergraduate students of computer science (the selected topics in this book can be lectured in 60 to 80 hours)
- Undergraduate students of computer engineering and electrical engineering
- Professionals in the electronics industry

After grasping the system concepts, readers can proceed to study more topics on computer hardware, system software, and networks.

Organization of This Book

This book has ten chapters. The first four chapters cover fundamental computer principles. Chapter 5 continues the discussion of intermediate level topics and Chapter 6 describes microprogrammed CPUs. Chapter 7 discusses superscalar machine principles. Chapter 8 covers vector and multiple-processor machines. Chapter 9 is devoted to processor design case studies and virtual machines. Finally, Chapter 10 teaches stack machine principles and the design of a virtual stack machine. Every computer science major should read this chapter before graduation. A brief description of each chapter is given below.

Chapter 1 introduces the history of computers, hardware components, software components, application programs, computer simulation, and the program design language to describe logic flow. After learning the basics of disk files and commands, readers are ready to run a program on an IBM PC.

The second chapter discusses number systems and basic mathematics in regard to computing. Topics include positional notation, radix, number conversions, integers, negative integers, floating points, packed decimals, and characters.

Chapter 3 introduces the stored program concept, instruction format, and basic computer principles. Topics include opcodes, addresses, instruction register, and instruction address register. A register transfer language is introduced to describe CPU operations — instruction fetch and operand execution. Other general topics include carry look-ahead adders, hardwired logic, microprogrammed logic, hybrid logic, and software interpretation.

Chapter 4 covers assembly language that is used to describe the internal operations at the target machine level. The purpose is to develop basic coding skills. Because of its popularity, the Pentium processor is used as a tool to describe instruction executions in a computer. Topics include assembly language syntax, machine ops, pseudo ops, basic addressing modes, looping, and macros.

Chapter 5 covers the common design features of a central processor. General topics include addressing modes, indexing, subroutine linking, interrupts, I/O structure, I/O programming, software polling, direct memory access, memory mapped I/O, and cycle stealing.

Chapter 6 focuses on the design of a microprogrammed CPU using segment base registers. The execution of microcode is overlapped with target instruction fetches. Topics include microcode engine, encoding, sequence control, conditional branch, and unconditional branch. Via a single adder, discussion is given to the algorithms for unsigned multiply, signed multiply, unsigned divide, and signed divide, as well as floating point operations.

Chapter 7 covers all the look-ahead, look-aside, and look-behind features of superscalar machine design. A balanced system allows all the hardware components to operate in parallel. Selected topics include storage hierarchy, resource dependencies, one-clock shifter, multiplication trees for unsigned or signed numbers, pipelined CPUs, instruction queues, instruction caches, data caches, decoupled pipes, and virtual memory.

Chapter 8 discusses vector and multiple-processor machines. The multiple-processor machine class consists of multistation systems, multiprocessing systems, and computer networks. Selected topics include processor-to-processor communications, intertask messages, protocols, local area networks, and wide area networks.

Chapter 9 focuses on processor design case studies. Examples include the IBM mainframe, Power PC, Alpha, Itanium, and the reduced software solution computer. At the end of chapter, we introduce virtual machines and the JAVA engine.

The final chapter continues the discussion on stack machine design. Essential topics include postfix notation, operator stack, operand stack, S-ops, and the design of a virtual stack machine.

Acknowledgements

I am forever grateful to my teachers, particularly C. L. Sheng, Martin Graham, Ivan Frisch, Arthur Gill, and Paul Morton. They taught me how to face challenges and endure. Andy Grove, a colleague at Fairchild in 1967, was kind enough to send me the technical manuals on Pentium. This book, in part or whole, was reviewed by many individuals. My students, including Diller Ryan, Zetri Prasetyo, Kurt Voelker, Ihab Bisha, Tam Tran, and Delora Sowle, were all helpful. I salute all the reviewers who helped me shape the manuscript to its final form. Some of their names and affiliations are: Alan Beverly (Ziatech), Dave Braun (Cal Poly), Wesley Chu (UCLA), Jim Gray (Microsoft), Elmo Keller (Cal Poly), Steve Luck (Hitachi), Miroslaw Malek (Humboldt U., Germany), Frederick Petry (Tulane), Cornel Pokorny (Cal Poly), C. Ramamoorthy (U. of California, Berkeley), and Charles Summers (Telesoft International). The artwork for the figures was done by Long T. Nguyen. Gerald Papke and his project team at CRC Press deserve recognition.

John Y. Hsu
San Luis Obispo

About the Author

John Y. Hsu received his B.S.E.E from National Taiwan University (1955-59); his M.S.E.E. (1963-64) and Ph.D. (1967-69) from the University of California, Berkeley specializing in computer system hardware and software. He is currently a professor of computer engineering at California Polytechnic State University in San Luis Obispo. In the academic year of 1979, he was a visiting research professor at National Taiwan University. He has held many industrial job titles, such as computer architect, project engineer, and senior software specialist. In addition, he has done over 10,000 hours of consulting work for companies including Federal Electric/ITT, ILLIAC IV, III in Taiwan, CDC, IBM, etc. He is the author of *Computer Networks: Architecture, Protocols and Software,* Artech House, 1996. Dr. Hsu is a member of IEEE and ACM.

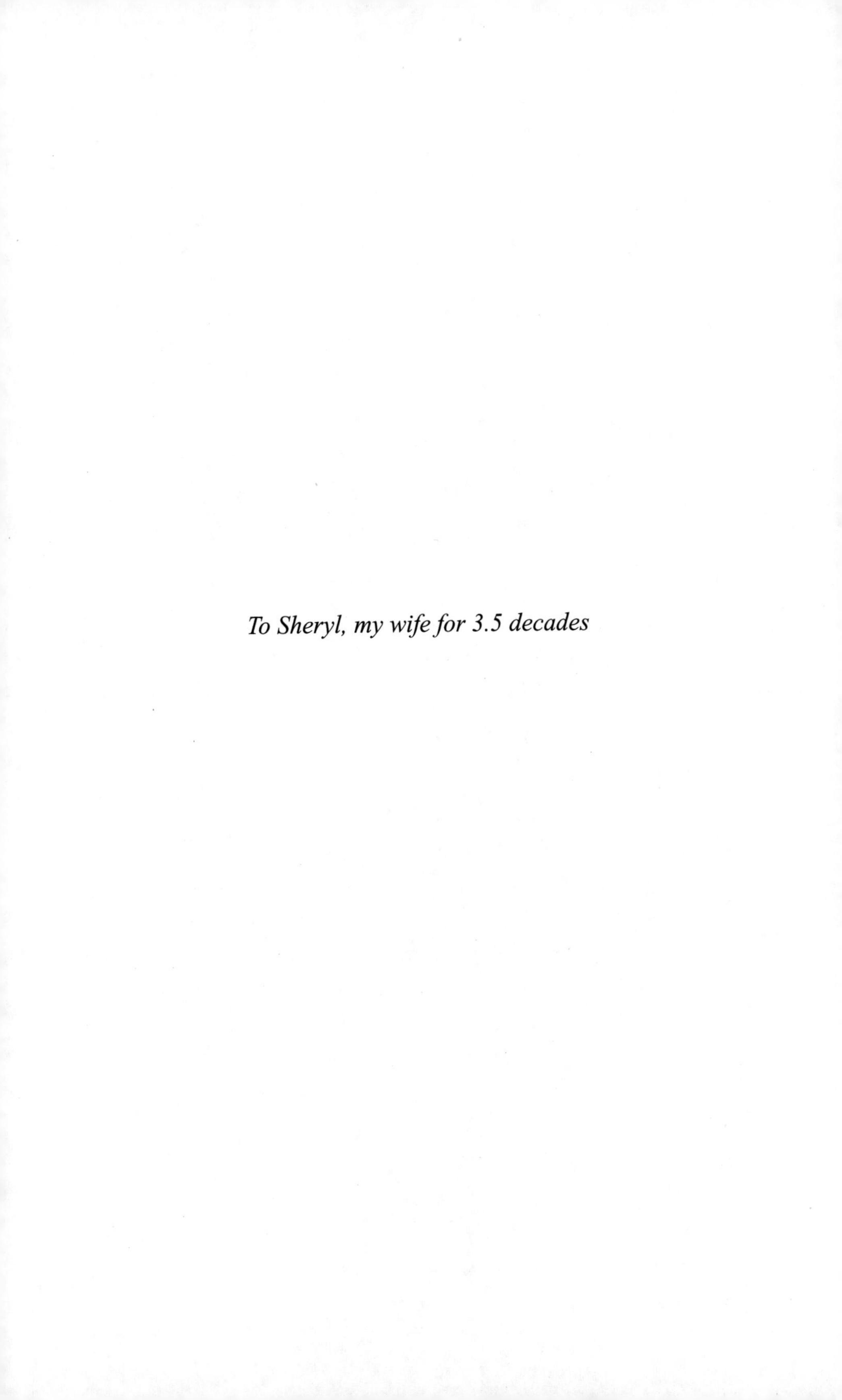

To Sheryl, my wife for 3.5 decades

Acronyms and Abbreviations

∧	exponential, concatenate
A	address; auxiliary
acc	accumulator
ACIA	asynchronous communications interface adaptor
ACM	Association of Computing Machinery
adc	add with carry
AIX	advanced UNIX
ALU	arithmetic and logic unit
AM	access method
ANSI	American National Standards Institute
ASCII	American standard code for information interchange
b	bit, binary digit
B	byte
BCD	binary coded decimal
bit	binary digit
BP	base pointer
bps	bit per second
Bps	byte per second
BU	bus unit
C	carry
CACM	communications of the ACM
CATV	Community Antenna Television or Cable TV
CC, cc	condition code
CD	compact disc
CF	carry flag
CISC	complex instruction set computer
com	communication; commercial
CPI	clocks per instruction
CPU	central processing unit
CS	code segment
CU	control unit
D	decimal; delay; direction; displacement
DARPA	Defence Advanced Research Project Agency
db	decibel (decimal bel); define byte
dc, DC	define constant; direct current; dynamic code
Disp	displacement
div	divide
DMA	direct memory access
DMC	differential Manchester code
dopd	destination operand
DRAM	dynamic random access memory
DS	data segment; define storage

EA effective address
EBCDIC extended binary coded decimal for information interchange
EC external code
EISA extended industry standard architecture I/O bus
e-mail electronic mail
endm end macro
endp end procedure
ends end segment
ENOR exclusive nor
EOR exclusive or
Epb errors per bit
EPROM erasable programmable read-only memory
EPT external page table
equ equate
ES extra segment; external symbol
EU execution unit
Exp exponent, exponential
F flags register
FAT file allocation table
ff flip-flop
fixed fixed point
float floating point
ftp file transfer protocol
g, G giga: 10^9 to measure speed or 2^{30} to measure memory size
gbps gigabit per second
GC global code
ghz gigahertz
GPR general purpose register
GR general register
hex hexadecimal
hz hertz
IA-32 Intel architecture-32 bits
IA-64 Intel architecture-64 bits
IAR instruction address register
IC integrated circuit
ID instruction decoder; identifier
IEEE Institute of Electrical and Electronics Engineers
IH interrupt handler
Int interrupt
Internet inter-networking
I/O Input/Output
IOR I/O register
IP instruction pointer; internet protocol
IR instruction register
ISA instruction set architecture; industry standard architecture I/O bus
ISDN integrated services digital network
ISR interrupt service routine
IU instruction unit

k, K kilo: 10^3 to measure speed or 2^{10} (1024) to measure memory size
km kilometer
LAN local area network
LHS left-hand side
LRU least recently used
LSB least significant bit
LSI large scale integration
LT literal table
m, M mega: 10^6 to measure speed or 2^{20} (1,048,576) to measure memory size; meter; memory; more; multiplicand
MAN metropolitan area network
MAR memory address register
mbps megabit per second
MB megabyte
mBps megabyte per second
MC machine check; manchester code
MDR memory data register
MFLOPS million floating point operations per second
MIMD multiple instruction multiple data
MIPS million instructions per second
MMU memory management unit
mop machine op; micro operation
MOS metal oxide semiconductor
MOSFET MOS field effect transistor
MRU most recently used
ms milli (10^{-3}) second
MSB most significant bit
MSI medium scale integration
mul multiply
MVS multiple virtual storage
µ micro
µIR micro instruction register
µPC micro program counter
µs micro (10^{-6}) second
µIR micro instruction register
µPC micro program counter
NAM network access method
NBS National Bureau of Standards
NIST National Institute of Standards and Technology
NOS network operating system
ns nanosecond (10^{-9})
O overflow
Op opcode; operation code; operator
opd operand
org origin, organization
OS operating system
P peta: 2^{15} to measure memory size; parity; program; page number
PAS physical addressing space

PB	petabyte
PC	program counter; personal computer; program check; printed circuit
PCI	peripheral component interconnection bus
PDL	program design language
PE	processing element
PEM	processing element memory
PF, PFN	page frame number
PFT	page frame table
PN	page number
pop	pseudo op
PPU	peripheral processing unit
PROM	programmable ROM
PSW	program status word
PT	page table
PTBR	page table base register
ps	picosecond (10^{-12})
R	register, real address
RAM	random access memory
RCL	rotate with carry left
RCR	rotate with carry right
Reg	register address field
RHS	right-hand side
RISC	reduced instruction set computer
ROL	rotate left
ROM	read-only memory
ROR	rotate right
RPM	revolutions per minute
RS	reset-set; recommended standard
RssC	reduced software solution computer
RTL	register transfer language
S	stack; sign; source
SAR	shift arithmetic right
sbb	subtract with borrow
SC	static code
SCSI	small computer system interface
SDRAM	synchronous dynamic random access memory
Seg	segment
SF	stack frame
SHL	shift logical left
SHR	shift logical right
SIMD	single instruction multiple data
SISD	single instruction single data
SMP	symmetric multiple processor
SOC	system on chip
Sop	stack machine op; sum of product
sopd	source operand
SP	stack pointer
SR	status register

SRAM	static random access memory
SS	stack segment
SSI	small scale integration
ST	segment table
SVC	supervisor call
T	tera: 2^{12} to measure memory size
TB	terabyte
TLB	translation look-aside buffer
TO	time-out
TOS	top of stack
Trans	transactions
UART	universal asynchronous receiver/transmitter
ULSI	ultra large scale integration
UNIX	UNICS (Universal Information and Computing Service)
USB	universal serial bus
USRT	universal synchronous receiver/transmitter
V	volt, virtual
VA	virtual address
VAS	virtual addressing space
VC	virtual circuit
VLSI	very large scale integration
VM	virtual machine; virtual memory
VS	virtual storage
VSM	virtual stack machine
WAN	wide area network
WORM	write once read-only memory
WWW	World Wide Web
Z	zero

Contents

Chapter 2 Number Systems

Chapter 3 Basic Computer Principles

Chapter 1

Introduction to Computers

1.1 PROLOGUE

A computer is an electronic machine for the purpose of computation. As far as design is concerned, there are analog computers and digital computers. An analog signal may vary continuously, while a digital signal is represented by many digits. For example, the slide rule is an analog computing device while the Chinese abacus is a digital computing device.[46] Both are mechanical devices for computing, but there is a fundamental difference in concept. A slide rule has three rulerbars made of bamboo or plastic as shown in Figure 1.1a.

The upper and lower bars are fixed in position but the middle one can slide both ways. Decimal numbers are carved on all the bars. To add 1.2 to 1.1, we slide the middle bar to the right and align its reference point 0 to 1.1 on the upper bar. From the 1.2 mark on the middle bar, its aligned position on the upper bar indicates 2.3 as the result. That is, by adding the distance of 1.1 and 1.2, we obtain the sum of 2.3 on an analog scale. The distance is an analog signal that is linearly proportional to the real numeric number.

In contrast, an abacus is used to perform decimal arithmetic with many fixed columns and movable beads divided into two partitions. Each bead in the upper partition carries a weight of five and each bead in the lower partition carries a weight of one. Therefore, as shown in Figure 1.1b, the abacus displays 123,456,789. The number has nine decimal digits: the left-most digit is one and the right-most digit is nine. Adding 1.2 to 1.1 means adding beads on each of the two columns. Therefore, the result is 2.3 on a digital scale.

Modern computers are designed for computation as well as information retrieval. The term computer architecture means the structural level design of a computer that includes the layout of instructions and registers as seen by the software developer. A computer architect defines the instruction set and the register set of a computer. After the instruction set is defined, a computer may be built with different hardware components so that its speed varies. Nevertheless, all the instructions execute the same

(a)

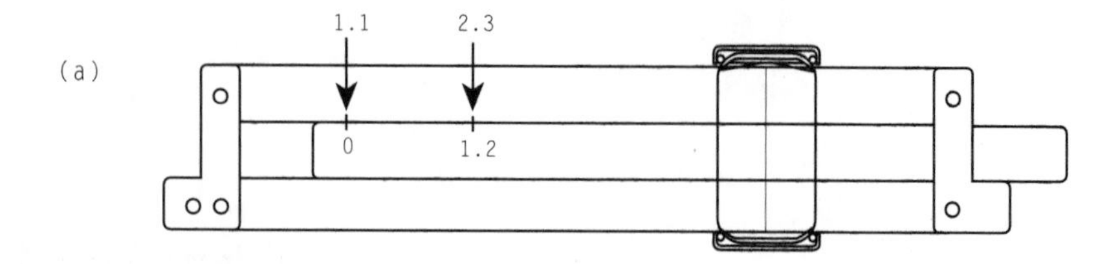

(b)

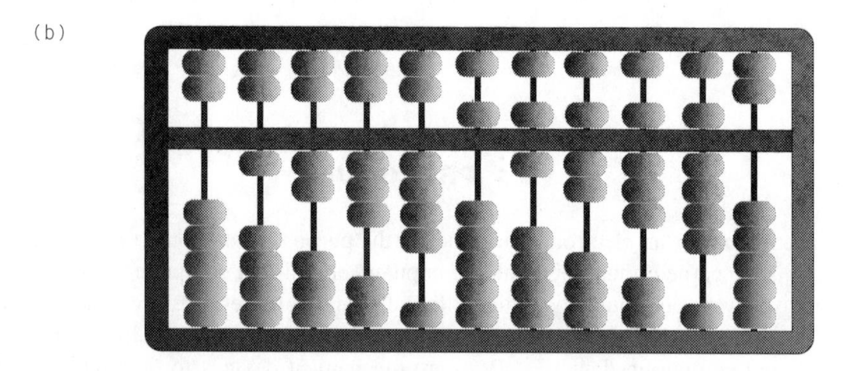

Figure 1.1 Mechanical computing devices: (a) slide rule and (b) abacus.

way on a functional basis if machines share the same architecture. We say that hardware components are designed to execute software, or, software is written to drive hardware. In that regard, the design issues of hardware and software are closely related. To begin, we introduce digital signals, the history of computers, hardware components, software components, and system software tools. PDL (program design language) is used to describe the logical flow of software or hardware.

1.1.1 Analog vs. Digital

An analog electric signal is a voltage that may vary with time on a continuous basis as shown in Figure 1.2a. The vertical axis is the coordinate of voltage, and the horizontal axis is the coordinate of time. The voltage waveform in the box is amplified in Figure 1.2b.

Adding a sample signal of 1.2 v (volts) to another sample signal of 1.1 v, we obtain a sum of 2.3 v in an analog computer. Because the voltage is proportional to its amplitude value, it is an analog signal. But in a digital computer, both 1.2 and 1.1 are represented by a group of ones and zeros and the two bit strings are not the same. Adding 1.2 to 1.1, we obtain 2.3 as the sum, which is comprised of a different combination of ones and zeros. That is to say, the number is always represented in bits (binary digits), and each bit can be one or zero. This book covers the design issues of digital computers.

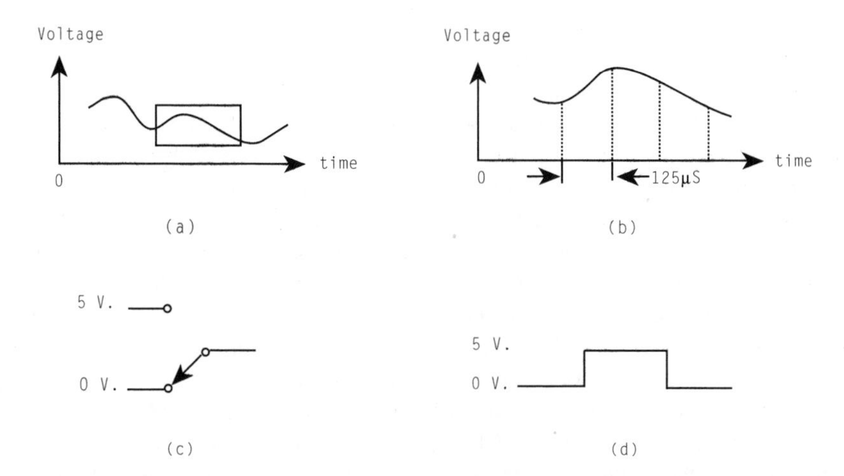

Figure 1.2 Signal waveforms and switch: (a) analog signal, (b) amplified amplitude, (c) binary switch, and (d) digital signal.

1.1.2 Digitizing

Voice is an analog signal which may be converted to the form of an electric voltage. A voice can be digitized so that it can be transmitted and processed by a digital computer. To begin, we take the sample amplitude of a voice signal at fixed time intervals. As shown in Figure 1.2b, 8000 samples are taken per second, and the time period between two consecutive samples is computed below:

$$\text{Sampling period} = 1 / 8000 \text{ sec.}$$
$$= 125\ \mu\text{s (micro sec.)}$$

Via electronic circuits, we can convert the amplitude of each voice sample into a string of binary digits, in other words, analog to digital (A–D) conversion or digitizing. If we transmit the binary digits to another station fast enough before the time interval expires, the receiving station can restore the signal from digital to analog (D–A) at the same fixed time intervals. That is, after the voice is digitized, a computer can do message handling such as send, receive, record, transfer, etc. Another example is the digital camera. After a picture is taken, it is digitized and processed by a computer process known as image processing. The future trend is to digitize all information so it may be processed by a computer.

A binary digit or bit has two states, one or zero. A bit string is a sequence of ones and zeros. Based on the bit patterns and the ways to define them, we obtain different information. A bit may be stored on disk or in an electronic circuit. If we use two voltage levels, 5 v for 1 and 0 v for zero, a binary switch is a bistable device to store one bit as depicted in Figure 1.2c. If the switch position is down, its output is 0 v; if the switch is in the up position, its output is 5 v. If we hold down the switch for a while and then flip the switch up for a while and down, its output is a combination of a low voltage, followed by a high voltage and a low voltage as depicted in Figure 1.2d. Inside the computer, the waveforms look just like this except the switching speed is much faster.

A computer has three major hardware components: a CPU (central processing unit), internal memory, and input/output (I/O) devices. The CPU is a hardware device to process data. The terms internal memory, central memory, or memory are all synonymous. The memory interacts with the CPU as it provides a temporary storage for the CPU during computation. As a matter of fact, the memory contains millions or billions of binary switches. All the instructions and data look alike as bit strings in memory. Based on physical appearance there is no difference between the two. An instruction tells the CPU what to do, for example, find data and perform an operation on it. If instructions and data are arranged correctly in memory, the instructions execute correctly on the CPU as expected. Generally speaking, an instruction is active as it tells the CPU what to do. In contrast, data are passive because they are the result of executing an instruction. By grouping instructions and data together, we obtain a computer program, otherwise known as software. Writing software means writing computer programs; the person who writes programs is called a programmer or coder.

1.2 HISTORY OF COMPUTERS

Computer architecture means the design of a computer system at the structural level that generally includes hardware and software. The computer development effort has gone through five generations. Each generation was characterized by some sort of hardware break-through along with some architectural improvements. Consequently, each generation has produced some changes, such as smaller size, lower cost, and substantial performance increase. Some of the major events in computing are briefly introduced in Table 1.1.[73]

1.2.1 First Generation Computers

The first generation computers were made of vacuum tubes. For example, the ENIAC was built between 1943 and 1946 by the Moore School of the University of Pennsylvania. The machine weighed 30 tons with more than 19,000 vacuum tubes, 1500 relays, etc. as shown in Figure 1.3.

Even though ENIAC had only 20 words of internal memory and required manual operations for setting up a program on a hardwired plugboard, it was the first electronic digital computer in history. The instruction sets of first generation computers were small, 16 or less. A computer program contains instructions and data, called machine executable code or code for short. Therefore, machine instructions, executable code, and machine code are all synonymous. Preparing computer instructions means programming or coding, and there is great demand for good coders.

Table 1.1 Major Events in Computing

Year	Description
1945	Stored program concept
	First computer bug — a moth found in a relay
1946	ENIAC (electronic numerical integrator and computer)
1947	Transistor
1951	Core memory
	Microprogramming
1952	The A-0 compiler by Grace Hopper
1953	Magnetic drum, disk memories
1957	FORTRAN programming language
1958	Integrated circuit
1961	Time sharing computer system
1963	ASCII (American standard code for information interchange)
1964	IBM 360, the first third generation 32-bit mainframe using integrated circuits
	Mouse (not in use until 1973)
	CDC6600, the first supercomputer (60-bit)
1965	PDP-8, the first minicomputer (12-bit)
1968	Structured programming constructs
1970	Floppy disk
1971	E-mail
1972	Intel 8008, the first 8-bit microprocessor
	Digital calculator
	Programming language C
1973	Ethernet
	Large scale integration (LSI) with 10,000 transistors on one chip
1974	Four-kilobit MOS (metal oxide semiconductor) memory chip
1975	RISC (reduced instruction set computer)
	Laser printer
1976	CRAY 1, the first 64-bit supercomputer
1977	Apple II, an 8-bit personal computer
1981	IBM PC
1982	PacMan game — man of the year, *Time Magazine*
1984	CD-ROM (compact disk-read only memory)
	Very large scale integration (VLSI) with 1,000,000 transistors on a chip
1987	16-megabit MOS memory chip
1988	Fiber optics as a transmission medium
1990	WWW (World Wide Web)
1995	JAVA programming language
1996	Pentium, a 32-bit microprocessor
2000	Itanium, a 64-bit microprocessor

Figure 1.3 ENIAC (courtesy of Charles Babbage Institute, University of Minnesota).

In first generation computers, there were no programming tools and a programmer had to prepare the instructions by toggling switches on a hardware panel. Subsequently, UNIVAC I and IBM 704 were developed in the early 1950s, and eventually evolved into second generation computers.

1.2.2 Second Generation Computers

Starting in the late 1950's, second generation computers emerged. These machines commonly used discrete transistors and magnetic core memories. Their processors had powerful instruction sets along with many architectural features. At the same time, progress was being made in system software development. For example, the operating system (OS) was developed as a set of control programs running alongside a user application program in the computer. In addition, high-level programming language compilers, such as COBOL and FORTRAN, were developed to make programming much easier. A compiler is a software tool that translates a program into some form of machine code. The popular second generation machines included the IBM 7000 series and the CDC 6000 series. It is interesting that CDC 6600 and 7600 were the fastest computers in their era, but are now out-ranked by the performance of personal computers (PC).

1.2.3 Third Generation Computers

In 1964, while every competitor still dwelled on the second generation development, IBM made a monumental decision to abandon its old 7000 product line and push its 360 product line, a family of third generation computers. The IBM 360 systems, which later became 370, mainly used integrated circuits (ICs) in the system and magnetic cores as central memory. An IC chip contains many transistors, so the size of a circuit board is smaller. Some are fast and some consume little power. Without dispute, ICs were the basic building blocks of third generation computers. IBM also pioneered the concept that if a program executes on a lower model, it can also execute on upper models. That is, all models are upward compatible because they all share the same architecture, or ISA (instruction set architecture). However, hardware components are different from model to model in terms of speed, size, and cost.

The IBM 360 computer was designed with many architectural innovations. It pioneered the concept of having 16 general purpose registers, hexadecimal notation to represent a binary number, twos complement arithmetic, powerful interrupt structures, and sophisticated Input/Output structures. Most importantly, its OS (operating system) was completely reliable. These topics will be explored in subsequent chapters. Due to this strategic move, IBM made a quantum jump in computer design, took the lead, and never looked back.

1.2.4 Fourth Generation Computers

As technology moved forward, fourth generation computers were characterized by using very large scale integration (VLSI) circuits and semiconductor memories. It is amazing that 5.4 million transistors could be packed on a single chip as small as a fingernail.[66] If we pack the electronic circuits of the entire central processor on one chip or several chips, we obtain a microprocessor. Due to the short distance between two hardware circuits on a chip, it is possible to operate the chip at a higher speed. As a consequence, a fourth generation computer usually has a fast microprocessor with a large amount of memory using VLSI circuits. That is, every new computer system will use microprocessors in its design, regardless of whether it is a mainframe or a supercomputer. However, the large systems are characterized by using more sophisticated hardware, many processors, and high-capacity memory chips, along with a very reliable operating system. Based on this notion, any PC using a microprocessor chip fits nicely into the low end of a fourth generation computer, while the IBM 360/370 systems have been evolved into its 390 series at the high end. Figure 1.4 shows the printed circuit board with dual microprocessors. Circuits are laid out on the left-hand side as add-ons. There are more than 350 copper pins, some of which are signal lines to communicate with other devices.

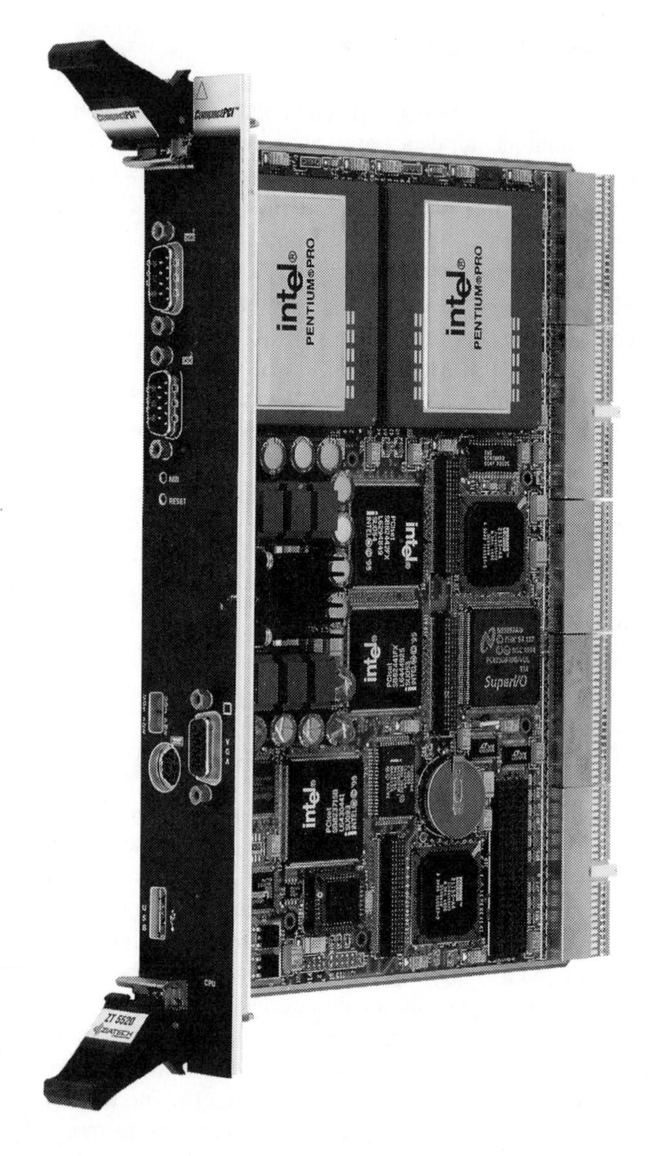

Figure 1.4 Microcomputers: printed circuit board with dual processors (courtesy of Ziatech).

1.2.5 Fifth Generation Computers

It is hard to define a true fifth generation computer in regard to technological breakthroughs. However, there was a fifth generation computer project in Japan which worked on the design of appropriate software running on a cluster of high-speed processors in order to perform a specific common goal.[44] Such a system requires extensive parallel computations, and the basic idea was to divide a large problem into pieces. Each processor could execute a program to solve one piece of the big problem. Such systems would be useful for library searches, global weather predictions, airplane wing design, global database searches, playing chess, etc. But such machines have been in the works since the second generation, and the proposed computer was just a special-purpose multiple processing system for parallel computations. The notion of interconnecting multiple computers, however, could be extended to designing a computer network. A general purpose computer network requires intricate system software to support interprocessor communications between any two computers on the globe.

Since high-speed links are used to transmit data over long distances, after extensive research, fiber optics will most likely become the future transmission media. Using a fiber wire, up to hundreds of billion bps (bits per sec) can be transmitted between two computers. As far as architectural changes are concerned, interconnecting over several million computers using fiber optics is a hardware breakthrough. Writing system software to support such applications over a global network is also a breakthrough. Therefore, the design of a computer network over several million different computers using fiber optics for data communications qualifies as a fifth generation computer. It should be mentioned that a computer network needs good system software to ensure reliability and performance. It took more than a decade for designers just to agree on network standards, e.g., physical interface, protocols, design approaches, etc. More reference materials can be found in a different book by this author. [24]

1.3 HARDWARE COMPONENTS IN A COMPUTER SYSTEM

The simplified block diagram of a computer system is depicted in Figure 1.5.

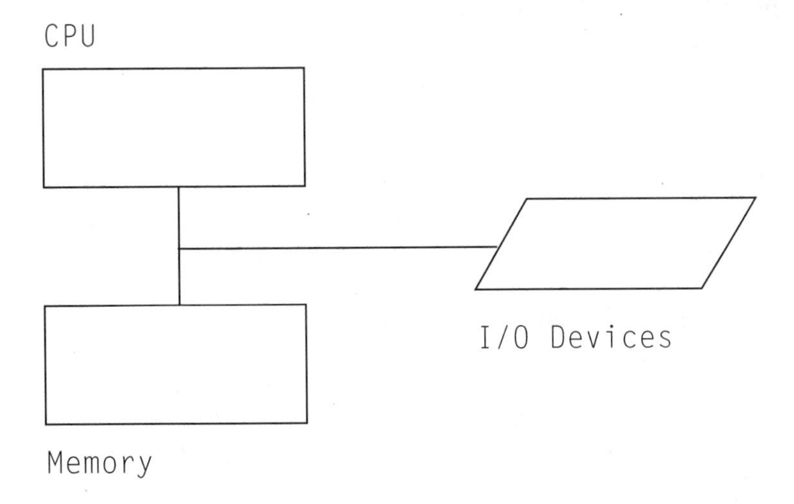

Figure 1.5 The simplified block diagram of a digital computer.

The hardware components include a CPU, memory, and input/output devices. In simple terms, the CPU is the central processor where data are processed. The memory is used to store the intermediate results during the execution of a program. That is, the CPU is similar to a processing factory and memory is the storage house. When the program executes on the CPU, it may want to read data from an external device for input or want to store data on an external device for output. Thus, we see that an input device provides data to the CPU while an output device receives data from the CPU. I/O devices are used by the CPU to communicate with the outside world. A magnetic disk or tape is classified as an I/O device serving the purpose of external memory. A program on disk must be loaded into central memory first before its execution can begin.

1.3.1 Central Processing Unit

The CPU is the brain of a computer and its design is quite intricate. Before the advent of microprocessor chips, a CPU may have cost several million dollars. Nowadays, a PC employs a microprocessor chip that has over ten million transistors on it and the cost is one thousand dollars or less. As speed is concerned, the chip is faster than a third generation mainframe.

A CPU is designed to execute a program, or, a program executes on a CPU. A program is a collection of machine instructions and data in binary form. Before a program can execute on the CPU, the instructions and data must first be stored in memory. As a program executes, each instruction is loaded into the CPU one by one and executed by hardware. In the chapters to follow, we will discuss the basic principles of a computer and the innovative features to make it run faster. The last chapter will teach how to write a simulator for a computer.

1.3.2 Memory

Memory, internal memory, and central memory are all synonymous. The memory system is the second most important component in a computer because it interacts with the CPU all the time during program execution. Memory is divided into basic addressable units called bytes, and each eight-bit byte is called an octet. As each byte in memory is assigned an address, the address must be passed to memory in order to fetch what is contained in the byte. The bit setting of a byte means its content (i.e., the value of a character variable) in a high-level programming language.

1.3.2.1 Memory Address vs. Content

The address of a location in memory is completely different from its value or content in that location. A variable occupies a memory location whose address never changes, but the content at that memory location may change from time to time. That is, a variable may occasionally be set at a different value. The memory location is similar to a mail box at the post office. Each box has a number assigned as its address or PO box number. Anything in the box is its content or value, such as letters. The content in the box may change from time to time, but its box number, i.e., the address will never change. The memory addressing concept is also similar to the address assigned to a house in a city. The person living in the house may change but not the house address. The memory address is used to find an instruction or data in memory. Therefore, if a memory operation is issued during execution, the address is transmitted to the memory system along with the memory request.

The basic addressable unit in memory is a byte of eight bits. Because a computer uses a binary number system, the memory size is measured as 2^N bytes. There are some memory sizes worth mentioning, as shown below:

- One KB (kilobyte) means 1024, or 2^{10} bytes.
- One MB (megabyte) means 1,048,576, or 2^{20} bytes.
- One GB (gigabyte) means 1,073,741,824, or 2^{30} bytes.
- One TB (terabyte) means 2^{40} bytes.
- One PB (petabyte) means 2^{50} bytes.

Nowadays, microcomputers tend to have 64 or 256 MB memory, while a mainframe supports a memory of 4 GB. When a CPU retrieves an instruction or a piece of data from memory, it must issue a memory read cycle. When a CPU decides to store data into memory, it issues a memory write cycle. The memory read and write cycles or operations are performed, back and forth, as the program executes. There are two major types of memories: cores and semiconductors, and both are quite interesting, as introduced below.

1.3.2.2 Core Memory

The invention of core memories was a breakthrough in the early 1950s.[15] A magnetic core has a diameter of 20 to 30 mil (10^{-3} inches), and the direction of the magnetic flux is shown in Figure 1.6a. If the direction is counterclockwise, it is defined as one; otherwise it is zero.

There are three electric wires, X, Y, and sense, through each core. The X and Y wires are used to select the core, and the sense wire is used to read the content of the core. The magnetic core has a property such that the sum of electric current flowing through the core on X and Y must be strong enough to switch the magnetic flux from one direction to the other. Therefore, on each wire X or Y we select the strength of the current in such a way that the current on one single wire will not be able to switch the flux, but the sum of currents flowing on two wires will switch the flux provided that the original flux is in the opposite direction. Thus, in a read operation, coincident currents flow on both X and Y to switch the flux towards zero as shown in Figure 1.6b. According to the right-hand rule, the current needs to flow from left to right on X, and from bottom to top on Y. At the same time, the sense wire will detect an induced voltage whose magnitude depends on the total amount of flux change in the core. If the bit contains a one, the induced voltage is big because the magnetic flux is completely reversed 180 degrees from one to zero. If the bit contains a zero, the induced voltage is small because the flux change towards the same direction is not

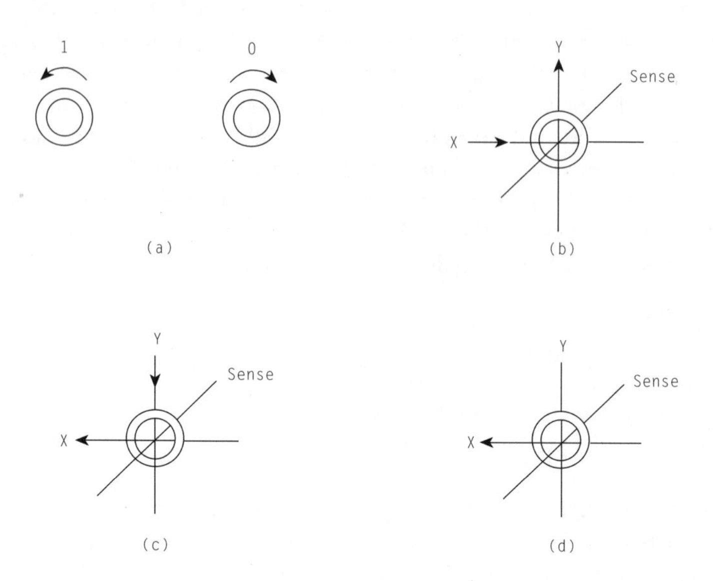

Figure 1.6 Magnetic cores: (a) magnetic flux direction of a bit 1 or 0, (b) read a bit 1 or 0, (c) write a bit 1, and (d) write a bit 0.

big enough due to the saturation of its magnetic flux. We have only one sense wire in the core, yet it picks up the induced voltage due to the change of magnetic flux surrounding it. This inductive coupling technique is also used to read or write a magnetic disk or tape.

The trouble comes after the core is read and the bit is cleared to zero, that is, the magnetic flux is switched to the zero direction. If the original bit is one, it is necessary to restore the original bit via a restore operation. In design, the hardware logic does the restoring operation by writing the original bit back regardless of its state. If the original bit is one, coincident currents will flow in the opposite direction to switch the flux back from zero to one as shown in Figure 1.6c. If the original bit is zero, the current will only flow on X and not on Y to avoid a magnetic flux switching as shown in Figure 1.6d. In that case, the direction of its flux remains a zero. Since a restore operation is always necessary after a memory read, it slows down the speed substantially. A core memory has the following attributes:

- Reliable
- Non-volatile
- Can sustain high temperatures
- Large size
- Fairly fast
- Fairly expensive

Core memory was fast but fairly expensive as used in second and third generation computers. Its major drawback was its large size and power consumption; one advantage was its non-volatility. In the 1970s, if a system crashed, we could always turn off the computer and boot a program to print out the core contents. This core dump reveals the history of executions that provides clues to a possible cause of computer crash. In third generation computers, core memory was replaced by semiconductor memory. However, due to core memory's tolerance of high temperatures, it still finds some military applications. In the 1970s, IBM had the vision to use semiconductor memories in some of its low end 370 series computers. At that time, because of its slow speed and low yield, it didn't succeed. However, after a series of breakthroughs in technology, semiconductor memory finally became the dominant force because of its speed, high density, and low cost. Today, all fourth generation computers and beyond use semiconductor chips as their internal memory.

1.3.2.3 Semiconductor Memory

A simplified diagram of a one-bit semiconductor memory cell is depicted in Figure 1.7. The cell is a bistable device, comprised of a semiconductor. The detailed circuits are omitted in the diagram, and the four signal lines are for address selection, read command, write command, and data. Note that the data line is bi-directional for both read and write operations. For a read operation, the selection and read lines are activated with a high voltage. After a while, the input bit stored in the cell shows up on the data line. For a write operation, the output bit is placed on the data line, and meanwhile the selection and write lines are activated with a high voltage.
If a single memory IC chip whose size is smaller than a fingernail has a storage capacity of 16 megabytes, then eight such chips would make a 128MB memory bank with the following characteristics:

- Reliable
- Volatile
- Cannot withstand high temperature
- Small size
- Fast
- Inexpensive

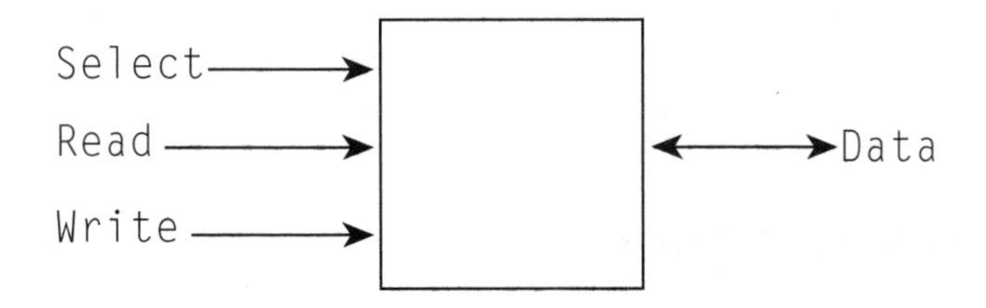

Figure 1.7 One-bit semiconductor memory cell.

There are two types of RAM (random access memory) chips: DRAM (dynamic RAM) and SRAM (static RAM). They are for general purpose applications that require both read and write operations. As shown in Figure 1.8, eight memory IC chips are soldered on a PC (printed circuit) board, which is 5.25 inches by 1.25 inch, known as a SIMM (single in-line memory module) strip. Depending on the cost, each chip may have a storage capacity of 4MB or 16MB, so the strip has a storage capacity of 32MB or 128MB but the wiring layout remains the same. A DIMM (double in-line memory module) has memory IC chips on both sides of the strip, so its capacity is doubled. A DIMM of 64MB or 256MB can be plugged into the same motherboard slot of a PC. According to the revised Moore's Law, the number of transistors on a chip doubles about every 18 months.[53] Hence, a DIMM strip with 16 128MB chips will soon be available with a storage capacity of 2 gigabytes.

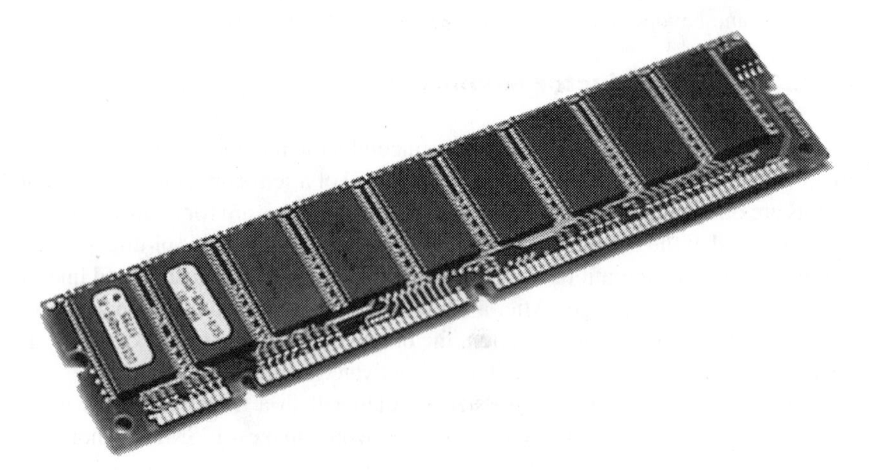

Figure 1.8 Single in-line memory module.

The second popular type of semiconductor memory is the ROM (read only memory) chip that can perform read only operations. In a ROM chip, all the bits are permanently set by the manufacturer so they can not be altered by a user. In a telecommunication switching system, software is usually placed in ROM. We say that the system is embedded for real-time applications.

The third type of semiconductor memory is the EPROM (erasable programmable ROM) chip where a user can change the bit pattern before production. The flexibility that allows design changes after errors are detected makes it very attractive in building prototypes. In relative terms, the EPROM chips are more expensive than ROM and RAM chips.

1.3.3 Input/Output Devices

All the I/O devices can be grouped into three categories: input only, output only, or both input and output. Input devices include the keyboard, mouse, joystick, etc.

Output devices include the display, printer, film making devices, etc. Disks, tapes, and laser disks are classified as I/O devices, but they are really external mass storage devices. A disk or tape supports both input and output in that we can read from the device as well as write to the device.

1.3.3.1 Magnetic Disk

There are two types of magnetic disks: hard disks and floppy disks. The basic principle for each type is the same. The disk surface is coated with magnetic materials to store non-volatile bits just like core memories. A disk spins on a fixed axis at a speed from 3000 to 5000 RPM (revolutions per minute). Figure 1.9a shows an IBM MicroDrive, a 340MB hard disk scarcely larger than a U.S. quarter. As shown in Figure 1.9b, the disk surface is divided into many concentric rings. Each ring is called a track, i.e., a circular path with the same radius.

(a)

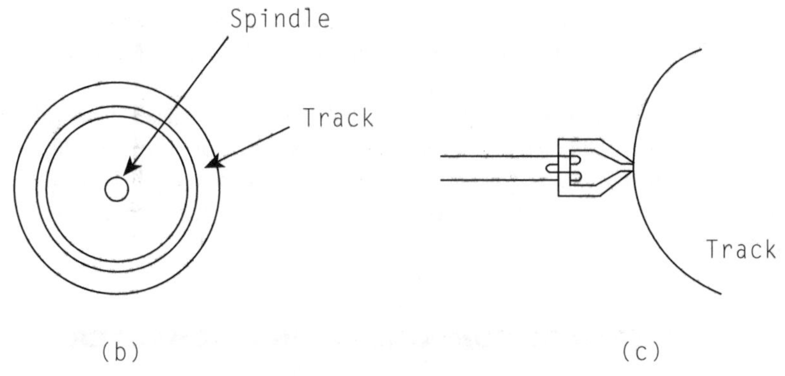

(b) (c)

Figure 1.9 Magnetic disk: (a) an IBM MicroDrive, courtesy of IEEE Spectrum & IBM, (b) concentric tracks, and (c) coil through a magnetic head.

Usually, one read/write head is mounted on a single mechanical arm that can move between the innermost track and the outermost track. The arm moves to select one track on a disk. The head is made of magnetic materials, and we say it is flying on the disk surface to read or write the track. As shown in Figure 1.9c, the head has a cut or gap in the front to pick up magnetic flux, so current is induced through a winding coil. That means that, while the disk is spinning, the magnetic field changes in the head so as to induce current in the coil. During a disk write operation, an electric current flows through the coil to generate a magnetic field that changes the magnetic flux on the track. Because the direction of current changes, the disk surface is magnetized like many tiny magnets aligned along the track with a property such that two adjacent magnets are in opposite directions. During a disk read operation, the magnetic head can pick up the magnetic flux change through its opening cut. Again due to the change of magnetic field or inductive coupling, an electric voltage is generated in the coil. After amplifying and reshaping, the voltage waveform can be further decoded into ones and zeros. A disk read operation does not change the data on the track, so there is no restore operation.

There are some differences between a hard disk and a floppy disk. First, a hard disk usually spins all the time while a floppy disk spins only as required. Second, a hard disk pack can not be removed easily without using special tools, so it can spin much faster. Third, the hard disk has a higher storage density, of approximately two billion bits per square inch. In contrast, a floppy disk is a piece of flexible plastic covered with magnetic material and wrapped in a paper bag or plastic case. Moreover, a floppy disk can be removed from the drive by hand. The term file refers to a collection of data identified by a name. If the file is on disk, it is a disk file. Generally, a floppy of 1.44 MB is big enough to store a whole book. Both hard and floppy disks are external or intermediate storage devices in that they are slower than the central memory. In a broad sense, they are called I/O devices to store programs and data during execution.

Magnetic tapes are similar to disks as far as magnetic recording principle is concerned. There are, however, some differences. First, a reel of tape can store a trillion bits, but it has a much slower speed. Secondly, it is a sequential access device that needs a rewind operation before it can be read again. However, because of its relatively inexpensive cost, it is occasionally used to store the contents of a hard disk as back-up. In the case of a hard disk crash, the files on the tape can be loaded back into the system to recover.

The third kind of disk is a CD (compact disk), which is a laser disk with data stored on its surface. The read operation is performed by a laser, and the disk has a very large data capacity. Some laser disks are writable, for example WORM (write once read-only memory). That is to say, the disk can be written once, and after that it is read only.

1.4 SOFTWARE COMPONENTS IN A COMPUTER

Let us examine some of the software components regarding program development. All software components can be divided into two classes: system and application. The number one system component is the OS (operating system), namely, a set of control programs running with other programs in memory. Other system software components include tools for software development. The application class of software consists of all the programs other than system software. Software tools are classified as system software because they are used to develop programs, systems, or applications. Note that system software tools execute just like any other application programs under the supervision of an OS.

Any program, including the OS, may reside on a floppy disk, hard disk, or optic disk after development. The loader in the system brings the program from disk to memory before its execution can start. The OS performs two major functions: supervision and service. It is correct to say that the OS supervises the execution of other programs in a computer. If a program tries to do something incorrect, the CPU hardware and OS work as a team to detect such a wrong move. As a consequence, the program in execution is forced to terminate without causing further damage. This happens a lot during program development. The terms OS, control programs, supervisors, and executives are all synonymous. The portion of OS routines that resides in memory all the time is also called the nucleus or kernel. The OS also provides services to other programs. For example, if a program needs to interact with an I/O device, it asks the help of the OS by issuing an I/O request system call. To perform program development, we need to introduce system software tools, namely editors, assemblers, compilers, linkers, loaders, and debuggers. In some computer systems, the system software tools and OS are bundled as one package.

1.4.1 Boot an Operating System

The bulk of an OS may reside on the hard disk or floppy disk and a cold boot means to turn on the power. When that happens, a routine at a fixed location in memory is executed. As a consequence, a small piece of the OS, i.e., the boot routine, is loaded into memory. After receiving the control, the boot routine executes to bring the rest of the OS into memory. A warm boot is slightly different in that pressing certain keys passes control to the boot routine in memory that is not damaged yet. This also means that the OS in memory is partially crashed due to some programming errors. If the warm boot does not work, the OS is completely dead and a cold boot is necessary. In either case, a software routine, when executed, boots the OS from a disk. A computer system usually has its OS stored in low memory as shown in Figure 1.10.

1.4.1.1 System Prompt

After booting is completed, the OS is in control and it displays a short message on the left side of the screen followed by a blinking cursor. This short message is called the system prompt, which may comprise a single character such as >, or a few characters, like C:>, on a PC. The prompt signals that the OS is ready to receive a command from the keyboard and the blinking cursor shows the last character just

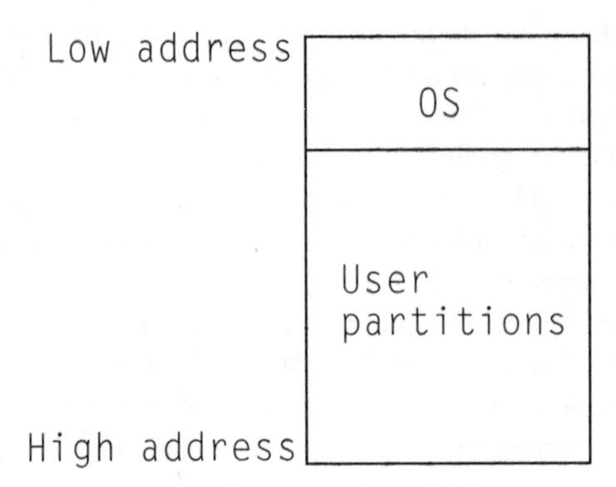

Figure 1.10 Operating system in memory.

typed in. The character C in the prompt denotes a disk drive ID (identifier) that is used to store temporary results when no other drive is explicitly specified. The C drive, therefore, is the default drive. Other systems may display a single character prompt: % or $ followed by a blinking cursor. After seeing the system prompt, a user can type a command on the keyboard followed by the <cr> (carriage return) or enter key. Note that <cr> is a meta-symbol with letters enclosed in a pair of angular brackets. The command is usually the name of an executable program on disk. After receiving the command, the OS does the following:

1. Reads the name of an executable program from the keyboard
2. Loads the program from disk into memory
3. Passes control to the program and executes

The program name serves as a command to the OS. System programs, i.e., routines, are referred to by names just like applications. For example, the cls (clear screen) command tells the OS to clear the screen, the help command asks the OS to display the help menu of a particular command, etc.[79] If a piece of software provides a service in the system, it is a system utility program. Both cls and help are program names of two system utilities that usually reside in memory. Before the execution of any program, a user must inform the OS of his intention. If the program is not in memory, the OS loads the program into memory first before its execution. While running, the program is under the supervision of the OS. When the program terminates, control is returned to OS.

Modern PCs use window-based operating systems in which a graphic user interface (GUI) runs on top of an OS. That is, a window-based OS is a superset which contains extra software to handle the graphic user interface.

$$\text{Window-Based OS} = \text{OS} + \text{GUI}$$

After booting is completed, the system displays many graphic symbols as icons on the screen, and each one represents an executable routine in memory. The label of an icon represents a program name. Since the mouse is an input device, a user can move the mouse to an icon and double click the mouse button to signal that the OS should start the execution of this routine. As a consequence, a window is displayed on the screen. If the current routine in execution requests a keyboard input, it then displays its own prompt in a window so the user can press keys to enter input. In concept, a prompt is just a message indicating that the current routine in execution is expecting a keyboard input from the user.

The design of an OS is complicated, and some of the routines can be written in a high-level programming language while others are written in assembly language.

1.4.1.2 File Name

A file system in the OS supports many files on disk or on tape. Each file may have a name of up to eight letters, and it may have an optional extension name of up to three letters. A period (.) needs to be inserted between a file name and an extension name. An extension name is used to provide extra information about the file. Some examples of file names are given in Table 1.2.

Table 1.2 Simple File Names

File Name	Description
pgm1.asm	program 1 in assembly language
pgm2.c	program 2 in programming language C
pgm1.obj	program 1 in object form
pgm1.exe	program 1 in executable form

A user can change the file name via the ren (rename) command on the keyboard as follows, ren pgm1.exe hello, where pgm1.exe is the old file name, and its new name becomes hello. After renaming the file, a user would enter hello instead of pgm1.exe on the keyboard in order to execute the program.

1.4.2 Editor

Recall that all the system software tools are used to develop a program, system, or application. The development job consists of many job steps, and each job step is an independent unit of computation. Since any program must be in executable form on disk, it takes many job steps to accomplish this goal. After writing a program in a programming language on paper, the first job step is to store the program on disk by executing an edit (editor) that is usually bundled with the OS. The editor allows a user to input a program via the keyboard. The program input from the keyboard is the source code. Thus, a disk file with a source code is also called the source file or

Figure 1.11 Execution of an editor.

source module. By means of an editor, a user not only can create a source program on disk by typing on the keyboard but can also make changes to the file while typing. In sum, an editor can read input from the keyboard, accept commands to make changes, and save the file on disk. The execution of an editor is shown in Figure 1.11. The editor reads input from the keyboard, makes changes as commanded by the user, and generates pgm1.asm as output.

The term processor is quite popular. Hardware people refer to it as a hardware processing unit, and software people refer it as a processing program. Hence, a word processor is a word processing software program and a language processor is a compiler. A word processor is a powerful editor because it can edit text as well as graphics. We can execute a word processor to prepare a large program because a mouse is an input device to select an edit command in a menu with no need to memorize. Because the data output can not contain any format control information such as page margins, numbering, line spacing, etc,, it is necessary to inform the word processor via a special command that the output file contains pure text.

For example, the commercial word processor WordPerfect (WP) is a powerful application program to edit a large programming file. After seeing the system prompt, we type the following command:

wp pgm1.asm

where wp is the name of an executable file, followed by a space as the separator, followed by pgm1.asm, the name of the file that contains the assembly source. If pgm1.asm is a new file, WP will create the file from scratch. If pgm1.asm is an existing file, WP allows a user to make editing changes in the file. Either way, the source program in assembly language is stored in a disk file named pgm1.asm. Running WP to prepare a program is one job step. After exiting from WP, the system prompt shows up again on screen and the system is ready to execute the next job step, e.g. assembly.

1.4.3 Assembler

An assembly language is a low-level programming language which may be used to write system software as well as application programs. An assembler is a software tool that translates a program from assembly language source to object code that is very close to machine code. Because each assembly language statement is usually translated into one machine instruction, assembly code programming also means machine code programming. Thus, while a compiler is a high-level translator, an

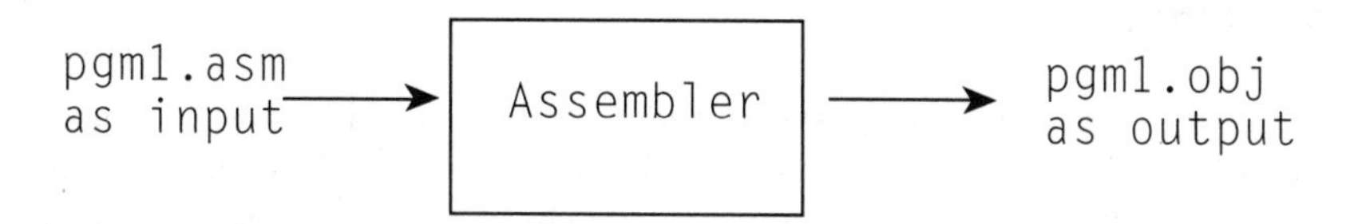

Figure 1.12 Execution of an assembler.

assembler is a low-level translator. Nonetheless, both are programming tools to translate programs.

The execution of an assembler is shown in Figure 1.12. The software reads an assembly program as input and translates it into some object code in an output file newly created on disk. The input file pgm1.asm contains the source code, and the output file pgm1.obj contains the object code generated by the assembler. After translation, the original source file remains intact without being modified.

1.4.4 Compiler

Most application programs are written in high-level programming languages for reasons of convenience and readability. Some of the programming languages are C, C++, Fortran, COBOL, JAVA, etc. A compiler is a software program that translates another program from a high-level language to object code. Thus, a compiler is like

Figure 1.13 Execution of a compiler.

an assembler except that it has the intelligence to parse and translate a high-level programming language statement into one or more machine instructions. In general, a compiler is a system utility or programming tool executing in the user partition just like any other program under the supervision of the OS.

The execution of a compiler is shown in Figure 1.13. Suppose that a program written in programming language C is in the disk file named pgm1.c. The C compiler reads the pgm1.c file as input, parses and translates each statement in the program, and outputs the object file named pgm1.obj on disk. That is, after a program is translated by an assembler or a compiler, an output file is created on disk to contain the object code that is close to the target machine code, but is not exact. The terms source code, source file, and source module are all synonymous. Likewise, object file also means object code or object module. In practice, it is more expensive to purchase a source program because the code can be easily modified to save the development cost.

1.4.5 Linker

There are many language translators, therefore it is a good idea to standardize the format of an object file. In other words, the output generated by the C compiler should not be different from the output generated by an assembler. For performance reasons, part of an application program is written in C and the other part can be written in assembly. Two translators are executed to generate two different object files. If a standard format is enforced, a linker program can link the two object files into one executable file as shown in Figure 1.14.

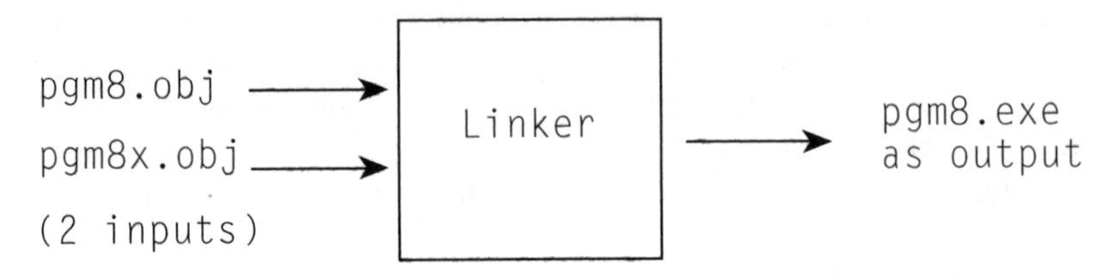

Figure 1.14 Execution of a linker.

The C program in the disk file is named pgm8.c, and the assembly routine is in the file named pgm8x.asm. As the C compiler translates the C program into pgm8.obj, the assembler translates the assembly routine into pgm8x.obj. Then, the two files, pgm8.obj and pgm8x.obj, are read as input to the linker that combines them into one load module named pgm8.exe. In fact, a load module is an executable file that contains almost 100% machine code, but not quite. Because an object file has a different format than an executable file, a linking step is necessary even though there is only one object file. That is to say, the linker links one or more object files into one executable module. After linking, all the object files remain intact, and some compilers can generate an executable file directly to bypass the linking step.

1.4.6 Loader

Only an executable file may be loaded into memory before its execution. The loading function is performed by a routine in the OS known as the loader. Since the loader is executed often, its code should be compact and efficient. Every time a system prompt appears on the command line, a user may enter either a system command or a user command via the keyboard. A user command is identical to an executable file name. If the file is not in memory, the OS must initiate the loader routine to load the program from disk to memory. The loading function is implicit in that we merely type in the executable file name without entering the loader name on the command line. That is, the execute step really consists of two job steps, first loading the code and then executing the code as shown in Figure 1.15.

The load and go system concept means that after loading, the OS passes control to the code to execute. Note that the loader routine does not generate any output file and it is implemented whenever an executable file needs to be loaded from disk to

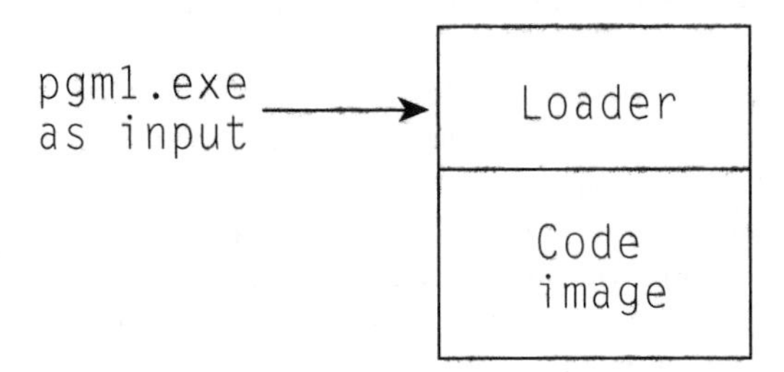

Figure 1.15 Load and go.

memory. Usually after loading an executable file into memory, the loader modifies some of the memory addresses in the code according to the exact memory locations. In other words, the executable file on disk is a little different from its code image in memory. From a system viewpoint, the address modification job is a nuisance, but it is present in most systems. If special hardware and software features are provided, as discussed in later chapters, we can implement an executable file that requires no address modifications after loading.

1.4.7 Debugger

When a program first executes on the computer, it may not work correctly as expected. Sometimes, the errors in the program are easy to detect. Other times, the errors are very difficult to detect. Under such circumstances, we rely upon a programming tool called the debugger. Debugging really means finding the design errors in hardware or software. For fifty years we have faced the same problems of software debugging.[37] Some bugs may take months or years to locate; others are impossible to locate. The execution of a debugger is shown in Figure 1.16.

In fact, the program debug requires the loader to bring in the executable file pgm1.exe. The debug program interacts with the human user at the terminal. That is to say, the test program executes under the control of a debugger, and a user can obtain computation results from his program through a debug command, step by step. Using logical deductions, a user can find errors in his program under the debug program. It is now easily understood that an editor, assembler, compiler, linker, loader, and debugger are all software tools to assist program development.

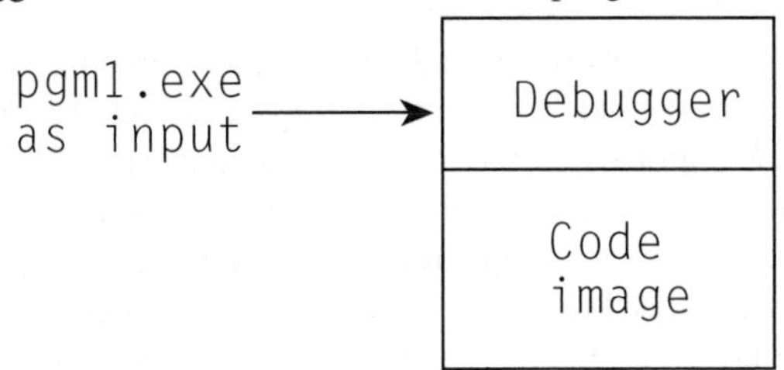

Figure 1.16 Execution of a debugger.

1.5 APPLICATIONS OF COMPUTERS

When an application program executes on a CPU, it performs a particular function, such as word processing, spread sheet, accounting, banking, web site search, and stock market trading, to name a few. Programming tools are also application programs, and they are used to develop other applications. An application program or software tool runs along with the OS in a computer. An application program is usually stored in a disk file that must be loaded into the computer via an OS command. If program-to-program communications do not exist, we have a group of single computer applications. Otherwise, we have network computer applications that continue to grow rapidly.

1.5.1 Single Computer Applications

If an application program executes on one CPU without communicating with other computers, it is a single computer application. A popular example is the word processor (i.e., word processing application program) that is used to prepare reports, documents, and books. A powerful word processor not only handles the text but also the graphics. Since a disk file is used for both input and output, a user can store his work on disk and fetch it later. As far as applications are concerned, there are many different files, some classified by file structure and others by information type. For example, if the file contains information in the form of machine instructions, it is an executable file. If the file contains data information, it is a data file. A word processor can read a data file from an input storage device, make modifications to the file, and store the file back on an output storage device.

Another popular single computer application program is a spread sheet, used to prepare quotations, transactions, and reports for small businesses. Other single computer applications include banking, billing, accounting, tax preparation, and game playing. Furthermore, a computer can be used as a teaching tool for dancing, singing, spelling, driving, etc. If we connect several computers together through telephone lines, we have a computer network. Any user program running over a network in order to achieve a particular function is a network computer application as discussed in the next section.

1.5.2 Network Computer Applications

The demand for network computer applications is immense and the field of network development is open and exciting. As the trend dictates, any single PC will have a connection to a global computer network of some sort. That is, all computers will be interconnected using either wire or wireless technology. If so, a person can call a teleconference, play ma-jong, bridge, or video games, watch movies, chat, vote, pay bills, reserve tickets, request quotations, trade stocks, etc. Using wireless technology, a cellular phone that has a computer inside can be used to exchange electronic mail (e-mail). Furthermore, a police officer can use a portable computer to validate a driver's license in a minute and a broker can trade stocks on an airplane in real-time. The advent of computers has indeed set the third industrial revolution in motion and the computer has become a household commodity like other appliances. We know for sure that the computer has improved our daily life, and its impact far exceeds our expectation. As the field for developing network computer applications is so exciting, it is important to know how to develop a program on a computer.

1.6 HOW TO PROGRAM A COMPUTER

There are two necessary steps to program a computer. First, a programming language must be studied, and second, a program must be written in that language. While a simple program can be entered via the keyboard, a large program usually requires many design phases. The program design may be written on paper first. After many design reviews, the program is entered into the computer and a test is performed. High-level language programming seems easy, but assembly coding is more intuitive because there is nothing to hide. That is to say, a good understanding of internal operations in a computer enables a programmer to produce elegant code.

As far as hardware is concerned, a computer operates on one level, the target machine level. The set of target machine instructions is what the system developer sees. As far as software is concerned, a computer can operate at one of two levels, that is, with an OS or without. When a system is first developed, its system reference manual describes the set of instructions and architectural aspects so the system developers can write an OS for the machine. There is also a hardware reference manual that contains the hardware information or specs in regard to machine maintenance.

1.6.1 Raw Machine

A raw target machine means that no operating system exists in its memory, as shown in Figure 1.17.

Figure 1.17 Raw machine without an OS in memory.

This is true when a CPU's hardware is first developed. The first order of business is to write some system software tools. Second, the tools are used to write an OS for the target machine. A program running without an OS is called a stand-alone program because it has its own system software to handle the control functions normally provided by an OS.

Nevertheless, a raw machine is a valuable tool to a group of students who would like to write an OS from scratch for such a machine. If a university has a mainframe computer that provides general computing services, students must request raw machine block time to do system software development. Allocating machine block time means that the whole computer is turned over to the student group so that no one else can use the computer. In industry, machine block time is allocated for special groups. For example, the real-time system project group often requests machine block time to debug their system work without interference from others.

Writing an OS is just like writing any other program except it is much more complicated. However, if a skeleton version of the OS is developed first, it can be executed as a base, and more system components can be added to it, step by step, with the aid of software tools.

1.6.2 Machine with an Operating System

After an OS is developed, a user program can execute on top of it as shown in Figure 1.18.

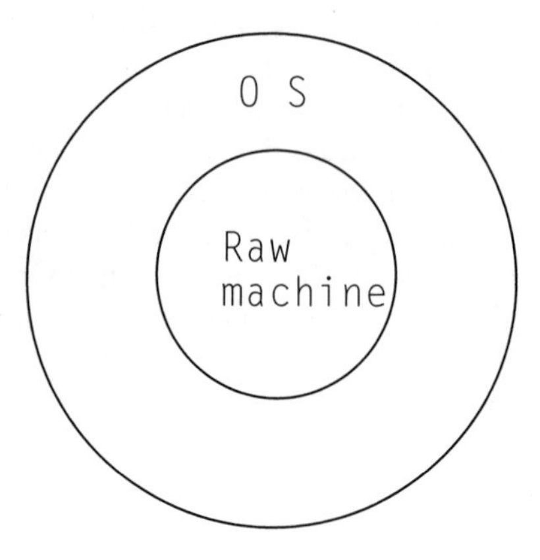

Figure 1.18 Machine loaded with an OS in memory.

In Figure 1.18, the inner ring represents hardware or the raw machine, and the outer ring represents the OS, i.e., the computer has operating system support. An application program running on top of the OS really means that it executes under the supervision or control of an OS. In fact, the OS executes on the CPU hardware just like any other program, such as the editor, compiler, word processing program, etc. The OS watches the execution of other programs and provides services. In particular, if a user program needs a service, e.g., data management, it calls upon the OS to perform the service on its behalf. The concept of levels sounds confusing when we say that the user program runs on top of the OS. In reality, a user program runs under the supervision of the OS. That is to say, the OS resides in memory and it provides services to a user program in addition to supervision. If anything goes wrong with a user program, the OS immediately takes control. Hence, the OS is nicknamed the supervisor, executive, or control program. Remember that both the OS and user programs take turns executing on the CPU at the target machine level. The OS runs on the raw machine to provide control functions and system support, i.e., services. Generally, the OS prepares the tasks to execute a user program, supervises the program while in execution, and handles the tasks after the program terminates. While a program is running, most instructions execute directly on the CPU, but there is one instruction, i.e., system call, that passes control to the OS. When that happens, the OS interprets the request and performs a service for the user program. After that, the OS usually returns control to the user program so as to resume its execution.

Note that with few exceptions, the OS executes on the CPU along with other user programs. All the programs including the OS execute at the target machine level. After the OS is developed, there is a boot record on disk. After turning on the power, a small piece of the OS in the boot record is loaded into memory. The execution of the boot routine brings the rest of OS into memory. More often, an OS is bundled with all the system software tools as one product. In other words, the package includes an OS, editor, a few compilers, an assembler, linker, loader, debugger, etc. After editing, translating, linking, and loading, the developed program executes on the CPU under the supervision of OS. The development effort is divided into many job steps, as shown in Figure 1.19.

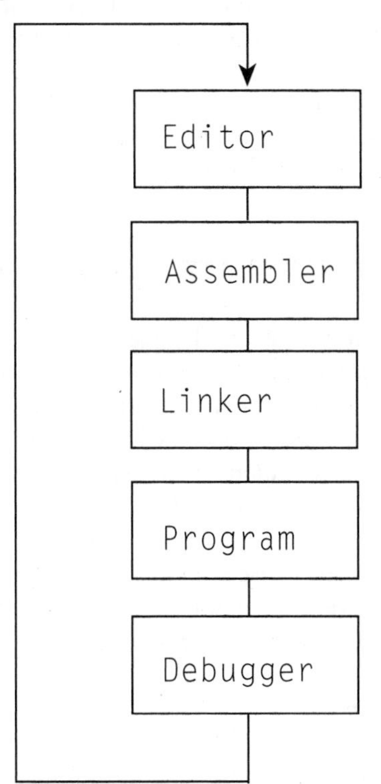

Figure 1.19 Program development steps.

For example, when writing a program in assembly language we have the following job steps:

1. Execute an editor, which reads the data input from the keyboard, makes changes,and stores the assembly source code in a file named pgm1.asm on disk.
2. Execute the assembler, which reads in a source file named pgm1.asm and generates an object file named pgm1.obj. If any errors exist in the program, the assembler prints error messages, and we must go back to step 1 to make corrections.
3. Execute the linker to generate an executable file named pgm1.exe. If any errors are detected during linking, we must go back to step 1 to make corrections.
4. Execute the program pgm1.exe to test its functions. If easy errors are found, we go back to step 1 and try again.
5. Execute the debug program to find the tough bugs in pgm1.exe. After locating the bugs, we go back to step 1 to start over.

Job steps 1 and 2 must be repeated until an object module is generated by the assembler. After loading, the program pgm1.exe executes on the CPU in step 4. If an easy error is detected during its execution, we go back to step 1. After making the changes, we go through the steps and try again. If the errors are hard to find, we must rely upon the debug program in step 5. That is, we execute the target program under the debugger in order to locate its programming errors. This development loop is repeated until the program is fully debugged. All the job steps, however, execute on the computer sequentially, one at a time.

1.7 SIMULATE A COMPUTER

An OS may be written in assembly language, high-level language, or a combination of both. For example, part of the OS may be written in programming language C and part of it may be written in assembly. After writing an OS for a new machine, testing the software can start before the new hardware becomes operational. Thus, system simulation software is written on another computer, known as the host, that may belong to the previous generation. The new machine is said to be the target machine. This notion is important so both hardware and software design teams can work in parallel. In the history of computing, the OS of the Burroughs 5000 was developed before its hardware was built.[2] Nowadays, software simulation is performed from the OS function level to the logic gate level before the chip is built.

1.7.1 Testing an OS

In case the new machine is not ready but the old machine is available with system software tools such as an editor, compiler, assembler, linker, and debugger, we can modify an existing compiler and assembler such that they generate the object code for the new machine. As a result, the modified compiler is called a cross compiler and the modified assembler is a cross assembler. Both are cross translators running on the host machine so as to generate object code for the target machine. The term cross means that the software tool runs on a computer that is different from the target machine. A cross compiler or assembler is often used to write a new OS. This approach also requires a computer simulator of the host that can test all the instructions in an OS for the new machine.

A computer simulator is a software program that can test all the functions of a target computer at the instruction level. Since the new machine is not operational, we can debug its OS under the simulator. Writing a computer simulator requires a thorough understanding of the hardware specifications of the new machine. As all the functions of hardware components of the new machine can be simulated on a host, we are able to find bugs not only in the new OS, but also in the new hardware.

1.7.2 Data Management Services

The OS provides both control functions and services to the user programs. An OS can supervise the execution of many user programs in a computer during the same time interval, and each user program occupies a partition in memory. As far as software development is concerned, the data management services are also tools for a user. For example, a user may want to create and manipulate the files on disk. The OS supports a file system that is comprised of a file directory, files, and I/O routines, and a user can enter a command and the OS will take action accordingly.

A floppy disk has both sides magnetized, and each side is divided into many concentric rings. Each track or ring is divided into many sectors. A sector is the basic unit to be accessed on a disk. The I/O routine reads a sector from a disk or writes a sector onto a disk. The directory on the disk describes all the files including their disk locations.

1.7.2.1 Directory

A typical PC disk has four logical areas: boot record, file allocation table (FAT), directory, and files, as shown in Figure 1.20a.

The boot record contains a set of PC instructions that gets loaded into memory when the power is turned on. The execution of the boot record brings the rest of the OS into memory. Each entry in FAT contains the disk address of a file extent (i.e., a portion of the file). The directory is an upside-down tree as shown in Figure 1.20b. That is, the directory has many nodes arranged in a linked list. Each tree node or leaf is a descriptor containing information to describe a file or a subdirectory. The rectangular node is a file descriptor containing attributes of a file, such as file name, extension name, creation date, starting address of the file, file type, etc. The circular node is a directory descriptor containing information about a subdirectory at a lower level. Between two adjacent directories, the upper node is the parent and the

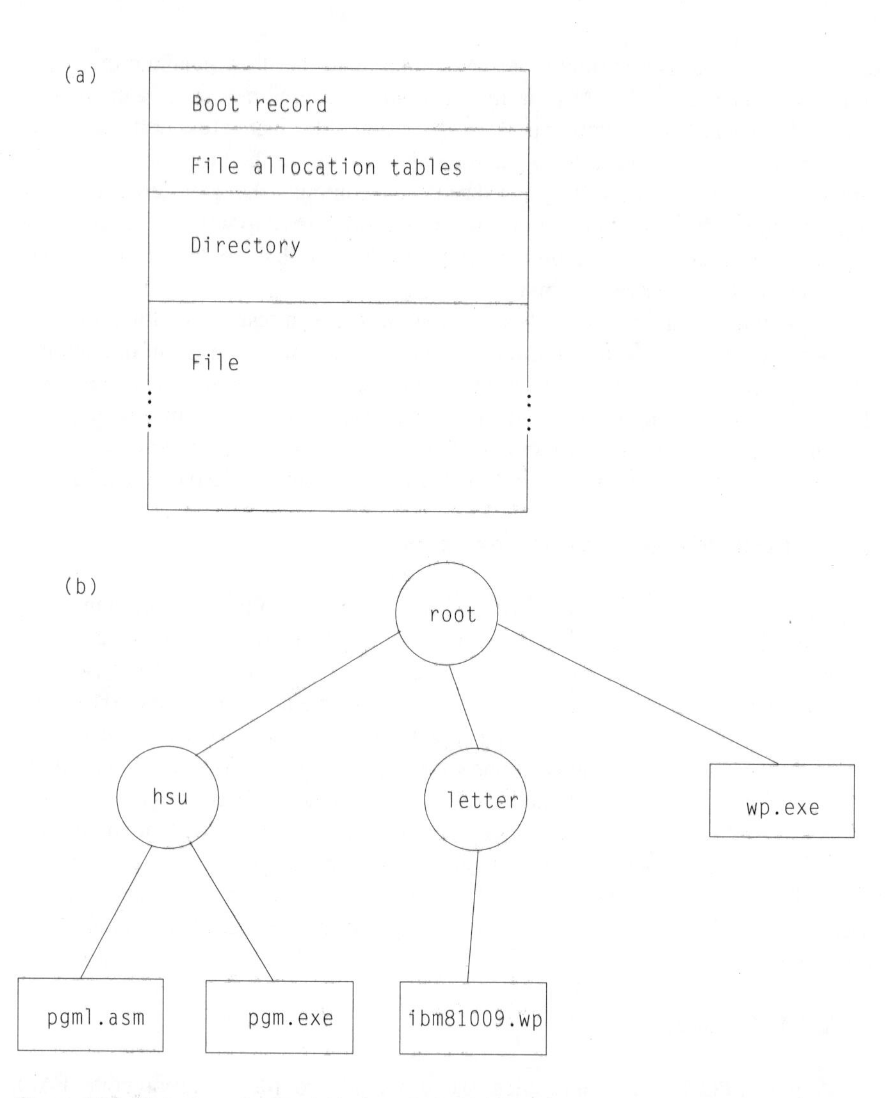

Figure 1.20 Disk space: (a) four logical areas and (b) the tree structure of a file directory.

lower node is the descendent. Thus, the lower directory is a subdirectory and it may contain either file or directory descriptors. The top node is the root directory that usually resides at a fixed location on the disk. From the root, many nodes below can be accessed and each lower node is either a file descriptor or a directory descriptor. The system directory is an internal file in which the OS can search for a particular directory or file. In fact, the file directory is separated from the data files that may be scattered all over the disk.

This file directory resembles a file cabinet that has many drawers, and each drawer has many folders. Each folder contains many files, e.g., letters and memos. Each directory on a disk, hard or floppy, has many tree nodes, and each node describes one file or another directory at a lower level. The OS maintains the information of the current directory in memory. When a user issues a command on the keyboard to access a directory or file, the OS searches for the directory to locate the object on disk.

1.7.2.2 Path Name

Because the directory is a tree structure, in order to access any directory or file, a user supplies a tree path name for the OS to traverse the tree. A full file path name may start with the root symbol, a sequence of directory names, and a file name. In MS-DOS, a backward slash (\) represents the root, and the same symbol (\) is used as the delimiter to separate fields in any path name.[79] Each directory node contains information about its file descriptors or directory descriptors. Some file system commands can be used to manipulate files or directories. In MS-DOS, three special symbols are chosen to represent the root directory, the current directory, and the previous directory, as shown in Table 1.3.

Table 1.3 Special Names Used in a Directory Tree

Symbol	Description
\	The root symbol is a backward slash (\) in MS-DOS, but a for ward slash (/) is used for UNIX.
.	The current directory symbol is a single period (.) in most systems.
..	The parent or previous directory symbol is double period (..) in most systems.

The data management routines in an OS use the path name to traverse the tree. In the path name, a delimiter is used to separate two adjacent directories, and the delimiter is usually the root symbol. The first symbol in a path name tells where to start: the root, the current directory, or the previous directory. The two path names for directory search and file search are quite similar. If the last name identifies a directory, it is a directory path name. If the last name identifies a file, it is a file path name. If no directory name is given before a file name, the default is the current directory. A special symbol (*) is a wild card that matches any file name or file extension name. Some of the path names to identify a directory or a file are listed in Table 1.4.

1.7.2.3 File System Command

A user can issue a command to access a directory or a file. Some of the popular file system commands are shown in Table 1.5. If only an argument is given on the line without a command name given, the previous command is assumed to be the default.

Table1.4 Path Names in a Tree Directory

Name	Description
\ch4\chap4.wp	\ is the root directory, ch4 is the directory name below it, and chap4.wp is the file, where .wp means a WordPerfect generated data file.
\ch4*.wp	Same as above except the * is a wild card to match any file, That is, any file with wp as an extension name.
\csc215\pgm1.*	Any file named pgm1 under the directory csc215 with any extension name.
..\letters\ibm81009.txt	The directory called letters is at the current level, and the referenced file is ibm81009.txt dated 10-9-1998.

Table 1.5 File System Commands

Name	Description
<Ctrl>-c	Cancel the current program in execution
<F3>	Repeat previous command
A:	Switch to A: as the default drive
cd \ch4	Change directory to ch4 under the root
..	Change to previous directory
wp	Change to wp under the current directory
..\wp	Change to wp at the same level as the current directory
cls	Clear screen
copy pgm1.asm test.asm	Copy file pgm1.asm to test.asm
del pgm1.asm	Delete file pgm1.asm
del pgm1.*	Delete all the files named pgm1
deltree \csc215	Delete the entire tree named csc215 under the root, i.e., all the files and subdirectories in this directory are deleted
dir	Display the current directory with one entry per line
dir/w	Display the current directory with many entries per line (wide option)
diskcopy a: a:	Disk copy from A: to A: Follow instructions of the prompt to swap disks in and out on drive A:

Name	Description
format a:	Format drive A:
help edit	Help menu of the edit command
md csc215	Make directory csc215 under the current directory
print pgm1.asm	Print file pgm1.asm
ren pgm1.exe hello	Rename pgm1.exe as hello
rd csc215	Remove directory csc215 provided that all its files are deleted as a prerequisite
type pgm1.asm	Type (display) file pgm1.asm
undelete pgm1.asm	Undelete file pgm1.asm

1.8 PROGRAM DESIGN LANGUAGE

The acronym PDL stands for program design language which is used to describe the logic flow of software design or hardware design. PDL is plain English plus structured constructs, and the language is case insensitive. The statement is readable by means of a keyword (i.e., keyword symbol) that provides semantic meaning explicitly. To improve readability, some keywords are written in uppercase, such as IF, THEN, ELSE, etc. The GOTO statement passes control to a label with a colon suffix as defined in column one. The semicolon serves as the end delimiter of a phrase, clause, or statement. A system routine usually starts with an uppercase letter, such as Attach, Detach, IH (interrupt handler), etc. A comment statement may be enclosed in braces or written as a sentence with a semicolon in column one as shown below.

```
...     {This task is reentrant.}
; —— This task is reentrant.
```

Certain syntactic rules should be observed. As spaces are delimiters, they are used to improve the readability of the program. PDL consists of six structured constructs as listed below:

1. IF - THEN - ENDIF
2. IF - THEN - ELSE - ENDIF
3. CASE - ENDCASE
4. DO-WHILE - ENDDO
5. REPEAT - UNTIL
6. DO - ENDDO

1.8.1 Short IF

The short IF construct provides a basic means to change the program flow under a condition. The short IF construct can be used to implement any other construct, i.e., the constructs 2–5 listed above. The flow chart of short IF is shown in Figure 1.21a.

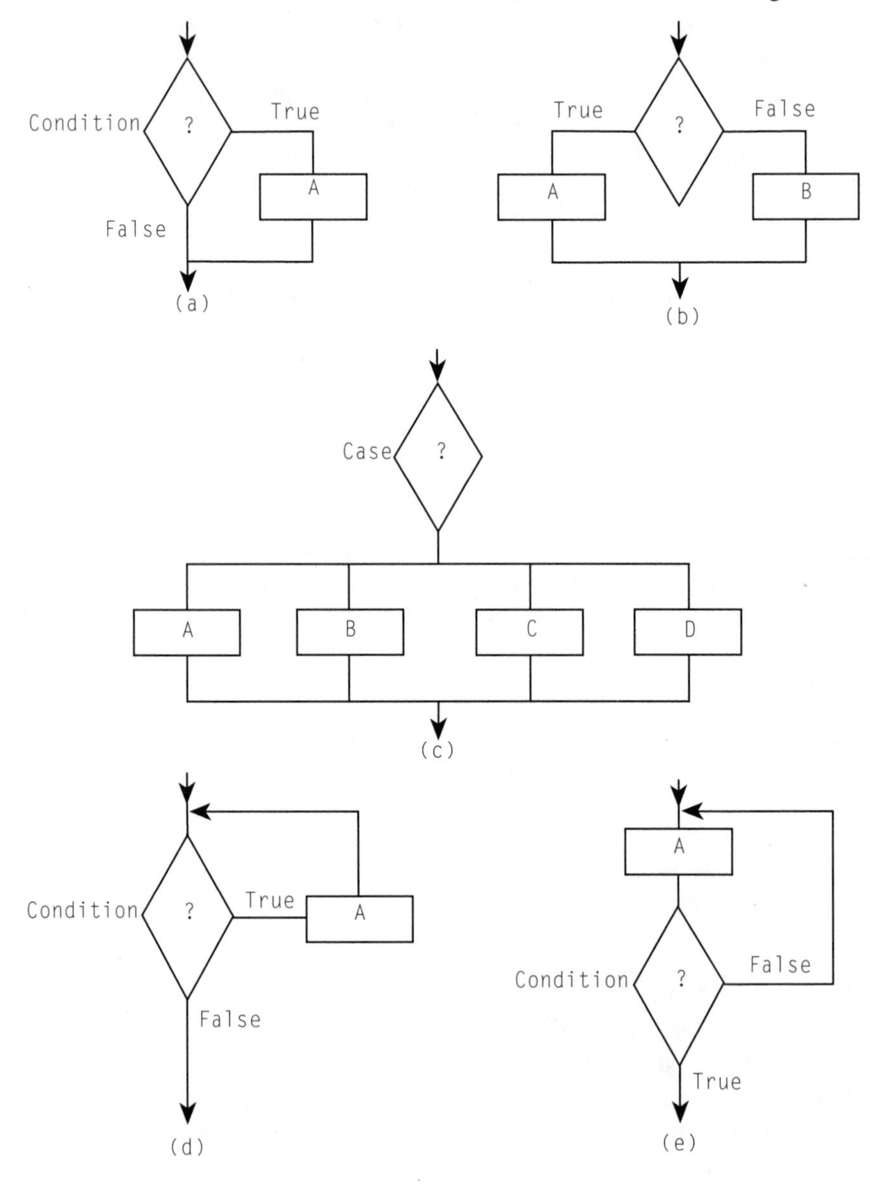

Figure 1.21 Structured programming constructs: (a) short IF, (b) long IF, (c) case, (d) do-while, and (e) repeat-until.

The diamond shaped box is the decision box that evaluates a condition. If the condition is true, control is passed to the branch labeled true; otherwise, control is passed to the other branch. Each branch indicates the direction of logic flow based on two possible outcomes, true or false. In that regard, a condition really means the evaluation of a logical expression. The rectangular shaped box A denotes an operation that may be one statement or a block of statements. If the result is true, we execute the operation box; otherwise the next statement after the construct is executed. The short IF is the fundamental programming construct which provides the conditional test capability to a loop that comprises a group of statements which may be executed repeatedly if the condition is true.

1.8.2 Long IF

The long IF construct provides two possible branches, the true clause and the else clause, as depicted in Figure 1.21b. In the then clause we have operation box A, and in the else clause we have operation box B. Only one box will be executed, determined by the outcome of the logical expression. After evaluation, if the condition is true, we execute operation box A, i.e., the then clause. Otherwise, we execute operation box B, that is, the else clause. After executing either box, the next statement after the construct is executed. In either a short or long IF statement, its then clause and else clause may contain another IF. If so, we have a nested IF that is often used to test a series of conditions. With nested IFs, we evaluate conditions one by one, and depending on the outcome we execute the designated branch based on the condition. This leads us to discuss the CASE construct as follows.

1.8.3 Case

Figure 1.21c depicts the CASE construct. A case condition is tested at the beginning of the construct and one of many possible outcomes is true. Based on the outcome, the clause designated as the individual case for that particular outcome is executed. After the particular case A, B, C, or D is executed, control is passed to the next statement after the construct. This is a little different than language C, where a break statement is needed for the case to skip.[74] The CASE construct can be rewritten as a nested IF, but as a CASE construct, it provides convenience and readability.

1.8.4 Do-While

The DO-WHILE construct is shown in Figure 1.21d. At the beginning of a loop, we evaluate a logical expression. If the result is true, we execute the operation box A, which is the body of the loop. After executing the loop, we go back to the beginning of the loop and evaluate the logical expression again. The loop executes repeatedly as long as the logical expression is true. Note that in the loop, it is necessary for the condition to be changed from false to true; otherwise we have an infinite loop. It should also be mentioned that the body of a DO-WHILE loop may not execute at all if the condition fails the first time.

1.8.5 Repeat Until

The REPEAT-UNTIL construct is slightly different from the DO-WHILE as shown in Figure 1.21e. It provides a different way to write a loop — we evaluate the logical expression at the end of the loop instead of at the beginning. As a result, the operation box A in a repeat loop executes at least once.

1.8.6 Do-Enddo

There are two types of DO blocks: repetitive and non-repetitive. The DO construct with a STEP clause executes a loop many times, as determined by a controlled variable. The STEP clause has a control variable, initial value, final value, and step value separated by keywords. An example is given below:

```
DO i = 0 to 4 STEP 1,    {Default step value is 1.}
...
ENDDO;
```

The initial value is zero, the final value is four, and the step value is one, the default. The loop executes five times, controlled by the variable i. Following each loop execution, the variable i is incremented by the step value. When the variable i exceeds the final value four, the loop exits. If the step value is omitted, its default value is one.

A DO block may be non-repetitive if there is no STEP clause at the end. That is, the block or group of statements executes only once. The keyword ENDDO improves the readability of a program as shown below:

```
DO;
...
ENDDO;
```

The PDL alone is adequate to describe the logic flow of most algorithms. However, we need RTL (register transfer language), as will be explained in Chapter 4, to describe the movement of bits at the hardware register level. That is to say, if PDL is coupled with RTL, we can describe the internal operations of a computer.

It is our intention to discuss assembly language coding at both the target machine and microcode levels. Such languages use keyword symbols like period, comma, left/right parentheses (), square bracket [], angular bracket <>, etc. That is, they are terminal symbols in the language, e.g., RTL, assembly, or microcode. Be that as it may, the syntax of a target language can be described by a meta-language. Ideally, the meta-language has its own set of symbols called meta-symbols that do not belong to the target language. In other words, the meta-symbols are used to describe the syntax or an object in a target language. Unfortunately, the symbols on a keyboard are limited, so some symbols in the target language are also used as meta-symbols for the sake of convenience. For example, the set of <, >, [,], and | in RTL are used in the meta-language. That is, the angular bracket means class of so <value> means class of value, e.g., 0, 50, or 100. The square bracket means an option and | means or. Some meta-symbol notations are listed in Table 1.6.

Table 1.6 Meta-symbol Notations

Notation	Description
<data type>	A class of data type.
[<dup>]	An option that is an integer declared as a dup (duplicate) factor.
[B|W|D]	An option that is a keyword B (byte), W (word), or D (double).
[L<length>]	An option that is a keyword L followed by a class of length.

1.9 SUMMARY POINTS

1. An analog computer uses analog signals with various voltage levels, while a digital computer uses signals with only two voltage levels.
2. Computer architecture means the design of instructions and registers as seen by the system software developer.
3. First generation computers were made of vacuum tubes.
4. Second generation computers were mostly made of discrete transistors and magnetic core memories.
5. Third generation computers were mainly made of integrated circuits and core memories.
6. A fourth generation computer has many microprocessor chips inside along with a very large memory using VLSI semiconductor circuits. That is, every computer is comprised of one or more microprocessors, regardless of whether it is a PC, mainframe, or supercomputer.
7. The hardware components in a digital computer include the CPU, memory, and I/O devices.
8. A microprocessor chip may contain over ten million transistors.
9. A memory cell is a bistable device that is used to store a bit, one or zero.
10. The OS supervises the programs running on the CPU during the same time interval, and it also provides services to a user.
11. Editor, assembler, compiler, linker, loader, and debugger are all programming development tools.
12. A general purpose computer may operate at only two levels, either with an operating system or without.

13. After seeing the system prompt followed by a blinking cursor, a user can type on a keyboard a command that is the name of an executable file.
14. A system command from the keyboard needs a <cr> (carriage return) at the end so the OS knows when to take action.
15. After reading the name of a program from the keyboard, the OS loads the program from a disk if it is not already in memory, and control is passed to the program to begin its execution.
16. The load function provided by the OS is implied because the name of the loader program is not typed on the command line.
17. There are three job steps to develop a software product, i.e., edit, assemble or compile, and link.
18. If programming errors are hard to locate, it is necessary to execute a debugger.
19. All system software for the target machine can be tested under a computer simulator running on the host.
20. Data management services are also tools provided by the OS.
21. Physically, the disk space is divided into tracks and sectors, but logically, it is divided into boot record, file allocation tables, directories, and files.
22. A file system includes a directory, files, and I/O routines.
23. The directory is designed as an internal file through which the OS can find a file or another directory.

PROBLEMS

1. What are the attributes of a third generation computer?
2. What are the attributes of a fourth generation computer?
3. Exactly how many bytes are there in one KB, MB, GB, TB, or PB?
4. What are the attributes of a core memory?
5. What are the attributes of a semiconductor memory?
6. What is a disk file?
7. Name three input devices to a computer system.
8. Name five basic programming tools in a computer system.
9. What is a system prompt?
10. Name the two operating levels of a general purpose computer system.
11. What is a compiler?
12. Describe the actions taken by the OS after a user enters a command from the keyboard.
13. After seeing the system prompt, a user can request that the OS execute an executable program by typing in the name of the program on the keyboard. If the program is not in memory, the OS must load the program in first. Why is this loading function implicit to the OS?
14. Explain briefly how to develop the system software for a new machine before its hardware is developed.

15. What are the main functions of a computer simulator?
16. In your opinion, what should be classified as a fifth generation computer system?
17. On a PC, what are the four logical areas on a disk?
18. Each node in the file directory can be a file descriptor or a directory descriptor. What is the difference between the two?
19. Turn on the power to boot a window-based IBM PC from the hard drive. Pop the DOS window by clicking the icon sequence Start-Programs-Command Prompt as described below:
    - A. After moving the mouse, place the cursor on the Start icon at the lower left corner of the screen and click the left button on the mouse to open the Start menu.
    - B. After moving the cursor on to the Programs command in the menu, click again to see its menu.
    - C. After moving the cursor on the Command prompt in the menu, click again to see the DOS window pop up. Press the left mouse button to drag the mouse to any command in the menu, and release the button so as to execute the command.
    - D. The blinking cursor tells you to enter a command followed by a <cr> from the keyboard. Type edit /? to get help on edit (editor) or diskcopy /? to get help on diskcopy.
    - E. To terminate DOS, click the exit command or the X button on the upper right corner of the window. If the system crashes for whatever reason, you must reboot by turning the power off and back on.
    - F. To copy a disk on drive A, enter the command diskcopy a: a: Remember to always place the source disk on your right and the destination disk on your left so that you will not make mistakes after working long hours.
    - G. After the session, click the Start-Shutdown sequence to turn the machine off.

CHAPTER 2

Number Systems

2.1 BASIC MATHEMATICS

As all digital information is represented by binary digits, the mathematical backbone of designing a digital computer is Boolean algebra. Before learning instructions and data, it is necessary to know the different number systems, number conversions, and arithmetic operations performed by a computer. To accomplish this goal, we must review some basic mathematics and definitions.

2.1.1 Integer Part vs. Fraction Part

Definition 1: If X is a real number, positive or negative, [X] is defined to be the greatest integer $\leq$ X, where $\leq$ stands for less than or equal to.

For a positive integer X, after truncating the fraction part of the number, we obtain its integer part as the answer. A positive number in square brackets represents an integer but, for a negative number X we need to subtract one from the negative integer to get the correct result. The square bracket notation represents the floor function and some examples are given below:

1. [7.45] = 7
2. [3.14159] = 3
3. [3.00] = 3
4. [-5.6] = -6
5. [-6.00] = -6

Definition 2: If X is a positive number, {X} is defined to be the fraction part of X. Therefore, we obtain,

$$\{X\} = X - [X]$$

Therefore, a positive number in curly brackets denotes a fraction, and some examples are listed below:

$$
\begin{aligned}
1.\ \{7.45\} &= 7.45 - [7.45] \\
&= 7.45 - 7 \\
&= .45 \\
2.\ \{8.00\} &= .0
\end{aligned}
$$

2.1.2 Modulus Concept

Definition 3: Assuming that / is the arithmetic divide operator, and + is the add operator which has lower precedence than /, the Euclidean division algorithm for positive integers is shown below:

$$
\begin{aligned}
&A / B = Q + R / B \\
&\text{where } 0 \le R < B \\
&\quad Q = [A / B] \\
&\quad R / B = \{A / B\}
\end{aligned}
$$

Q is the quotient, and R is the remainder.

Definition 4: Integers A and B are said to be congruent of modulus N where N is an integer, if there exists an integer k, such that

$$A - B = k * N$$

where the - sign denotes the subtract operator, and * denotes the multiply operator. Symbolically, we have

$$A \equiv B \quad \text{mod } N$$

Note that either A, B, or both can be negative integers; and more examples are given below:

$$
\begin{aligned}
7 &\equiv -1 \quad \text{mod } 8 \\
7 &\equiv -1 \quad \text{mod } 4 \\
7 &\equiv -1 \quad \text{mod } 2 \\
3 &\equiv 0 \quad \text{mod } 3
\end{aligned}
$$

If the positive number N is equal to 2M where M is a positive integer, we have some very interesting examples as follows:

$$
\begin{aligned}
15 &\equiv -1 \quad \text{mod } 16\ (2^4) \\
16 &\equiv 0 \quad \text{mod } 16
\end{aligned}
$$

$$0 \equiv -16 \mod 16$$
$$8 \equiv -8 \mod 16$$

Given 16 bits or 32 bits, the integer representations of 15 and -1 in a computer are different. If we use twos complement notation to represent a negative number (as discussed later) the lower 4 bits (mod 2^4) of each of the two numbers are identical. Likewise, this is true for 16 and 0, 0 and -16, 8 and -8, etc.

A high-level programming language may support either a mod function or a mod operator. For example, in C, the mod operator is %, and the arithmetic expression (A % B) really means: dividing A by B, we obtain a remainder as the answer. We say that A and the remainder are congruent mod B.

2.2 POSITIONAL NOTATION

If B is a positive integer, then any integer A may be written in a positional notation as follows:

$$A = C_n * B^n + C_{n-1} * B^{n-1} + \ldots + C_1 * B^1 + C_0 * B^0 = \sum_{i=0}^{n} C_i * B^i$$

where Ci is the coefficient of the ith term under the condition that $0 \le C_i < B$, and B is referred to as the base, or radix. In a decimal system, the base B is 10. If a number is written as a decimal, then all the coefficients are readily visible from its notation. For example, the number 221 in decimal form has the following meaning:

$$221 = 2 * 10^2 + 2 * 10^1 + 1 * 10^0$$
$$= 2 * 100 + 2 * 10 + 1$$

Each digit in the number carries a weight, and the right-most digit carries a weight of one because any base to the power of zero is defined to be one. The second digit from the right carries a weight of 10, and the third digit carries a weight of 100.

2.3 NUMBER SYSTEMS

To generalize the concept, any positive integer of base B may be written in a positional notation abbreviated as

$$C_j\, C_{j-1} \ldots . C$$

If the base B is omitted in the above notation, the default base is assumed to be 10, unless otherwise specified. A number of base 10, is said to be in decimal form. In the above notation, C_0 is the least significant digit, and C_j is the leading non-zero digit. The number 221 in decimal form is different from a number 221 of base 8 or base 16 as shown below:

$$\begin{aligned}
221_{10} &= 2 * 10^2 + 2 * 10^1 + 1 * 10^0 = 221 \\
221_8 &= 2 * 8^2 + 2 * 8^1 + 1 * 8^0 = 145 \\
221_{16} &= 2 * 16^2 + 2 * 16^1 + 1 * 16^0 = 545
\end{aligned}$$

It is important to know that the positional notation is used to evaluate a positive integer of any base, and its decimal value serves as a reference point for number conversions.

2.3.1 Binary

The number system used in a computer is binary, or base two. In a binary number, each coefficient C_i in the notation carries a weight of 2^i. Any piece of information, machine instruction or data, is represented by a bit string of ones and zeroes. While many different data types are used in a computer, they are all bits; only the way they are used creates the difference between them. We may use four bits to represent an unsigned integer whose value is always positive; that means the four-bit number really represents five bits with the leading bit zero hidden as a positive sign. Since the total number of combinations is 16 (2^4), we can represent a total of 16 positive numbers ranging from 0 to 15. Each coefficient is indicated by a one or zero. Obviously, a four-bit zero represents zero. To represent one, logically we need to add 1 to 0000 to get 0001. To represent two, we add 1 to 0001. Since the binary digit can not be more than one, the sum bit becomes zero and a carry bit is added to the next bit as shown below:

```
     0001
+)      1
  -------
     0010
```

Given any binary number, we can use its positional notation to compute its value in decimal form. In the next example, the decimal value of a given binary number is computed.

$$\begin{aligned}
1001\ 0001_2 &= 1 * 2^7 + 1 * 2^4 + 1 * 2^0 \\
&= 128 + 16 + 1 \\
&= 145
\end{aligned}$$

In a simple CPU, an adder is used to perform arithmetic operations between two binary numbers A and B. The adder receives input A and input B and generates S, the arithmetic sum. A 16-bit adder means that the adder can add two operands of 16 bits each in one adder cycle, while a 32-bit adder can add two 32-bit operands in one adder cycle. The block diagram of a one-bit full adder is shown in Figure 2.1.

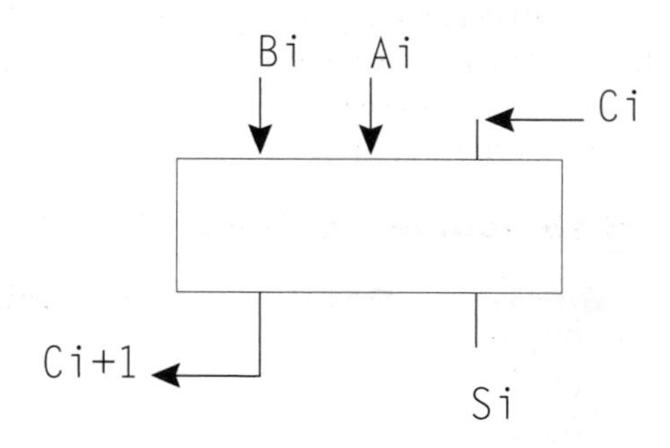

Figure 2.1 One-bit full adder.

The one-bit full adder has three inputs, Ai, Bi, and Ci, and two outputs, Si and Ci+1, as described below:

Ai is bit i in A.
Bi is bit i in B.
Ci is the carry input for bit i.
Si is the sum output for bit i.
Ci+1 is the carry output for bit i or the carry input for bit i+1.

A one-bit full adder can add three input bits, two data bits, and one carry input bit, so the result depends on the combination of all three bits. If an adder can add only two input bits, it is a half-adder. To get a simple idea, picture Si and Ci+1 as two output bits denoting the sum and carry after the add operation. Thus, S0 carries a weight of 2^0, and C1 carries a weight of 2^1. Together they constitute a two-bit unsigned integer. If all three input bits, A0, B0, and C0, are ones, the output should be 11 in binary form.

2.3.2 Octal

If the base is eight, the number is said to be in octal form. Octal digits range from 0 to 7, and there is no 8 or beyond. Because the base is different, a number in octal has a different value than the same digits in decimal. How do we say 221 in octal? The answer is: two, two, one octal that means two hundred, twenty, and one in octal. But the octal hundred has a different definition than the decimal hundred and its actual value is 64 in decimal. The positional notation is used to compute the decimal value of a number of any base, and the result provides a reference point for number conversions.

2.3.3 Hexadecimal

If the base is 16, the number system is in hexadecimal form or hex for abbreviation. In a hex number system, we need 16 different symbols to represent all the hex digits. Conveniently, we have 10 decimal digits at hand, so we need six more. One thought was to use the first six letters of the alphabet from A – F. A hex number A means 10, B means 11, C means 12, D means 13, E means 14, and F means 15. The first 16 numbers should be memorized as shown in Table 2.1.

Table 2.1 First 16 Numbers of Base 10, 2, 8, and 16

Decimal	Binary	Octal	Hexadecimal
0	0000	0	0
1	0001	1	1
2	0010	2	2
3	0011	3	3
4	0100	4	4
5	0101	5	5
6	0110	6	6
7	0111	7	7
8	1000	10	8
9	1001	11	9
10	1010	12	A
11	1011	13	B
12	1100	14	C
13	1101	15	D
14	1110	16	E
15	1111	17	F

Each four-bit unsigned integer has a value ranging from 0 to 15. Because 16 means hexadecimal, it is quite common to replace the base 16 with x (i.e., hex or hexadecimal). Therefore, 221_{16} can be written as 221x. It is simple to convert a binary number to octal or hex. A binary number is usually written in hex on modern computers because the bit pattern can be easily visualized as explained below.

2.4 NUMBER CONVERSION

After understanding the basic concept of positional notation, we can convert a positive integer A into a number of any base B. One method of conversion is the divide algorithm:

Given a number N;
REPEAT
Dividing N by B, we obtain a quotient and a remainder;
The remainder is the next digit of base B starting from
  the right-most end;
Set N to be the obtained quotient as the new number;

UNTIL the number N is 0;

This algorithm is explained in the below. Recall that the positive integer A of base B can be represented as

$$A = C_j * B^j + C_{j-1} * B^{j-1} + \ldots + C_0 * B^0$$

The positive integer A has a decimal value, even though it may be written in binary or any other base. Our challenge is to convert A to a number of any base B and determine all the coefficients in its new positional notation. In the case that the integer A is written in decimal form, we can easily perform the arithmetic divide operations in decimal. Thus, dividing A by B, we obtain both the quotient and the remainder as follows:

$$A \equiv C_0 \mod B$$
$$Q1 = [A / B] = C_j * B^{j-1} + C_{j-1} * B^{j-2} + \ldots + C_1$$
$$R1 / B = \{A / B\} = C_0 / B$$

where Q1 is the quotient, and R1 or C_0 is the remainder. Dividing the quotient Q1 again by B, we obtain the quotient Q2 and the remainder R2, or C_1. Repeat this process until the quotient is zero, all the remainders or coefficients, and place them from right to left in sequential order as follows:

$$C_j\, C_{j-1} \ldots C_2\, C_1\, C_{0\ B}$$

Let us convert 145_{10} to a number of base eight:

```
8 | 145     1 = C0
  |_____
8 |  18     2 = C1
  |_____
      2       = C2
```

After dividing, step by step, we obtain a total of three remainders as shown below:

$$145_{10} = 221_8$$

Interestingly enough, using the convert by divide algorithm we can convert a positive integer A in binary to a number of any base, change each of the coefficients from binary to a text character, and display them on the screen.

2.4.1 Convert a Number from Base B to Base B^n

If B is two and n is three or four, we have a special case of converting a binary number to octal or hex. Because 2^3 is 8 and 2^4 is 16, we can apply a special rule. As a matter of fact, it is quite easy to convert a positive integer among binary, octal, and hex. To generalize the case, we choose B as any base which is positive, and a positive integer A may be represented by the following positional notation:

$$A = C_j * B^j + C_{j-1} * B^{j-1} + \ldots. + C_1 * B^1 + C_0$$

Dividing A by B^n, we obtain a remainder

$$0 \leq C_{n-1} * B^{n-1} + \ldots. + C_1 * B^1 + C_0 < B^n$$

Now, starting from the rightmost digit, if we collect digit n in a group and evaluate its value based on its positional notation within the group, we obtain a digit of base B^n. This concept can be extended to the next n-digit group, until the leftmost end. Note that the leftmost group may have less than or equal to n digits as shown below:

$$\underbrace{C_j C_{j-1} \ldots C_{j-n+1}}_{\leq \text{ n digits}} \underbrace{C_{j-n} C_{j-n-1} \ldots C_{j-2n+1}}_{\text{n digits}} \ldots. \underbrace{C_{n-1} \ldots C_1 C_0}_{\text{n digits}}{}_B$$

Applying the positional notation within each group, we compute the value of the n digits in the group as follows:

$$(C_j * B^{n-1} + C_{j-1} * B^{n-2} + \ldots + C_{j-n+1}) \; (\ldots) \; (\ldots) \; \ldots$$
$$(\ldots) \; (C_{n-1} * B^{n-1} + C_1 * B^1 + C_0) \text{ of base } B^n$$

For example, we convert 145_{10} to binary, then to octal and hex.

2	145	$1 = C_0$
2	72	$0 = C_1$
2	36	$0 = C_2$
2	18	$0 = C_3$
2	9	$1 = C_4$
2	4	$0 = C_5$
2	2	$0 = C_6$
	1	$= C_7$

Therefore, we have

$$\begin{aligned} 145 &= 10\ 010\ 001_2 \\ &= 221_8 \\ &= 1001\ 0001_2 \\ &= 91_x \end{aligned}$$

Conversely, we can convert an hex number to binary, then to octal.

$$\begin{aligned} \text{CBD1x} &= 1100\ 1011\ 1101\ 0001_2 \\ &= 1\ 100\ 101\ 111\ 010\ 001_2 \\ &= 145721_8 \end{aligned}$$

To further understand the concept, we convert a number of base B to base B^n, and leave the positional notation inside each group without evaluation. For example, we convert a number from base 2 to 2^3 or 2^4, as shown below.

$$\begin{aligned} A &= 145_{10} \\ &= 10\ 010\ 001_2 \\ &= (1*2^7 + 0*2^6) + (0*2^5 + 1*2^4 + 0*2^3) + (0*2^2 + 0*2^1 + 1*2^0) \\ &= 221_8 \end{aligned}$$

We can apply the conversion by divide algorithm to verify the remainders. After dividing A by 2^3, we obtain

$$\begin{aligned} Q1 &= 1 * 2^4 + 0 * 2^3 + 0 * 2^2 + 1 * 2^1 + 0 * 2^0 \\ R1 &= 0 * 2^2 + 0 * 2^1 + 1 * 2^0 = C_0 \end{aligned}$$

Dividing Q1 again by 2^3, we obtain

$$\begin{aligned} Q2 &= 1 * 2^1 + 0 * 2^0 \\ R2 &= 0 * 2^2 + 1 * 2^1 + 0 * 2^0 = C_1 \end{aligned}$$

Dividing Q2 one more time, we obtain

$$\begin{aligned} Q3 &= 0 \\ R3 &= 1 * 2^1 + 0 * 2^0 = C_2 \end{aligned}$$

We further evaluate R1, R2, and R3 based on positional notation within the group as follows,

$$\begin{aligned} A &= 10\ 010\ 001_2 \\ &= C_2\, C_1\, C_{0\ 8} \\ &= 221_8 \end{aligned}$$

How do we convert a number from base seven to base eight? Recall that in the convert by divide algorithm, the dividend is a positive integer of any base. If we perform base seven arithmetic during the divide operation, we can find all the remainders of base eight directly. But base seven arithmetic is not intuitive, so the alternate way is to convert the number to decimal form first, and then divide the number by eight using decimal arithmetic and collect all the remainders.

$$\begin{aligned} 145_7 &= 1 * 7^2 + 4 * 7^1 + 5 * 7^0 \\ &= 49 + 28 + 5 \\ &= 82_{10} \\ &= 122_8 \\ &= 101\ 0010_2 \\ &= 52_x \end{aligned}$$

Finally, we want to convert a positive number from binary to base 10 and display the number on the screen. Follow the steps below:

1. Divide the number by 10, and collect the remainders. Even though the number is in binary form, the computer knows how to divide it by 10 using binary arithmetic.
2. After collecting all the remainders, we need to convert the decimal digits into external characters that can be displayed on the screen.

For example, let us print out the decimal integer 221 in memory. This integer is a binary number in the computer that may be the result of an arithmetic operation. After dividing the integer by 10, we obtain the first remainder one, and the quotient is 22. Dividing the quotient 22 by 10 again, we obtain a remainder two and the quotient is two. But the problem is that all the remainders collected after the divide are in binary and we need to convert them to external characters. Each character occupies one eight-bit byte and we elaborate the discussions on internal and external data in the next section.

2.5 DATA REPRESENTATION

As memory is usually divided into bytes, each byte is associated with an address. Because the address is in binary form, the total number of bits in an address also determines the size of memory. If the memory size is 4 GB, the total number of bits in its address must be 32. Thus, we have a total of 4 giga (2^{32}) combinations. The lowest address is 0000 0000 in hex and the highest address is FFFF FFFF (2^{32} - 1). To obtain this largest value intuitively, we can add one to it to make 2^{32} as seen by its positional notation. An address is needed to access data in memory and any piece of data in memory occupies one or more bytes.

There are two general data types for internal and external use. The internal data are for computations inside a computer, and the external data are mainly used for communicating with the outside world. In the following sections, we discuss the representations of positive integers, negative integers, characters, floating point numbers, and packed decimal numbers.

2.5.1 Positive Integers

An integer in memory has 16, 32, or more bits. An integer has four bytes while a short integer has two bytes as shown in Figure 2.2. The leftmost bit, or b15 (bit 15), indicates a sign and the least significant bit (LSB) is denoted as b0 (bit 0) or the rightmost bit. A sign bit zero means positive while a sign bit one means negative. Looking at the sign, we can tell whether the integer is positive or negative. In addition, the largest positive 16-bit integer is a zero followed by 15 ones representing $2^{15} - 1$.

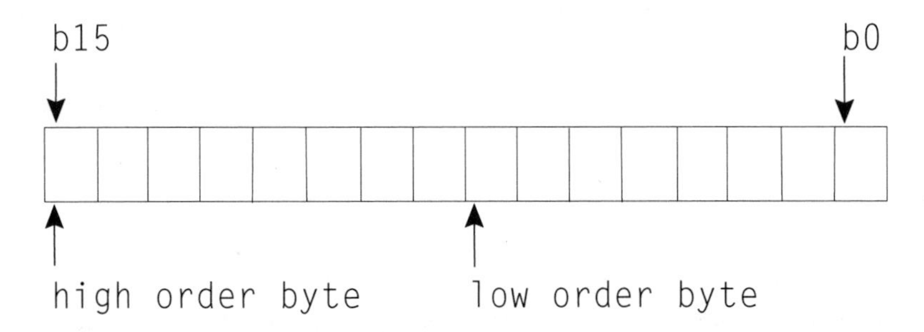

Figure 2.2 A 16-bit short integer.

If a 16-bit integer in memory has two bytes, it occupies two consecutive addresses, one for each byte. In a microcomputer, a memory word usually means two bytes but in a mainframe a word means four bytes. The low order byte in a word means the lower eight bits, while the high order byte means the upper eight bits. A 32-bit integer in memory occupies four bytes B3, B2, B1, and B0, as shown in Figure 2.3a. Byte 3 (B3) is the high order byte in the high order word and B0 is the low order byte in the low order word. There are two byte ordering methods in memory: the little endian computer and the big endian computer. In a little endian system, the low order word is stored before the high order word, and within the word, the low order byte is stored first as shown in Figure 2.3b. To store in memory first means that the low order byte has a lower address. The big endian system has the opposite order, as shown in Figure 2.3c. That is, the high order word is stored first followed by the low order word, and within the word, the high order byte is stored first. Note that a 32-bit positive integer has the largest value of 2^{31}-1, that is a sign bit zero followed by 31-bit ones. In order to get a larger value in an integer representation, we need more bits.

The next issue is how to represent a negative integer.

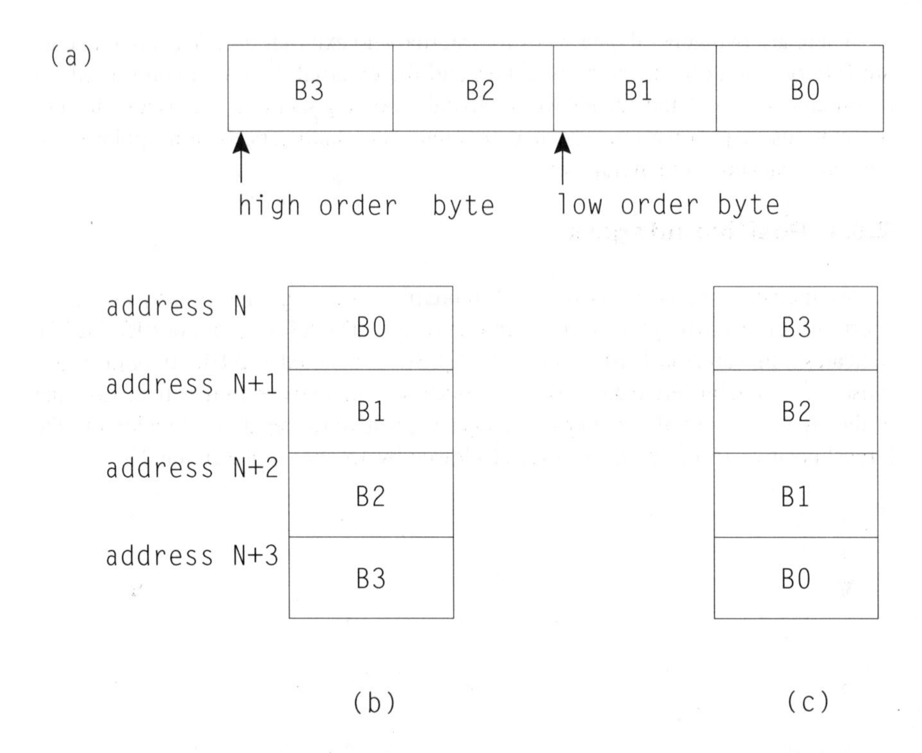

Figure 2.3 A 32-bit integer: (a) format, (b) byte ordering in the memory of a little endian system, and (c) big endian system.

2.5.2 Negative Integers

The evolution of representing negative integers in a computer was quite interesting. At first we chose a signed magnitude notation which was intuitive to humans, but awkward for hardware. Later, we switched to twos complement notation, which is good for machines but not easy for human users. For pedagogical reasons, we discuss signed magnitude, ones complement, and twos complement as follows.

2.5.2.1 Signed Magnitude

The signed magnitude format is shown in Figure 2.4. With 16 bits, a negative integer has a sign bit one followed by a 15-bit magnitude. To represent a positive value, we change the sign bit to zero. Therefore, in such a system, the sign bit provides the only difference. For example, the short integers +5 and -5 have the same magnitude, but their sign bits are different as shown below.

+5 0000 0000 0000 0101
-5 1000 0000 0000 0101

The spacing is used to separate the binary number into four-bit groups for easy comparisons. The signed magnitude notation was phased out after the introduction of second generation computers. Nonetheless, the concept itself has merit, and it is still used to represent a floating number or a packed decimal number.

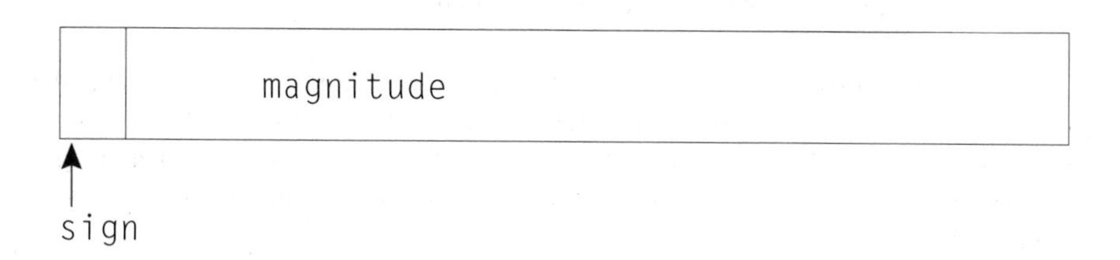

Figure 2.4 Signed magnitude format.

2.5.2.2 Ones Complement

We study the ones complement notation as a concept that was used by some computers.[5] To negate a number, we simply complement every bit in the number, positive or negative. The 16-bit +5 and -5 are compared below.

```
+5    0000 0000 0000 0101
-5    1111 1111 1111 1010
```

Interestingly, the ones complement notation has two drawbacks. First, there exist two different zeroes, a +0 and a -0. A +0 contains all zeroes, and a -0 contains all ones and both zeroes are valid. As a consequence, it requires extra hardware to convert a -0 to a +0 after an arithmetic operation. Second, it has the end-around carry problem. Given two integers A and B, the operation of (A - B) means adding A to the ones complement of B. The result is perfect if no carry is generated at the leftmost end. In the case that an end-around carry is generated, hardware logic must pull the carry bit back to the LSB position and add it again to the result. That is to say, it takes two add cycles to accomplish the operation. To further understand the concept, we subtract four from five, i.e., add -4 to +5. The sum contains all zeroes but an end-around carry is generated at the leftmost end. The hardware logic pulls the end-around carry bit to the right, and adds it one more time to make the result perfect.

```
     +5    0000 0000 0000 0101
+    -4    1111 1111 1111 1011
          ---------------------
         1 0000 0000 0000 0000
                             1 ← end-around carry
          ---------------------
     +1    0000 0000 0000 0001
```

2.5.2.3 Twos Complement

A negative integer in twos complement notation means taking ones complement first and adding one to it as shown below:

twos complement = ones complement + 1

The twos complement notation has two advantages. First, it does not have the end-around carry problem so it takes one add cycle to accomplish an add or subtract operation. Second, it has a unique representation of zero: +0 or -0. Given an integer zero of 16 or 32 bits, after complementing all zeroes and adding one to it, we still obtain zero.

Using signed magnitude, we negate an integer by flipping its sign bit. Using ones complement, we negate an integer by flipping all the bits in it. Negating a number in an easy way is essential in computer design. Can we negate an integer easily using twos complement notation? The answer is yes. By taking the twos complement of a positive integer, we obtain the negative integer. Taking the twos complement of a negative integer, we obtain the positive integer. Henceforth, to take the twos complement of an integer means to negate as shown in Figure 2.5. Logically, we tend to think that to convert a number from negative to positive we first subtract one, and then take the ones complement. However, this is the same as taking the ones complement first and then **adding** one.

Subtract one & take one's complement = Take one's complement & add one

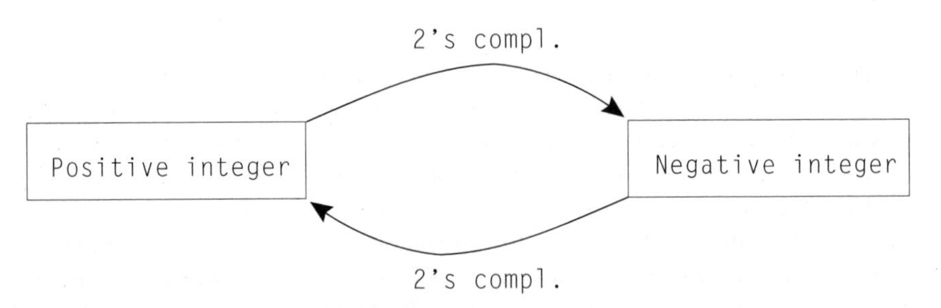

Figure 2.5 Take twos complement means to negate.

Using twos complement notation, we use one add cycle to subtract a number via the adder. This is possible if a one-bit full adder is used at the LSB position. As shown in Figure 2.6, a 16-bit carry rippled adder has 16 1-bit full adders provided that the carry output is the carry input at the next bit position. We can use the adder to subtract B from A where A and B are two integers, positive or negative. The trick is to feed the ones complement of B to one input of the adder and feed A to the other input. At the same time, we turn on the Cin (carry in) bit to add the twos complement of B to A. The Cin is denoted as C0 in the figure. In the case that A is chosen

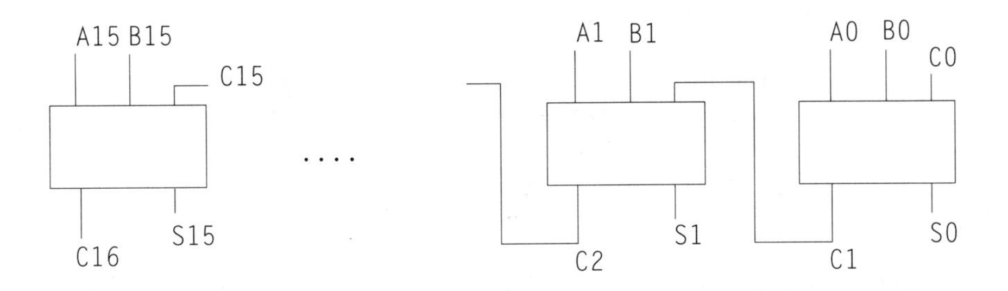

Figure 2.6 A 16-bit carry rippled adder.

to be zero, -B becomes the negated output. If there is an end-around carry generated, we simply discard the carry.

Before we perform examples of twos complement arithmetic, let us determine the bit patterns for +25, -25, +26, and -26. Using the convert by divide algorithm, we first convert each number to binary form. If the number is negative, we take the twos complement, that is, take the ones complement first and add one as follows.

```
      +25   0000 0000 0001 1001
            1111 1111 1110 0110   <- ones complement
+                             1
      ---------------------------
      -25   1111 1111 1110 0111   <- twos complement
```

```
      +26   0000 0000 0001 1010
            1111 1111 1110 0101   <- ones complement
+                             1
      ---------------------------
      -26   1111 1111 1110 0110   <- twos complement
```

A 16-bit adder can add two 16-bit numbers. Because it is easy to convert a number from binary to hex, we show the following two examples with all additions performed in decimal, binary, and hex.

Example 1. 26 - 26 = 0.

```
     Decimal      Binary                  Hex
      +26         0000 0000 0001 1010     001A
+     -26         1111 1111 1110 0110     FFE6
     -------      ---------------------   ----
        0         1 0000 0000 0000 0000   0000
                  ^
                  carry discarded
```

Example 2: 25 - 26 = -1.

	Decimal	Binary	Hex
	+25	0000 0000 0001 1001	0019
+	-26	1111 1111 1110 0110	FFE6
	-1	1111 1111 1111 1111	FFFF

Remember that a -1 in computer language has all binary ones, regardless of whether the integer length is 16 bits or 32 bits.

A third example clarifies the mod concept. Assume that a four-bit machine performs four-bit arithmetic using the twos complement notation to represent a negative number. That is, a signed integer in 4 bits. There are 16 (2^4) combinations for both positive and negative integers. The positive integers range from 0 to 7 with a sign bit zero, and the negative integers range from -1 to -8 with a sign bit one. What happens if we subtract 1 from -8 or add -1 to -8? We say that the number is busted because its sign changes and the result is +7 ,which is incorrect as proven below:

```
    -8      1000
 +  -1      1111
 ------   -------
     -     1 0111
           ↑
           carry discarded
```

Recall that -9 and +7 are congruent mod 16 from the mathematical point of view. In practice, if we examine the lower four bits of -9 and +7 represented in 16 bits or 32 bits using the twos complement, they look the same.

2.5.2.4 Serial Algorithm for Generating Twos Complement

A serial circuit may be designed to convert a binary number to its twos complement form assuming the bit stream arrives, bit by bit, in a serial manner starting from the LSB. As each bit arrives, we can convert the bit right away based on the algorithm provided below:

```
loop1:
IF the leading bit is 0,
THEN DO;
   Copy bit 0 down;
   Goto loop1; ENDDO;
ELSE copy bit 1 down; ENDIF;
loop2:
REPEAT
Complement all the subsequent bits arrived;
UNTIL the entire bit string is received;
```

This algorithm converts a binary number starting from the LSB. Search from right to left. If the bit is zero, we copy it down until we encounter the first bit which is not zero. Then copy the first bit one down and complement the rest of the bits. Note that this method works both ways, from positive to negative and vice versa.

We can apply this method to negate a number written in hex. As we search from the rightmost digit towards the left, copy down any hex digit zero, until we encounter the first digit that is not zero; subtract its value from 16 and jot down the answer. Subtracting the remaining digits, one by one, from 15, and jot down the answer until all the digits are converted. Some examples of hex number conversion are given below for practice.

16-bit number in hex	Twos complement in hex
0000	0000
FFFF	0001
0777	F889
0019	FFE7

To negate a 32-bit number, take its twos complement and add one to it. Therefore, given 32 bits, we actually divide all the integers into two groups: the positive number group with a sign bit zero, and the negative number group with a sign bit one. With 32 bits, the most negative integer in twos complement is -2^{31}.

2.5.3 Characters

In order for one computer to communicate with a keyboard, display, printer, or another computer in a network, it is necessary to establish an external code standard. This is, after we press a key on the keyboard, bits flow into the computer, and a system routine translates the bits into ASCII (American standard code for information interchange), pronounced as as-key. Each character is an ordered set of seven bits. Because the basic addressable unit in memory is a byte of eight bits, we place an ASCII character in a byte and clear its leftmost bit. Note that ASCII and the international alphabet set number 5 (IA5) are the same as proposed by the ITU-T (International Telecommunications Union-Telecommunication Standard Sector). The seven-bit ASCII character set is shown in Figure 2.7a where b0 denotes the LSB. The bit pattern of each code is specified by a two-dimensional matrix as shown in Figure 2.7b. The upper three bits, b6, b5, and b4 constitute a hex number ranging from 0 to 7 which is the column number and the lower four bits, b3, b2, b1, and b0, constitute a hex number ranging from 0 to F which is the row number. The entire set of ASCII is placed in a 16 x 8 matrix with 128 combinations. To obtain the seven-bit code of a character in the matrix, locate the character symbol first in the matrix, and concatenate its three-bit column number with its four-bit row number.

(a)

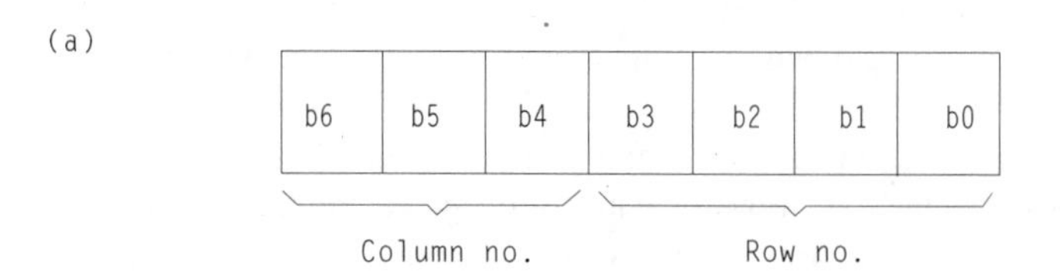

(b)

			b6		0	0	0	0	1	1	1	1
			b5		0	0	1	1	0	0	1	1
			b4		0	1	0	1	0	1	0	1
b3	b2	b1	b0		0	1	2	3	4	5	6	7
0	0	0	0	0	NUL	DLE	SP	0	@	P	`	p
0	0	0	1	1	SOH	DC1	!	1	A	Q	a	q
0	0	1	0	2	STX	DC2	“	2	B	R	b	r
0	0	1	1	3	ETX	DC3	#	3	C	S	c	s
0	1	0	0	4	EOT	DC4	$	4	D	T	d	t
0	1	0	1	5	ENQ	NAK	%	5	E	U	e	u
0	1	1	0	6	ACK	SYN	&	6	F	V	f	v
0	1	1	1	7	BEL	ETB	'	7	G	W	g	w
1	0	0	0	8	BS	CAN	(	8	H	X	h	x
1	0	0	1	9	HT	EM	)	9	I	Y	i	y
1	0	1	0	A	LF	SUB	*	:	J	Z	j	z
1	0	1	1	B	VT	ESC	+	;	K	[	k	{
1	1	0	0	C	FF	FS	,	<	L	\	l	|
1	1	0	1	D	CR	GS	-	=	M	]	m	}
1	1	1	0	E	SO	RS	.	>	N	^	n	~
1	1	1	1	F	S1	US	/	?	O	_	o	DEL

Figure 2.7 ASCII: (a) 7-bit format and (b) specifications.

For example, to find zero in the matrix, its column number is three and its row number is zero, so its code value is 30 in hex. Intuitively, the character nine should be 39 in hex. The alphabets are arranged similarly so that the first A has a value of 41 in hex. To find Z, the last in a set of 26 alphabetical characters, we add 25 in decimal or 19 in hex to the value of A. Consequently, the value of Z is 5A in hex. The arithmetic operations in binary, hexadecimal, and decimal are shown below:

Alphabet	Binary	Hex	Decimal
A	0100 0001	41	65
	+ 0001 1001	19	25
Z	0101 1010	5A	90

To obtain the value of a, we simply add 20 in hex to A. That is to say, a has a value of 61 in hex, and z is 7A. In ASCII, a code from 00 to 1F in hex are mainly designed for control purposes as explained below:

Code 00 - 0F	Code 10 - 1F
NUL (null)	DLE (data link escape)
SOH (start of header)	DC1 (device control 1)
STX (start of text)	DC2 (device control 2)
ETX (end of text)	DC3 (device control 3)
EOT (end of transmission)	DC4 (device control 4)
ENQ (enquire)	NAK (negative acknowledgement)
ACK (acknowledgement)	SYN (synchronization)
BEL (bell)	ETB (end of block)
BS (back space)	CAN (cancel)
HT (horizontal tabulation)	EM (end of medium)
LF (line feed)	SUB (substitute)
VT (vertical tabulation)	ESC (escape)
FF (form feed)	FS (file separator)
CR (carriage return)	GS (group separator)
SO (shift out)	RS (record separator)
SI (shift in)	US (unit separator)

The rest of the code is self explanatory, except for SP (space) and DEL (delete). The ordering of each character in the ASCII set is called the collating sequence. In telecommunications, we sometimes transmit an escape character followed by another character to make a 16-bit special code known as the escape sequence. Some systems may just use the ESC character, while others use the DLE (data link escape) character instead, but the basic concept is to change an eight-bit code to a 16-bit code.

There are too many PCs in the world, and they all use ASCII, the de facto standard. If any piece of information is coded in ASCII, it is said to be in textual form. There are other types of external codes for information interchange. As an example, the IBM mainframes use the eight-bit EBCDIC (extended binary coded decimal interchange code), pronounced E-B-C-dick. The automatic teller machines use the six-bit transcode. Others use 16 bits or 24 bits, as required to transmit Chinese characters.

2.5.4 Floating Point Numbers

In scientific applications we have real numbers, each of which has an integer part and a fraction part separated by a radix point. The radix point is equivalent to a decimal point in a decimal system or a binary point in a binary system. In practice, the radix point does not need to be fixed. In fact, it can be floated anywhere in the number, as long as we can adjust its exponent properly. We use the positional notation to evaluate the value of its integer part as well as its fraction part. For the sake of simplicity, we assume that number A has only a fraction part and its positional notation is shown below:

$$A = .C_{-1} * B^{-1} + C_{-2} * B^{-2} + \ldots + C_{-n} * B^{-n} = \sum_{i=-1}^{-n} C_i * B^i$$

Each coefficient in the notation has a negative subscript, and its power of base B is also negative. In a short notation, we write down a radix point, followed by the coefficients in the positional notation, followed by a base, or radix. Several examples are given below:

$$.11_{10} = 1 * 10^{-1} + 1 * 10^{-2}$$
$$.11_2 = 1 * 2^{-1} + 1 * 2^{-2}$$

Therefore, .11 in decimal tells us that the first coefficient to the right of the decimal point carries a weight of one tenth (10^{-1}), and the next digit to its right carries a weight of one hundredth (10^{-2}). The .11 of base 10 may represent .11 dollars, or 11 cents. In contrast, .11 in binary tells us that the bit to the right of the binary point carries a weight of one half (2^{-1}), and the next bit to its right carries a weight of a quarter (2^{-2}). Therefore, .11 in binary means 75 cents. Given a finite number of bits, some decimal fractions can not be represented perfectly in binary just as 1/3 can not be represented perfectly in decimal for the answer is .3333333.... However, the number 1/3 has a perfect representation .1 of base 3. We can further show that the decimal .1 also needs an infinite number of bits to represent its value in binary.

Let us convert the decimal .1 to binary using the positional notation. Recall that the convert by divide algorithm converts a positive integer from one base to another. Converting the fraction part from one base to another, we can still use the convert by divide algorithm except for two things. First, the divisor is 2^{-1} instead of 2^1. Second, after dividing, we obtain a quotient which is a real number consisting of two parts: an integer and a fraction. Because the divisor is a fraction, a real number divide needs to performed and the integer part is the next bit after the binary point. Since the fraction part represents the value of the remaining bits after the binary point, it needs further division. Keep dividing the fraction and collect the integer part of the quotient, bit by bit, until the fraction reaches zero. Each bit obtained after the divide operation is laid from left to right after its binary point to comprise the answer. It is interesting to see that dividing the fraction by 2^{-1} really means multiplying it by 2^1 and the product is a real number consisting of an integer part and a fraction part. The integer part can be a one or zero. Therefore, to convert a fraction from base 10 to base 2, we modify the conversion by divide algorithm into conversion by multiply algorithm.

2.5.4.1 Conversion by Multiply Algorithm

REPEAT;

Multiplying the fraction by two, we obtain a real number consisting of an integer part and a fraction part;

Take the integer part as the next bit and the remaining fraction part becomes the fraction for the next iteration;

UNTIL the fraction is 0;

Using the conversion by multiply algorithm, we obtain

```
                 . ← binary point
2 * .1 =  .2     0
2 * .2 =  .4     0
2 * .4 =  .8     0
2 * .8 = 1.6     1
2 * .6 = 1.2     1
2 * .2 =  .4     0
2 * .4 =  .8     0
2 * .8 = 1.6     1
etc.
```

We have an infinite loop because the bit pattern 1001 is repeated forever. In other words, it takes an infinite number of bits to convert the decimal .1 to a perfect binary as shown below:

$$.1_{10} = .0001100110011001\ldots_2$$

In the convert by multiply algorithm, the selected base can be 16 or any positive integer. To convert a fraction from decimal to hex, we multiply the fraction by 16, and the integer part is the first hex digit after the radix point. Multiplying the fraction part repeatedly, we obtain a 16-digit number, .1999 9999 9999 999A in hex. The last digit is changed from 9 to A due to rounding. This is because after looping 15 times, we still have a fraction of .6 left, which is greater than 50%. As a result of this, we add one to the last digit so its value is changed from 9 to A.

In scientific applications, a real number is represented in floating point format. After an arithmetic operation, the binary point can float at any bit position so the number also contains an exponent part. The computer either employs a hardware floating point unit to support floating point arithmetic or it has special software to simulate such operations.

2.5.4.2 Floating Point Format

The basic idea is to divide the floating point number into three fields: overall sign, biased exponent, and significand. The format is shown in Figure 2.8.
Most microcomputers use a biased exponent of base two. Other computers may use

Figure 2.8 Floating point format.

a different base, e.g., 16, for its biased exponent defined as the characteristic, and in such systems the significand is known as mantissa.[26,75] The total number of bits in the significand field determines the precision of a number. Assuming that a biased exponent base two is used, if we shift the significand one bit to the left, the value of the number is multiplied by two. If we shift the significand one bit to the right, its value is divided by two. However, if we shift the significand one bit to the left and decrease its exponent by one at the same time, we retain the same value of the number. By the same token, if we shift the significand one bit to the right and increase its exponent by one, the number remains unchanged. Thus, a real number may be represented in many different ways in floating point format because the position of its radix point may change after each operation. A simple example is shown below:

$$\begin{aligned} .5_{10} &= .10 * 2\ _2 \\ &= 1.0 * 2^{-1}\ _2 \\ &= .01 * 2^{+1}\ _2 \end{aligned}$$

The decimal .5 has many different representations in floating point format. In this particular case, a significand of few bits will be adequate to represent the number perfectly. We can shift the significand one bit to the left and decrease its exponent by one or shift the significand one bit to the right and increase its exponent by one. However, there is a unique way to represent a real number in normalized form. That is, the number is represented by an integer part and a fraction part, provided that the integer part is always a one. If this one bit is not stored in the format, the real number gains one extra bit in precision. One exception should be mentioned: a real zero contains all zeroes in its format.

How many bits are needed to represent a floating point number? The answer really depends on the type of applications. Sometimes we can not represent a real number perfectly in binary, therefore, the more bits the better. The Pentium processor supports the IEEE 754 standard.[28] That is, the 32-bit single precision, the 64-bit double precision, and the 80-bit extended precision. For each format, there is one overall sign bit to indicate whether the number is positive or negative. The remaining bits constitute the biased exponent and the significand, which all have different lengths as listed in Table 2.2.

Table 2.2 Attributes of a Floating Point Number

Attribute	Single	Double	Extended
Total length (bit)	32	64	80
Sign	1	1	1
Biased exponent	8	11	15
Significand	24	53	64
Max exponent (value)	+127	+1023	+16383
Min exponent	-126	-1022	-16382

The sign bit is zero for a positive number and one for a negative number. Given two floating point numbers, one is positive and the other one is negative with identical magnitudes; their sign bit is the only differentiating factor. It also means that the signed magnitude concept is applied here to represent a floating point number.

The design of an exponent is intricate. First, it is biased by adding a positive number, which is a zero followed by all ones and the total number of bits is the same as the exponent field. In single precision, the exponent field is eight bits long, and the bias is +127 or 01111111 in binary. Such a biased exponent is said to be in excess of +127.

Two questions are answered below:

1. Why is the exponent biased?

   If we represent the exponent part, positive or negative, in twos complement notation, the CPU hardware would have a hard time to comparing two numbers and determining which is bigger. Adding a bias to the exponent in twos complement means that we rearrange all the exponents in ascending order from the most negative to the most positive; all exponents are treated as unsigned integers. During arithmetic operations, addition or subtraction, we can not extract the siginificands of two numbers and perform the operation unless their exponents are equal. For this reason, it is necessary for the CPU to compare the biased exponents of the two normalized numbers as unsigned numbers and determine which is smaller. Then, the CPU hardware shifts the significand of the floating point number with a smaller exponent to the right and increases its exponent accordingly until its value reaches the same level as the larger exponent.

2. Why add +127, instead of the unsigned 128 as we did in third generation computers?

   If the exponent field is 8 bits long, flipping its leading bit means that we add an unsigned bias of 128 (2^8). The biased exponents are still in ascending order, but the most negative one is -128 and the most positive one becomes +127 as shown in Table 2.3.

Table 2.3 An eight-bit Exponent in Excess of Unsigned 128

Decimal value	Biased exponent
+127	1111 1111
0	1000 0000
-127	0000 0001
-128	0000 0000

The advantage of adding +127 is the ability to represent one more positive exponent at the high end rather than one more negative exponent at the low end. In scientific applications, one extra large positive exponent is indeed more meaningful than a negative one as shown in Table 2.4.

Table 2.4 An eight-bit Exponent in Excess of +127

Decimal value	Biased exponent	
+128	1111 1111	(overflow)
0	0111 1111	
126	0000 0001	
-127	0000 0000	(underflow)

Furthermore, the most positive exponent +128 is reserved for overflow detection and the most negative exponent -127 is reserved for underflow detection. An overflow condition indicates that the exponent of the result after a floating point operation is greater than +128 and can not be represented by the machine. An underflow condition indicates that the exponent of the result is less than -127 and again the machine just can not handle it. Another case is after subtraction, when we normalize the result by shifting its siginificand to the left and decreasing the exponent field. If the value of its biased exponent reaches all zeroes, we stop and declare underflow. Thus, given an eight-bit exponent we have all the valid exponents from -126 to +127 as shown in Table 2.2.

Also in the table the single precision significand is 24 bits long and the double precision is 53 bits, but the extended precision is 64 bits. This is because for a single or double precision number, we always normalize its significand to make sure that the number is in the form of one followed by a fraction as shown below:

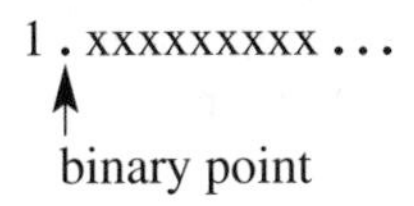

where x represents a bit in the fraction, and the leading bit one before the binary point is not stored in the significand. That is, the integer bit one is implicit, or hidden (so to speak) in the representation. Even though the integer bit one is not in the siginificand, the hardware knows of its existence. The reason for doing this is to achieve one extra bit accuracy. However, for extended precision, we have the integer bit explicitly specified in the siginificand.

The exponent field in a single precision format really contains a special code that is a twos complement notation in excess of 127, that is, uses twos complement notation to represent any exponent if it is negative and then add 127 to it. This particular excess chosen is not totally without debate. For example, if we multiply two very large positive numbers, hardware will handle the arithmetics for both the exponent and significand. As far as the exponent is concerned, it is necessary to add the two individual exponents together, but the result has one extra excess and it is necessary to subtract the extra excess from the result via another add cycle. This extra add cycle will not hurt the speed because both exponent and significand arithmetics can be performed in parallel, but nevertheless adds complexity. Detecting overflow and underflow conditions is quite complex.

Example 1: Convert the real number .1 in decimal form to single precision format.

Step 1: Find the 24-bit significand in the normalized number as shown below:

$$1.100\ 1100\ 1100\ 1100\ 1100\ 1101 * 2^{-4}$$

(binary point ↑ after the leading 1; exponent ↑ at 2^{-4})

Note that the spacing is for clarification purpose only, and the LSB is changed to one due to rounding. That is, after truncating the first 23 bits after the binary point, if the next bit in the remaining bit string is one, we add this one to the LSB of the significand as shown below.

```
                     the next bit
        ....  1100 ↓100  ....
  +              1           ←  rounding
  ---------------------
        ....  1101
```

Because the bit one before the binary point is omitted in the notation, we have the next 23 bits after the binary point as the significand. Note that the leading bit of the third hex digit is the LSB of its biased exponent.

Step 2: Find the exponent -4 in excess of +127.

	-4 in twos complement	1111 1100
+	127	0111 1111
	-4 in excess of 127 (123)	0111 1011

Step 3: By concatenating the sign bit zero, the eight-bit biased exponent, and the 23-bit significand, we obtain:

0 011 1101 1 100 1100 1100 1100 1100 1101

3 D C C C C C D (hex)

Example 2: Convert -.1 in decimal to single precision floating format.

Since the only difference is the sign bit, the answer is BDCC CCCD in hex.

Example 3: Convert .1 in decimal to double precision.

Recall that in Table 2.2, a double precision floating point number is 64 bits. Its exponent is 11 bits and the significand is 52 bits without showing the integer bit one before the binary point. Because the total number of bits of the sign and the exponent is 12 which is a multiple of 4, the 52-bit significand is shown as 13 4-bit groups as follows,

1001 1001 1001 1001 1001 1001 1001 1001 1001 1001 1001 1001 1010

that is, 9 9999 9999 999A in hex, where rounding is performed at the LSB. Next, the 11-bit exponent -4 in excess of 1023 is computed as

	-4 in two complement	111 1111 1100
+	1023	011 1111 1111
	-4 in excess of 1023 (1019)	011 1111 1011

which is 3FB in hex. Concatenating the sign bit zero, the biased exponent, and the significand, we obtain the 64-bit floating point number 3FB9 9999 9999 999A in hex.

Example 4: Convert the real number 178.125 in decimal form to single precision.

This real number has an integer part and a fraction part. We use the conversion by divide algorithm to collect the integer part and the conversion by multiply algorithm to collect the fraction part, so the obtained 24 significant bits are shown below:

1011 0010 . 0010 0000 0000 0000

The binary point is after the first eight bits. To normalize the number, we shift the bit string seven places to the right and increase the exponent by seven. Therefore, we have

$$1\,.\,011\ 0010\ 0010\ 0000\ 0000\ 0000 * 2^{+7}$$

The exponent +7 in excess of 127 is computed as,

+7	0000 0111
+ 127	0111 1111
+7 in excess of 127 (134)	1000 0110

Concatenating the sign bit zero, the biased exponent, and the significand, we obtain the 32-bit number

0011 0011 0010 0010 0000 0000 0000

which is 4332 2000 in hex. Note that in hardware design, there is no binary point, which is a concept built in the floating point arithmetic unit. In practice, the real number +0.0 contains all zeroes, regardless of its length in bits.

Recall that floating point arithmetic is ideal for scientific applications, and the significand length determines how many digits will be accurate in the result. For example, with 80-bit extended precision the significand is 64 bits so as to achieve an accuracy of 19 decimal digits as shown below:

$$2^{64} = 16 * (1024)^6$$

2.5.5 Packed Decimal Numbers

The packed decimal number concept was implemented on IBM mainframes for the purpose of business transactions. A typical number comprises a sign field, followed by 15 BCD (binary coded decimal) digits, and each digit has four bits ranging from 0 to 9. The only difference between a positive number and a negative number with identical magnitude is the sign. With 15 digits, we may use 13 digits to represent the dollar amount, and the last two digits represent dimes and pennies. Hence, a packed decimal number has a fixed decimal point in the digits, and after an arithmetic operation, the position of its decimal point is still fixed. A decimal number may be represented in two forms, packed and unpacked, as described below.

2.5.5.1 Packed Decimal Format

A packed decimal number of 18 digits on a Pentium processor is shown in Figure 2.9. The leftmost byte is the sign byte with its leading bit denoting the sign: zero means positive and one means negative. The sign bit is followed by seven trailing zeroes which are not part of the number. The sign byte is followed by 18 BCD digits, so the total number of bits is 80. Among the 18 digits, D17 is the leading digit, and the rightmost digit D0 is the least significant digit.

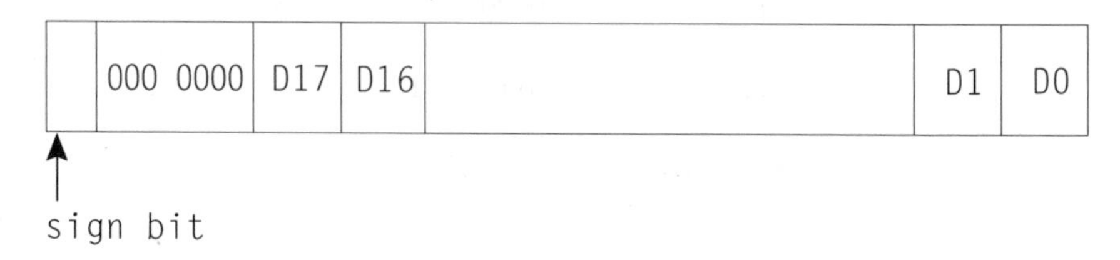

Figure 2.9 The 80-bit packed decimal format.

Packed BCD digits have applications in telecommunications. During the dialing phase, packed BCD digits are transmitted on the line as a telephone address, and their formats are conformed to an international standard.[80]

To support packed decimal numbers, the processor needs a special hardware unit to perform the BCD arithmetic.[43,55] Alternately, we may write software to convert the number to a 64-bit integer or an 80-bit floating point number with a 64-bit significand, perform the arithmetic, and convert the number back. A BCD arithmetic unit may be desirable in a pocket calculator. A BCD number is more readable than a binary number, but its hardware implementation is also more complicated. For example, when adding two BCD digits, we need to add six to the result if it is greater than nine as described below.

a. Algorithm to Add Two BCD Digits

IF there is a carry generated or the result is greater than 9,
THEN add 6 to the result;
ENDIF;

Example 1: Add nine and nine, and the sum digit should be eight.

```
     9   1001
+    9   1001
-------------
       1 0010
+        0110
-------------
     8   1000
```

Example 2: Add nine and three, and the sum digit should be two.

```
     9   1001
+    3   0011
-------------
         1100
+        0110
-------------
     2 1 0010
```

2.5.5.2 Unpacked Decimal Format

The unpacked BCD number is used as an intermediate step to print out packed digits in ASCII. Therefore, each unpacked digit occupies one byte, or eight bits and the leading four bits are all zeroes. Thus, the decimal number 215 in unpacked form is shown in Figure 2.10a. The leftmost byte contains the digit two, the next byte contains one, the least significant byte contains five, and each of the bytes has four leading zeroes. After obtaining the BCD digit in the unpacked form, we add a bias to it to make an ASCII character. According to the collating sequence in ASCII, the bias is 30 in hex. Therefore, after adding 30 we obtain the decimal number in ASCII characters or text as shown in Figure 2.10b.

2.5.6 Fixed Point Numbers

All the numbers in a computer can be grouped into three categories: fixed point, floating point, and characters. The integers and packed decimal numbers all belong

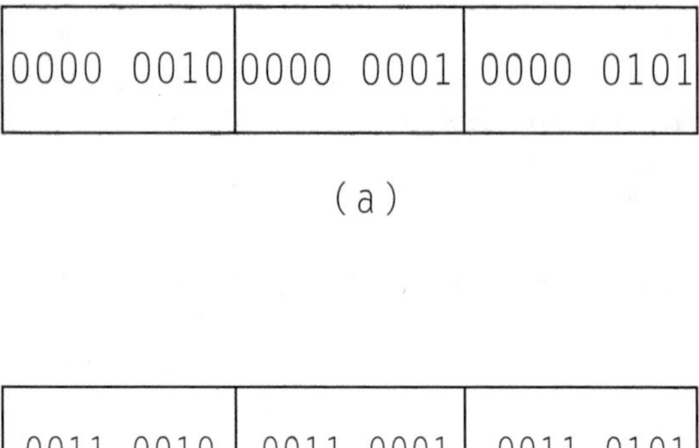

Figure 2.10 Unpacked decimal: (a) 215 in unpacked decimal and (b) 215 in ASCII.

to fixed point. After converting a binary integer, we know its radix point is to the right of the least significant digit. However, a fixed point number may also have a fraction with the understanding that the decimal point has a fixed position in the number. Assume that a decimal number has two fractional digits, we can represent this number in binary by scaling it up. Take 215.55 as an example, we multiply it by 100 to scale up to 21,555, which is a positive binary integer with a decimal point before the last two digits. Thus, we can convert a fixed point number to binary form, perform integer arithmetic, and convert the result back to decimal as long as we remember where the decimal point is. The following numbers are in fixed point formats with a decimal point before the last two digits.

$$215.55 = 21{,}555 * 10^{-2}$$

$$2155.50 = 215{,}550 * 10^{-2}$$

$$21555.00 = 2{,}155{,}500 * 10^{-2}$$

$$10.00 = 1{,}000 * 10^{-2}$$

$$100.00 = 10{,}000 * 10^{-2}$$

Multiplying two fixed point numbers, we need to adjust the radix point in the result.

Example 1: Multiply 215.55 by 10.00.

$$215.55 * 10.00 = 21{,}555 * 1{,}000 * 10\text{-}4$$

$$= 21{,}555{,}000 * 10^{-4}$$

$$= 215{,}550 * 10^{-2}$$

Example 2: Multiply 215.55 by 100.00.

$$215.55 * 100.00 = 21{,}555 * 10{,}000 * 10^{-4}$$

$$= 215{,}550{,}000 * 10^{-4}$$

$$= 2{,}155{,}500 * 10^{-2}$$

In our society, we use the decimal system mainly because we have 10 fingers. The computer system uses a binary system because it is easy to design such circuits. A decimal number with or without a decimal point may be represented as a binary integer or a packed decimal number as long as we remember where the decimal point is. A fixed point number covers a range of magnitude as limited by the total number of bits in the format. The total number of bits in a fixed point number determines its accuracy in terms of significant digits.

In contrast, a floating point number has an exponent field, so its radix point can be repositioned anywhere after each arithmetic operation. Thus, the number has a broader range of magnitude at the expense of accuracy. For example, if we have a total of 80 bits in the format, the significand field can only achieve 64 bits accuracy.

2.6 BIT STRINGS IN MEMORY

In a computer, instructions and data are all bit strings. Therefore, in a von Neumann type machine instructions and data all look alike and differ only in how they are used. The OS treats data just like an instruction until it is executed on the CPU. We conclude that a 16-bit string in a memory word may represent:

- A short positive integer where short means 16 bits in C
- A short negative integer in twos complement notation
- A 16-bit unsigned short integer
- Two ASCII characters
- Part of any floating point number
- Four BCD digits
- A 16-bit instruction
- Part of an instruction

2.7 SUMMARY POINTS

1. The integer representations of 7 and -1 are different in a computer using twos complement notation, but looking at the lower 3 bits (mod 2^3) of each number, they appear identical.
2. The number system used in a computer is binary or base two.
3. Any piece of information in a computer is an ordered set of bits.
4. If the base is eight, the number is octal.
5. If the base is 16, the number system is hexadecimal or hex for short.
6. An hex number is a way to represent a binary number.
7. The conversion by divide algorithm converts a positive integer from decimal form to another base.
8. A negative integer in twos complement notation means taking the ones compli-ment and adding one.
9. In order to communicate with a keyboard, display, printer, or another computer in a network, the industry has established an external code standard called ASCII.
10. In telecommunications, an escape character followed by another character is transmitted to make a 16-bit special control code as the escape sequence.
11. A fixed point number may have a fraction, and after an arithmetic operation, the decimal point is at a fixed position in the result.
12. By using the conversion by multiply algorithm, we can convert a decimal frac-tion into a fraction of any base.
13. A fixed point number can be in binary or packed decimal form.
14. A packed BCD number is comprised of a sign field followed by 15 BCD digits, and each digit has four bits ranging from 0 to 9.
15. A 16-bit string may represent an instruction, part of an instruction, a short inte-ger, two ASCII characters, four packed BCDs, two unpacked BCDs, part of a float, etc.

PROBLEMS

1. Using the twos complement notation, show that -9 and +7 have the same lower four bits, so that they are congruent mod 16.
2. Show that 1024 and zero are congruent mod 1024 by looking at the lower 10 bits.
3. Given 16 bits, convert the following positive decimal numbers to binary, octal, and hex.

   a. 1 b. 100 c. 255 d. 1023 e. 2^{15}-1

4 Using 16-bit arithmetic in twos complement notation, negate the above decimal numbers and write down the answers in binary, octal, and hex.

5. Negate the following numbers using twos complement notation. By inspection only, write down the hex answer directly in one step.

   a. CDEF b. 1234 c. 0FF0 d. FF00 e. 0001

6. Convert the following numbers of a given base to 16-bit binary:

   a. 221 base seven b. 221 base eight c. 221 base 16
   d. 215 base seven e. 215 base 16

7. If one byte is used to store an ASCII character with b7 (bit seven) equal to zero, write down the hex notation for each of the following characters enclosed in single quotes.

   a. `A′  b. `a′ c. `B′  d. `b′
   e. `Z′  f. `z′ g. `0′  h. `9′

8. Comparing `A′ and `a′ as two unsigned integers, which one is larger? Verify that the difference is 20 in hex.

9. Given a 64-bit double precision floating point number:

   a. What is the reason for adding a bias to the exponent?
   b. Why is the bias chosen as 3FF instead of 400 in hex?
   c. What is the exponent to indicate an overflow condition?
   d. What is the exponent to indicate an underflow condition?

10. By multiplying .1 by 16 successively, we can collect the integer part of the quo-tient after each iteration. Verify the number can be directly obtained as .1999 9999 9999 ... in hex.

11. Explain the rounding concept after converting a real number in decimal to a floating point number in binary.

12. Given a real decimal number +.1875:

    a. What is its single precision floating point representation?
    b. What is its double precision floating point representation?

13. What is the double precision representation for .1 in decimal?

14. In an 80-bit extended precision floating point number, the significand field is 64bits long. After normalization, it has the form of 1.xxxxxx ... where x denotes a bit, and the implied binary point is to the right of the leading bit one which is also shown in the format. The bias is the 15-bit 3FFF.

    a. Verify that the decimal +.1 has a biased exponent of -4 as shown below:

       3FFB CCCC CCCC CCCC CCCD

    b. What is the 80-bit representation for decimal -.1?

15. Omitting the sign, what are the packed and unpacked decimal number representations for 1875?

16. After omitting the sign, what are the packed and unpacked decimal number representations for 1875.00 if the decimal point is assumed to be at the rightmost end?
17. In a Von Neuman's computer, instruction and data are just bits, so what can berepresented by a 16-bit string in memory?
18. The empire of the little guy in the galaxy uses a floating point number format that has one overall sign bit, followed by a four-bit biased exponent field and a three-bit significand. The 4-bit biased exponent uses twos complement notation in excess seven. That is, if the exponent is positive, we add seven to it. If the exponent is negative, then first take the twos complement and add seven to it. All the exponents, unbiased and biased, are tabulated below where the symbol — indicates no representation.

Value	**Unbiased Exponent**	**Biased Exponent**
+8	—	1111
+7	0111	1110
+6	0110	1101
+5	0101	1100
+4	0100	1011
+3	0011	1010
+2	0010	1001
+1	0001	1000
0	0000	0111
-1	1111	0110
-2	1110	0101
-3	1101	0100
-4	1100	0011
-5	1011	0010
-6	1010	0001
-7	1001	0000
-8	1000	—

After normalizing a number to 1.XXX, we store the three fraction bits XXX after the biased exponent so the leading bit one is implied.

a. Determine the eight-bit representation of .1 in decimal with rounding. Compute the exact value of this representation that by no means is perfect.
b. What is the error in terms of percentage with rounding? Without rounding, what is the error percentage?

19. Study the mathematical theory behind twos complement notation. Let N1 and N2 be two positive n-bit integers of base two as shown below:

$$0 \leq Ni < 2^{n-1} \quad \text{for } i = 1, 2$$

To represent N1 and N2 in twos complement notation, we picture the bit pattern of the negative number as an unsigned integer as shown below:

$$-Ni = 2^n - Ni \qquad \text{for } i = 1, 2$$

Adding -N1 and -N2, we obtain perfect -(N1 + N2) in twos complement notation if the carry is discarded as follows:

$$-N1 + (-N2) = \underset{\substack{\uparrow \\ \text{carry}}}{2^n} + 2^n - (N1 + N2) = 2^n - (N1 + N2)$$

Prove mathematically, that the end-around carry must be added to the result in ones complement notation.

20. We can verify the format of a number by programming in C, or assembly. Find a PC or UNIX machine with a C compiler on it. We can edit a program, compile, and execute. After seeing the system prompt, enter a command to initiate a job step. Study the output and verify that 2's complement notation is used to represent a negative integer.

```
/* ___________________________________________

Description: This program displays the ones complement of a
32-bit integer in upper-case hex using two different approaches.
The following job steps were tested on a machine running UNIX.

1. vi test.c          {Run vi, the visual editor.
                       After editing, type ZZ to exit.      }
2. cc test.c          {Run cc, the c compiler which generates
                       a.out as the implied executable file.      }
3. cp test.c test.doc        {Copy test.c into test.doc. }
4. a.out >>test.doc          {Run the test program and append
                              the output to test.doc.      }
5. lpr test.doc  {Generate a hard copy of
                  test.doc on the printer.    }

The test.doc file contains the C program and the output:
(1) The ones complement of zero is FFFFFFFF.
(2) The ones compliment of zero is still FFFFFFFF.
___________________________________________ */
```

```
#include <stdio.h>
void main( void)
{
int num;        /* ________________________________________
                The variable num is declared as a 32-bit integer.
                We hard code the integer in the program.  But with
                minor modifications, we may enter an integer from the keyboard.
                ________________________________________________ */

num = 0;
printf( "(1) The ones complement of 0 is %X. \n", ~num);
                /* ________________________________________

                The symbol ~ means ones complement, and %X indicates
                the format control to upper-case hex.  The \n indicates a
                new line, i.e., a carriage return character in ASCII.
                ________________________________________________ */

printf( "(2) The ones compliment of zero is still %X. \n", -num - 1);
                /* ________________________________________

                (twos compliment = ones compliment + 1) is the same as
                (ones compliment = twos compliment. - 1)
                ________________________________________________ */

}
```

21. Run the C program below to verify that 0.1 in decimal conforms with the 64-bit double precision format:

```
                /* ________________________________________

                Description:  In C, the real constant .1 in decimal is
                stored in double precision format by default.  If you don't
                get the correct output on your computer, use the pointer
                concept as shown below:
                {double *num1ptr, num1 = .1;
                num1ptr = &num1;
                printf( "The decimal .1 is represented by %X %X in hex. \n",
                *num1ptr, *num1ptr + 1);}
                ________________________________________________ */

#include <stdio.h>
void main( void)
{

printf( "The decimal .1 is represented by %X %X in hex. \n", .1);

}

/*The decimal .1 is represented by 3FB99999 9999999A in hex.*/
```

CHAPTER 3

Basic Computer Principles

3.1 STORED PROGRAM CONCEPT

An instruction is a bit string that tells the CPU what to do. An instruction contains an op (operation code) and address information such as where and how to find the operand if there is one. The stored program concept was conceived by von Neumann stating that all instructions and data must be stored in memory before their execution can begin. Thus, instructions are retrieved from memory, one by one in sequence, interpreted, and executed. If the operation needs an operand in memory, the CPU issues a memory read request and waits for the arrival of the operand before the operation can start. Instructions and data look alike and programmers can modify the address field or the whole instruction by performing arithmetic or logical operations on them in the execution unit. This chapter discusses the design of an instruction and its execution on the CPU.

3.1.1 Instruction Length

An instruction may have a fixed length or a variable length. A fixed length means a uniform length in that all the instructions have the same length. For example, each instruction on a PDP-10 has 36 bits,[49] but on the POWER (performance optimized with enhanced RISC) PC or MIPS, each instruction has 32 bits.[16,35] The Burroughs B5000 stack machine has 12-bit instructions known as syllables. On such a machine, no address is coded in the syllable of add or subtract instructions. Most computers support variable length instructions, and VAX is an extreme case that has instruction from 1 to 21 bytes.[51]

The 16-bit instruction format on a Burks machine from the early days is shown in Figure 3.1a.[4] Each instruction has a four-bit op (operation code) followed by a 12-bit memory address. The reason that the instruction contains one address is because the first source operand implied in acc (accumulator) is also the destination. In consequence, only the address of the second source operand is coded in the instruction. The op tells the CPU what action to take between the acc and the memory operand. Take the add instruction as an example: the CPU retrieves the second source operand from memory and adds it to acc. Because the address is 12 bits long, its memory size is limited to 4096 (2^{12}) words, and each word contains 16 bits. An instruction may contain many fields, and each field has a group of bits to specify the operand in memory, or in the CPU.

Figure 3.1 Instructions: (a) simple format, (b) op and eop are adjacent, (c) op and eop are apart, and (d) op, eop1, and eop2.

3.1.2 Opcode vs. Extended Opcode

The op usually resides at the instruction's leftmost end, even though in theory it may be placed anywhere in the instruction. Every instruction has an opcode. The nop (no operation) instruction does nothing but it has an opcode. In some instructions, part of the address field is used as an eop (extended opcode). That is to say, the opcode has two levels, op and eop. More often than not, the op and eop are adjacent to each other, as shown in Figure 3.1b, but sometimes they may be apart as shown in 3.1c. If each op or eop has four bits, the op field provides 16 combinations. Assume that 15 combinations are used without an eop, so one combination is left to work in conjunction with an eop. The total number of combinations is increased from 16 to 31 as computed below:

$$15 + (1 * 16) = 31$$

For example, the op between 0 and 14 needs no eop. If the op is 15, then a four-bit eop is used to provide 16 more combinations. Often, the opcode field is extended to three levels: op, eop1, and eop2 as shown in Figure 3.1d. The eop1 provides the first level extension, and eop2 provides the second level extension. This encoding concept is used to design messages between two computers. The messages are for a program running on a remote computer to interpret and execute. That is to say, the message header contains an op with one or more eops and other information.

The CPU uses information in the address field to retrieve the operand. Because a 12-bit address has a value ranging from zero to 4095, the CPU can access any word in a memory of four KW. In a second generation computer and beyond, an instruction may have many address fields, each one providing information on exactly how to find the operand.

3.2 HARDWARE UNITS IN A CPU

A simplified block diagram shown in Figure 3.2 represents a CPU with two hardware units: IU (instruction unit) and EU (execution unit). Conceptually, a control path and a data path exist between the two units. The control path provides timing signals for control, and the data path supplies the data operand, perhaps retrieved from memory. The major function of IU is to fetch and decode the instructions. The IR (instruction register) is where the current instruction resides after being fetched from memory. After decoding the opcode, the IU generates control signals, e.g., to fetch data from memory, route data, and execute data. The EU (execution unit) is where that execution takes place. In a high-performance CPU, the IU and EU can be further decomposed into many units operating in parallel.

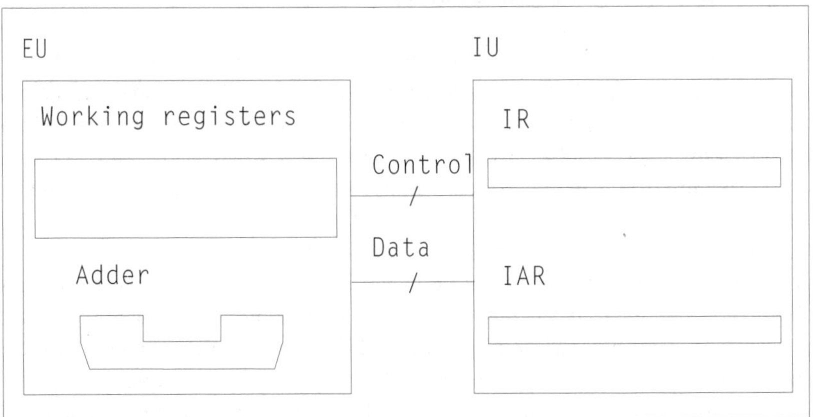

Figure 3.2 Central processing unit.

3.3 INSTRUCTION UNIT

The instruction unit is also known as the control unit (CU) that mainly provides timing signals to obtain, interpret, and execute instructions. Let us trace the logical flow of instruction execution. After an instruction is retrieved into the IR, the hardwired decoder generates timing signals for further actions based on the opcode. Several issues remain, such as, where does the IU find the instruction in memory? We say that the address of an instruction is kept in an IAR (instruction address register) which needs to be updated during each instruction retrieval cycle.

3.3.1 Instruction Register

The IR is hardwired in the IU and a decoder connected to it decides what actions to follow. The IU always fetchess the instructions one by one in sequence if there is no branch. A more sophisticated CPU may have an instruction queue or pipe so that a stream of instructions can flow into the queue. The leftmost end of the queue becomes an IR for the decoding logic positioned right on top. The IR has a fixed hardware location and an instruction must reside in the IR before its execution can start. There is a conceptual difference between hardware execution and software interpretation. For hardware execution, the IR is hardwired in the CPU. For software interpretation, the IR is a simulated variable in memory.

3.3.2 Instruction Address Register

The IAR points to the next instruction in memory. In a simple model, after retrieving an instruction in the IR, the IU increments the IAR by the length of instruction. In first generation computers, each machine instruction is one word long, so the IAR is a counter that contains a word address. If the instruction length is fixed, the IAR becomes a program counter.

3.3.2.1 IAR vs. Program-Counter

The program counter contains an unsigned integer as the address of the next instruction in memory. A program counter is incremented by an integer, provided that all the instructions have the same length. It is possible to design the IAR as a counter as long as the variable instruction length is 2^n where n is 1, 2, or 4. For example, after obtaining a 4B instruction into an IR, we can add 1 to the 3rd bit on the right of the counter.

It should be stressed that because all the instructions on a typical RISC (reduced instruction set computer) machine are four bytes long, its IAR is a counter as shown in Figure 3.3. As each instruction is four bytes, its address is an unsigned integer with two trailing zeroes (IA1 and IA0) implied. The program counter contains a 30-bit block number followed by 2 zeroes and each block has four bytes. To obtain a 32-bit byte address, we append two zeroes to the 30-bit block address so that the total number of bits is 32. In hardware design, the two trailing zeroes can be hardwired so to save two flip-flops. Thus, the 30-bit block address in the program counter really means a 32-bit byte address with the two trailing zeroes implied. After an instruction is obtained in the IR, the 30-bit program counter is incremented by one. The byte address is incremented by four (2^2).

Figure 3.3 Instruction address register as a 30-bit counter.

On a program counter, the instruction length may vary from 1 to 7 bytes, but the IAR can be designed as a counter provided that the IU increases its value right after interpreting a portion of an instruction. After the entire instruction is executed, the IAR contains the address of the next instruction in memory. For this reason, we will use program counter and IAR interchangeably in this book.

3.3.3 Memory Bus

Between the CPU and memory there exists a bundle of electric copper wires collectively known as the memory bus. As far as function is concerned, some wires are used to transmit a memory address, while others are used to provide data and control signals as shown in Figure 3.4. In a modern computer, the memory bus is etched on a printed circuit board so that single in-line memory modules (SIMMs) can be inserted into the connector sockets. The address bus is unidirectional: it is used to transmit a physical address from the CPU to memory. In contrast, the data bus is usually bidirectional, which means that in a write cycle the CPU places bits on the data bus, but in a read cycle the CPU receives bits from the data bus. A control signal is placed on the line to indicate which operation is to be performed, e.g., read or write. The memory may also provide a data available signal on the line to tell the CPU when the data are ready.

3.3.3.1 Address Bus vs. Data Bus

The number of wires in an address bus determines the size of the physical memory which is commonly measured in bytes. The physical memory size is defined as the PAS (physical addressing space). A 32-bit address is an unsigned integer that has a total of 2^{32} combinations so its PAS is four GB. The data bus width is defined as the number of bits to be transferred in one physical cycle. That is, multiple bits are transferred in parallel during a memory cycle, and each bit is placed on the line (wire) with a voltage. The data bus width of a low end controller's memory can be eight bits or 16 bits.

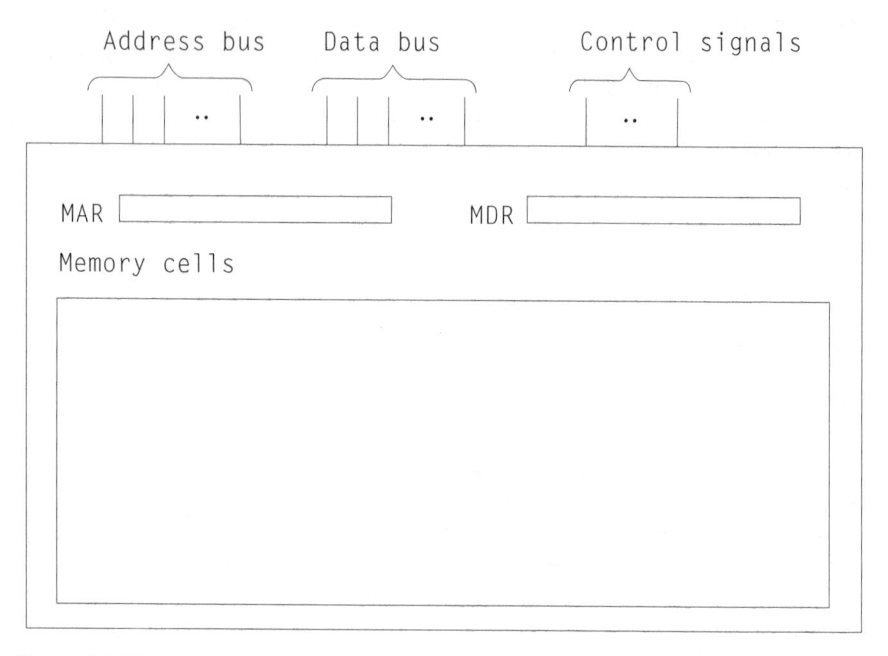

Figure 3.4 Memory system.

Assume that we have a family of computers. All the models share the same architecture (instruction set), but the data bus of each model may have a different width. For example, the Intel 8088 processor has a memory data bus of eight bits, while the 8086 has a data bus of 16 bits.[29] Therefore, it takes two physical memory cycles for an 8088 to fetch a 16-bit operand, but it takes only one physical memory cycle for an 8086 to obtain the same operand. In regard to system architecture, the 16-bit operand is obtained by one logical memory cycle for both machines. That is to say, an 8088's 16-bit logical memory cycle is comprised of two eight-bit physical memory cycles, and each cycle merely fetchess an eight-bit portion of the operand. An 8086's 16-bit logical memory cycle is one physical memory cycle. In layman's terms, it takes one memory dip to fetch the entire operand. Regardless of the different bus sizes, both machines support the same instruction set with a difference in speed or performance. In fact, the i386 uses a 32-bit memory data bus, the Pentium Pro uses a 64-bit data bus, and the Pentium IV uses a 128-bit data bus. That is to say, the CPU has logic to store the block on a chip and can access the operand in the block.

3.4 REGISTER TRANSFER LANGUAGE

The internal operations in a computer can be described by RTL (register transfer language). Recall that the structured constructs in PDL are used to describe the logic flow at the top level. The RTL is a tool to describe the logic flow at the hardware register level. A comment may be placed in brace bracket or begun with a semicolon (;) to give more details. Some RTL operator symbols are shown in Table 3.1.

Table 3.1 Operator Symbols in RTL

Operator	Description
←	Transfer (the source operand specified as RHS is copied to the destination operand specified as LHS.)
(?)	Grouping of operations
[?]	Address designator
[m .. n]	Address or index ranging from m to n inclusive
<? : ?>	Bit designator
+, -, *, /	Arithmetic Op: add, subtract, multiply, divide
.EXP.	exponential
^	Concatenate
.MOD.	Modulus
.NEG.	Negate
.NOT., .AND.	Logical Op: not, and
.OR., .EOR.	Or, exclusive or
.EQ., .NE.	Relational Op: equal, not equal
.LT., .LE.	Less than, less or equal
.GT., .GE.	Greater than, greater or equal
.SHL	Shift logical left
.SHR., .SAR.	Shift logical right, shift arithmetic right

3.4.1 Source Operand vs. Destination Operand

The RTL describes the movement of bits from one location to another location in a computer. The source operand refers to the bits stored at one location while the destination operand refers to where the bits are moved to, i.e., another location. Each operand may be in a register or memory that has a mnemonic name in letters followed by digits. If a register is designated for a special function, it may have a mnemonic to describe its function. For example, we have two registers in memory: the MAR (memory address register) and the MDR (memory data register). Both are hardware buffers that interface with the internal circuitry in memory. The EU has a working register set to store temporary results. Each hardware register is used to store data or an address during computation. To access an operand, we need to specify the symbolic name of a register or memory word with an address.

3.4.2 Address Designator

In order to access an operand that is a register or in memory, we specify a name followed by an address designator, namely a left square bracket, an address, and a right square bracket. As a simple example, if the register set is for storing addresses, the mnemonic may be A (address) and A[1] would means the second address register. Since a computer uses zero indexing, A[0] represents the first address register. Since the number of entries in a register set is much smaller than memory, a register address is usually in between 3 and 5 bits. Because the double period in a square bracket indicates a range operator, A[0..4] represents a set of five address registers, A0 to A4 inclusive.

3.4.2.1 Memory Address vs. Register Address

The mnemonic symbol for memory is M. At the machine level we may specify an address in hex, so M[100000] represents the operand at the 1 MB location. The default is memory so if M is omitted in the notation[100000] means M[100000]. For an operand in a general purpose register set named R, we specify its name with an address designator to access an element in the set. The address or index in the square bracket tells the position of the register in the set. Since the address in a computer always starts with zero, R[0] denotes the first CPU register, R[1] denotes the second register, etc. The address is either coded in the instruction or implied in the opcode.

An assembly language statement is a string of ASCII characters representing a message to the assembler. After assembling, the entire program becomes a block of instructions and data in binary form. An assembly statement is readable because it is in symbolic form. The notation R[1] is used in RTL, so in an assembly statement, we may use R1 (register 1) instead as an alias. That is to say, an alias is a special name for a register. For example, if R[1] is used to store a unsigned integer for tracking the number of times that a loop is executed, CX (count) may be used as an alias instead of R[1]. When we examine the code image in memory, its address is 001 in binary. In general, a memory word is 16 bits, which is the default size of an operand word.

The I/O controller has its own register set, the IORs (I/O registers). Each of which is assigned an I/O address that is separated from the register or memory addressing space. For example, IOR[3F8] means receive data register in a communication controller, and 3F8 is an I/O address in hex. The alias of this I/O register is Rxd (receive data register). In the case that the middle bits in an operand are of interest, we need to specify the bit string by spelling out the first bit and the last bit as explained next.

3.4.3 Bit Designator

A bit designator has a left angular bracket, a beginning bit position, a colon (:), an ending bit position, and a right angular bracket. It is used to specify a bit string in a register or memory. The first example is M[0] <31:0> which represents 32 bits in memory starting at address zero. In a second example, R[1] <7:0> represents the lower eight bits in R[1]. On a Pentium processor, because CX is the 16-bit count register, CL (count register — low order byte) represents its low order byte. That is, CL and CX <7:0> are the exactly same.

3.5 INSTRUCTION EXECUTION CYCLE

The execution of an instruction is divided into two phases: instruction retrieval and operand execution. Suppose the CPU has neither instruction pipe nor operand pipe so that all the operations are sequential. First, issue a memory read request to retrieve the next instruction in the IR. After decoding, if all the operandi are in the CPU, the operand execution phase starts. Otherwise, if one operand is in memory, one memory read request must be issued to retrieve the operand. If two operandi are in memory, then two memory read requests are issued to fetch the operandi. Each memory read request is called a memory dip, so the CPU must wait for the arrival of the memory operand.

3.5.1 Instruction Fetch

In order to understand the sequential hardware steps performed by the CPU, we use an add instruction as an example. The assembly language statement for add is shown below:

```
add    sopd2        ;add sopd2 in memory to acc
```

where add is the mop (machine opcode) for integer addition, and sopd2 is the symbolic name of the second source operand in memory. In the Burks machine for example, the assembler translates this statement into a 16-bit instruction as shown in Figure 3.1a of the previous chapter. This four-bit opcode is followed by a 12-bit memory address. Before execution starts, the IAR contains the memory address of the add instruction. First, the CPU fetches the instruction in the IR by passing the address through the IAR on the bus. After the instruction resides in IR, the CPU interprets and executes its opcode. As a result of its execution, the content of a memory location named sopd2 is added to the content of a register named acc (accumulator). Since the hardware location of the accumulator is implied, its address is not part of the instruction. Using RTL, we can specify the micro-hardware steps to fetch an instruction as follows:

Step 1: MAR $\leftarrow$ IAR {bus operation}
Step 2: MDR $\leftarrow$ M[MAR] {Memory read cycle}
Step 3: IR $\leftarrow$ MDR
Step 4: IAR $\leftarrow$ IAR + instruction length

In step 1, the CPU issues a memory read request by copying the address bits in the IAR on the memory address bus along with other control signals. After obtaining the address from the bus, the memory system reads the content bits at this address into the MDR. Then, the data bits in MDR are placed on the data bus for the CPU to fetch in the IR. Steps 1 to 3 represent a memory read cycle that is atomic or indivisible.

In other words, no other processors can perform a memory read operation to the shared memory at the same time. After decoding, the CPU also knows to increases the IAR by the instruction length, as shown in step 4.

3.5.2 Operand Fetch

If the operand is in the CPU, execution may start right after instruction retrieval. However, for this one-address add instruction, the CPU must issue a memory read cycle in order to obtain the operand whose address is specified as the lower 12 bits in the IR. In RTL, we have three micro steps.

Step 5: MAR ← IR <11:0> {bus operation}
Step 6: MDR ← M[MAR] {memory read cycle}
Step 7: T ← MDR

The temporary register T resides in the EU. When all the operandi are available in the CPU, operand execution may begin.

3.5.3 Operand Execution

The operand execution in the EU is specified as follows.

Step 8: acc ← acc + T {Integer add}

where acc denotes the accumulator. The EU feeds the bits in acc to one input bus of the adder, the bits in T (temp) to the other input bus of the adder, and clocks the adder output into acc. After the add, the accumulator contains the sum. At this point, the CPU goes back to step 1 and repeats the new cycle, that is, the instruction execution cycle is an infinite hardware loop. For convenience, we describe the instruction execution as follows,

add sopd2 ;acc ← acc + M[sopd2] {integer add}

After the semicolon, the RTL comment indicates that both instruction fetch and operand fetch are implied.

As the IAR is increased by the length of the instruction, the next instruction is fetched from memory, that is, instructions are executed in sequence until a branch instruction is executed. A branch instruction means a goto statement in a high-level programming language. Such an instruction has a memory address of the next instruction. Upon its execution, the CPU places the branch address into the IAR. As a result, the instruction at the new address will be fetched next. In other words, control is passed from one program address to another program address. A computer also supports conditional branch instructions. Upon executing such an instruction, the EU in the CPU performs a condition test. If the result is true, the new branch address in the instruction is placed into the IAR, otherwise there is no branch.

3.5.3.1 Program Termination

How does a program terminate? Recall that every computer has an OS running side by side with the user program. Therefore, in concept, the last instruction in the program should pass control back to the OS. In modern computer design, this is accomplished by executing an interrupt (int) instruction. As a result of its execution, the CPU relinquishes control from the current program to a specific location in memory where the OS resides. That is, after a program terminates, the OS resumes control and executes next.

3.6 EXECUTION UNIT

The EU (execution unit) is where operand executions are performed. A simple EU may consist of several working registers and an adder. If we incorporate the logical functions and shift functions into the adder, we obtain an ALU (arithmetic and logical unit). Note that in a high performance computer, the EU is composed of many hardware units, such as multiply, divide, shift, etc., using parallel logic. However, in order to illustrate the basic concept we limit our discussions to sequential operations. That is to say, the simple CPU has an ALU that can handle both unary and binary operations, e.g., add, subtract, multiply, divide, AND, NOT, OR, EOR, shift, etc.

3.6.1 Working Registers

The EU supports a set of working registers. Each register is an ordered set of flip-flops. Each flip-flop is a one-bit storage device. We show a simple D flip-flop in Figure 3.5a. The two inputs are for D (data) and Clk (clock). The two outputs are for Q and \Q. The circle to the right of \Q indicates a complement of Q. The output Q of this flip-flop always obeys the D input as long as there is a clock. The Clk signal is a rectangular voltage pulse that goes up from 0 v to 3.3 v and goes down as shown in Figure 3.5b. Note that the circle at the Clk input indicates that the flip-flop changes state at the trailing edge of a clock. That also means there is one-bit delay when a flip-flop changes state. If the clock frequency is 100 Mhz, its period is 10 ns representing the raw speed of a CPU. Remember two things: first, without a clock signal present as input, the flip-flop can not change state and second, a register has its output available in both true and complement forms.

In first generation computers, there was only one register, namely the acc (accumulator). For binary operations, the first source operand is in acc and the destination is also the acc. As far as programming is concerned, more working registers provide speed and convenience to a user. This is why a modern computer has 8, 16, or 32 working registers as shown in Table 3.2.

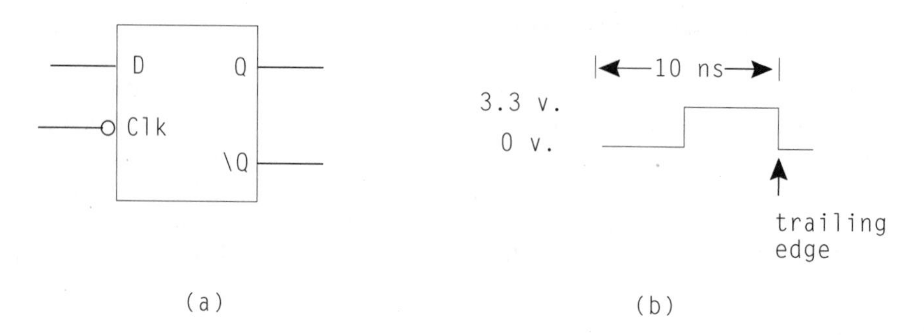

Figure 3.5 One-bit: (a) D flip-flop and (b) 10 ns clock.

Table 3.2 Working Registers in a Computer

Computer Model	Number of Working Registers	Number of Bits in Register	Number of addresses in an add instruction
Burks	1	16	1
PDP-8	1	12	1
VAX	16	32	2, 3
IBM360/370	16	32	2
CDC6600/7600	24	60, 24	3
CRAY 1	16	64, 24	3
B5000	0	48 (word)	0
MIPS	32	32	3
Pentium	14	32, 16	2
Itanium	128	64	3

All the computers listed above are well known in the history of computing. The Burks machine has a memory word of 16 bits with one address coded in the instruction. The PDP-8 is the first minicomputer with a 12-bit accumulator, so its add instruction contains one address. The VAX (virtual address extension) machine is a super minicomputer using 16 32-bit registers.[51] The IBM mainframe has 16 32-bit general purpose registers (GPRs).[26] The CDC mainframe has eight 60-bit operand registers plus 16 more 24-bit registers for storing addresses.[5] The CRAY 1, a supercomputer, has 8 64-bit operand registers, eight 24-bit address registers, plus many save registers as scratch pad. The Burroughs machine provides no working registers at all, and its add instruction only has opcode.[2] It is a stack machine in that all the operandi, source and destination, are implied on the stack, and each memory word has 48 bits.

The last three microcomputers in Table 3.2 are singled out because they have features dictating the future trend. The MIPS machine is a RISC (reduced instruction set computer) that has a total of 32 32-bit registers. The Pentium uses eight 32-bit registers for data or addressing plus six more 16-bit segment registers. The Itanium has 128 general registers and each register has 64 bits. The various computer classes are compared from a historical point of view, in Table 3.3.

Table 3.3 Design Features of Various Computer Classes

Class	Adder Width	Number of Registers	OS Protection
Micro	8	4	Poor
Mini	16	8	Poor
Mainframe	32	16	Very high
Super	64	8	High

The adder width is defined as the number of bits that can be added in one clock cycle. Generally speaking, a simple microcomputer has an adder of 8 or 16 bits with a less sophisticated I/O structure and operating system. Quite often, the OS on a microcomputer crashes while running an application or doing some development work. However, future microprocessors will have wider memory data buses, wider adders, and more reliable operating systems. Future mainframes or supercomputers will be designed as clusters of microprocessors with very sophisticated operating systems.

3.6.2 Look-Ahead Adder

The adder in an EU is comprised of many one-bit full adders. The faster the adder, the faster the CPU. The adder width dictates the CPU performance to a certain degree. In a family of mainframe computers sharing the same architecture, different models may have adders of different widths.[26] For example, the lowest model can add only eight bits at one time, so it takes four basic add cycles to add two 32-bit integers. A higher model can add 32 bits at one time, so it takes one add cycle to add two 32-bit integers. Regardless of their adder widths, all the models execute the same instruction to add two 32-bit integers. Because a lower model uses less hardware, its cost is low, but so is its speed.

Intuitively, the adder in an EU should have a width as same as its memory data bus. That is to say, an operand can be fetched in one physical memory cycle so the adder can perform its operation in one clock cycle. In general, supercomputers use 64-bit adders, the mainframes use 32-bit adders, and the microcomputers use 16-bit adders. Today, a modern microcomputer not only uses many 32-bit or 64-bit adders, but it also uses a 64-bit or 128-bit data bus.

The adder is the heart of a simple CPU. During an add operation, it is possible for a carry to propagate from the LSB to the MSB. In order to reduce the number of circuit delays, a carry look-ahead adder uses combinational logic that has many cascaded AND – OR gates to generate the carry output. The carry input at a particular bit position is expanded as a function of all the input bits at the previous stages. Consequently, the number of gate levels is reduced along with the number of circuit delays. The ultimate goal is to perform the add operation in one clock cycle. To illustrate this concept, we first design a four-bit carry look-ahead adder, then extend its design to 16 bits, and finally to 64 bits. The basic logical operators are explained below:

1. The parentheses ( ) have the highest precedence to group several operations.
2. The backward slash \ is the NOT operator so we can key in the symbol for logic simulations.
3. The period (.) or space is the AND operator.
4. The plus (+) is the OR operator.

The CPA (carry propagated adder) using carry look-ahead logic is composed of many one-bit full adders. At each bit position i, the three inputs are Ai, Bi, and Ci, and the two outputs are Si (sum i) and Ci+1 (carry i+1) whose logical equations are shown below:

Si = \Ai \Bi Ci + \Ai Bi \Ci + Ai \Bi \Ci + Ai Bi Ci
Ci+1 = Ai Bi + Bi Ci + Ai Ci

At bit 0 (LSB), the three inputs are A0, B0, and C0 and the two outputs are S0 and C1. Figure 3.6a, shows the logic of two functions as defined below.

G0 = (A0 B0) {Generate function zero.}
P0 = (A0 + B0) {Propagate function zero.}

The two-input AND gate provides G0 that is the generate function at bit zero. It means that a carry output is true if A0 is true, and B0 is true regardless of its carry input. The two-input OR gate provides P0, the propagate function at bit zero. It means a carry output is true if A0 or B0 is true, and its carry input (C0) is also true as an assist. Thus, a generate function does not need an assist from its RHS (right hand side) but a propagate function does. The carry output at bit 0 is the carry input at bit one (C1), whose logical equation is shown below:

C1 = A0 B0 + B0 C0 + A0 C0
= A0 B0 + (A0 + B0) C0
= G0 + (P0 C0)

We further define the following:

Pi = Ai + Bi {propagate function i for i = 0, 1, 2, ...}
Gi = Ai Bi {generate function i}

1. The Pi function propagates a carry from input to output if a carry input exists at bit i.
2. The Gi function generates a carry output, one or zero, regardless of its carry input at bit i.

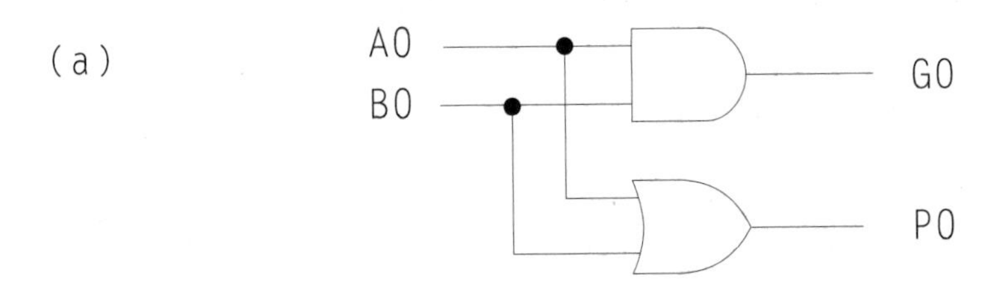

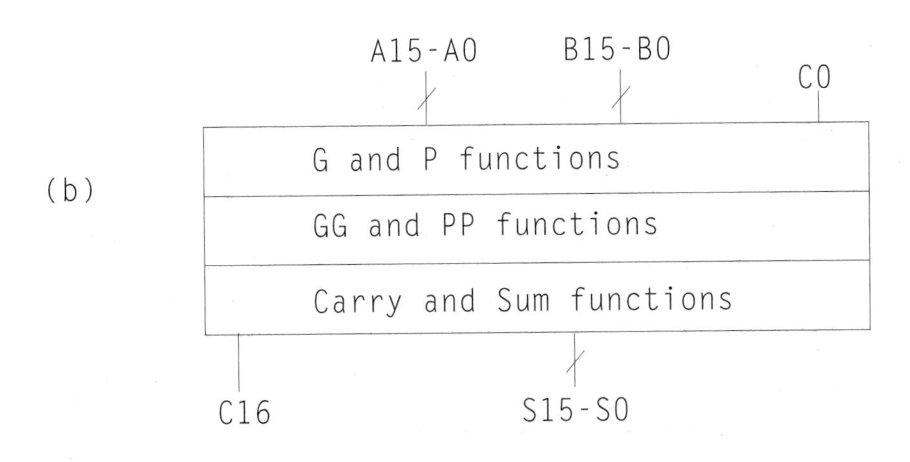

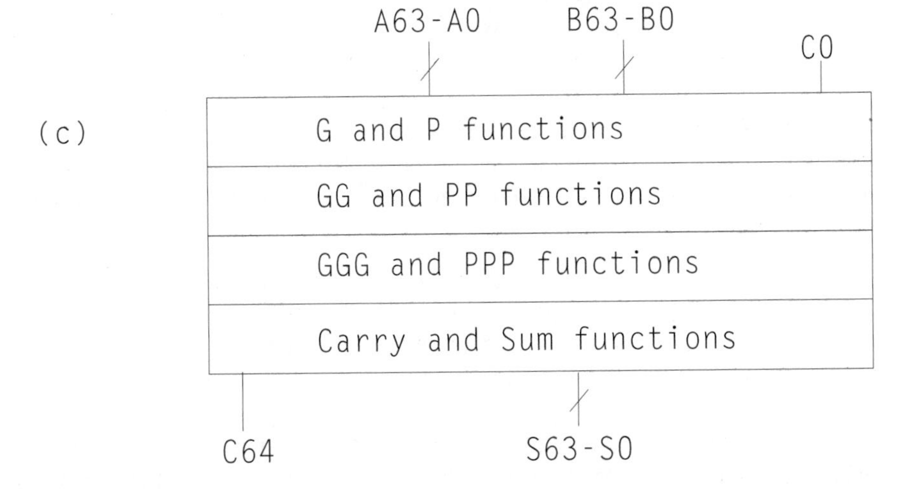

Figure 3.6 Carry look-ahead logic: (a) G0 and P0, (b) 16-bit adder, and (c) 64-bit adder.

We can use AND – OR logic to represent the carry outputs C2, C3, and C4, as shown below.

$$
\begin{aligned}
C2 &= A1\ B1 + A1\ C1 + B1\ C1 \\
&= A1\ B1 + (A1 + B1)\ C1 \\
&= A1\ B1 + (A1 + B1)\ (A0\ B0 + A0\ C0 + B0\ C0) \\
&= A1\ B1 + (A1 + B1)\ A0\ B0 + (A1 + B1)\ (A0 + B0)\ C0 \\
&= G1 + P1\ G0 + P1\ P0\ C0 \\
C3 &= G2 + P2\ G1 + P2\ P1\ G0 + P2\ P1\ P0\ C0 \\
C4 &= G3 + P3\ G2 + P3\ P2\ G1 + P3\ P2\ P1\ G0 + P3\ P2\ P1\ P0\ C0
\end{aligned}
$$

The C4 function is the sum of five products, so it is true if any of the five conditions listed below are true.

1. The generate function at bit three is true.
2. The generate function at bit two is true, and the propagate assist at bit three is also true.
3. The generate function at bit one is true, and the propagate assists from bit threeand bit two are true.
4. The generate function at bit 0 is true, and the propagate assists from bits three, two, and one are true.
5. The C0 or Cin (carry in) bit is true with the propagate assists from all four bits.

We can further expand the carries to level two and obtain a 16-bit carry look-ahead adder as shown in Figure 3.6b. We define the following propagate functions:

$$
\begin{aligned}
PP0 &= P3\ P2\ P1\ P0 \\
PP1 &= P7\ P6\ P5\ P4 \\
PP2 &= P11\ P10\ P9\ P8 \\
PP3 &= P15\ P14\ P13\ P12
\end{aligned}
$$

The double P symbol (PP) denotes the propagate function at level two. In general, PPi is a propagate function of the four-bit adder block i. For example, PP0 is the propagate function of the four-bit adder block zero. The generate functions of each block are defined as follows:

$$
\begin{aligned}
GG0 &= G3 + P3\ G2 + P3\ P2\ G1 + P3\ P2\ P1\ G0 \\
GG1 &= G7 + P7\ G6 + P7\ P6\ G5 + P7\ P6\ P5\ G4 \\
GG2 &= G11 + P11\ G10 + P11\ P10\ G9 + P11\ P10\ P9\ G8 \\
GG3 &= G15 + P15\ G14 + P15\ P14\ G13 + P15\ P14\ P13\ G12
\end{aligned}
$$

The double G symbol (GG) denotes the generate function at level two. For example, GG3 is true under four conditions or "ed" together. That is, if G15, the generate function at bit 15, is true, or if any generate function at the lower bit in the block is true, then it needs the assist from all its upper bits in the block to propagate the carry to the end. The four carry outputs of the adder blocks are derived below:

$$CC1 = GG0 + PP0\ C0$$
$$CC2 = GG1 + PP1\ GG0 + PP1\ PP0\ C0$$
$$CC3 = GG2 + PP2\ GG1 + PP2\ PP1\ GG0 + PP2\ PP1\ PP0\ C0$$
$$CC4 = GG3 + PP3\ GG2 + PP3\ PP2\ GG1 + PP3\ PP2\ PP1\ GG0 + PP3\ PP2\ PP1\ PP0\ C0$$

The double C (CC) means a carry output at a level two block. Comparing these carry outputs with the carry outputs of a single four-bit adder, they share the same schematic with different inputs. In fact, CC0 is C0, CC1 is C4, CC2 is C8, CC3 is C12, and CC4 is C16. What should be the equation for C15, the carry output at bit 14? Intuitively, we have,

$$C15 = G14 + P14\ G13 + P14\ P13\ G12 + P14\ P13\ P12\ CC3$$

If the CC3 term in the equation is not expanded, then C15 looks just like C3 using four-input AND and OR gates. As a challenge, we attempt to design a 64-bit look-ahead adder. After expanding the carry equations to level three, the new adder is shown in Figure 3.6c. The propagate functions are defined below:

$$PPP0 = PP3\ PP2\ PP1\ PP0$$
$$PPP1 = PP7\ PP6\ PP5\ PP4$$
$$PPP2 = PP11\ PP10\ PP9\ PP8$$
$$PPP3 = PP15\ PP14\ PP13\ PP12$$

The triple P (PPP) denotes a propagate function at level three. In general, PPPi is the propagate function of the 16-bit adder block i. For example, PPP3 is the propagate function of the 16-bit adder block three so the propagate functions of all the four-bit adder blocks in it need to be true. The PP15 (P63 P62 P61 P60) is the propagate function of the four-bit adder block 15. Similarly, the generate functions for the 16-bit adder blocks are as follows:

$$GGG0 = GG3 + PP3\ GG2 + PP3\ PP2\ GG1 + PP3\ PP2\ PP1\ GG0$$
$$GGG1 = GG7 + PP7\ GG6 + PP7\ PP6\ GG5 + PP7\ PP6\ PP5\ GG4$$
$$GGG2 = GG11 + PP11\ GG10 + PP11\ PP10\ GG9 + PP11\ PP10\ PP9\ GG8$$
$$GGG3 = GG15 + PP15\ GG14 + PP15\ PP14\ GG13 + PP15\ PP14\ PP13\ GG12$$

The triple G (GGG) denotes the generate function at level three, and GGGi is the generate function for the 16-bit adder block i that has a five-level circuit delay. We derive the carry outputs for each 16-bit adder block as follows:

$$CCC1 = GGG0 + PPP0\ CC0$$
$$CCC2 = GGG1 + PPP1\ GGG0 + PPP1\ PPP0\ CC0$$
$$CCC3 = GGG2 + PPP2\ GGG1 + PPP2\ PPP1\ GGG0 + PPP2\ PPP1\ PPP0\ CC0$$
$$CCC4 = GGG3 + PPP3\ GGG2 + PPP3\ PPP2\ GGG1 + PPP3\ PPP2\ PPP1\ GGG0 + PPP3\ PPP2\ PPP1\ PPP0\ CC0$$

We also have the following equations:

$$CCC1 = C16$$
$$CCC2 = C32$$
$$CCC3 = C48$$
$$CCC4 = C64$$

Note that each of the above carry output implementations has a seven-level circuit delay. The equation for C63, the bit before the MSB should intuitively resemble C15 or C3 as shown below.

$$C63 = G62 + P62\ G61 + P62\ P61\ G60 + P62\ P61\ P60\ CCC3$$

If we press the circuit delay per each gate down to 100 ps (10^{-12} sec), the add cycle can be done in one clock that is one nanosecond at one Ghz.

3.6.2.1 The Cin Input

It is important to understand that C0 is the Cin (carry in) bit that plays an ingenious role in twos complement arithmetic. The Cin bit serves as the carry input signal to the adder at bit zero before an add operation starts. In practice, the Cin bit can be set to zero, one, or the C (carry) flag generated from the previous adder operation as described below:

1. Cin is zero for an add operation.
2. Cin is one for a subtract (sub) operation. A subtract operation is accomplished by adding the twos complement of the subtrahend to the minuend. Thus, the minu-end is fed to one adder input, the ones complement of the subtrahend is fed to the other adder input, and meanwhile the Cin bit is turned on. In other words the twos complement is obtained by adding the ones complement and one.
3. Cin is set as the C (carry) output flag generated from the previous adder cycle. This design feature is used for implementing multiple precision arithmetic oper-ations. Take integer add as an example. If the adder bus is eight bits wide, we can add two 16-bit integers provided that the ADC (add with carry) instruction is available. That is, we write software to go through the eight-bit adder opera-tions twice. The first time we use an ADD instruction with a Cin zero to add the lower eight bits. After the operation, we obtain the low-order byte sum and its carry output (C8) is clocked into the carry

output flag in the status register. The second time, we use the ADC instruction instead to compute the high-order byte sum as the Cin bit is set to be the carry output flag generated from the previous eight-bit adder cycle. Consequently, the adder generates two sum bytes in two different cycles to provide the 16-bit result.

3.6.3 One-Bit Shifter

In some computers including the 8086, the logical and shift capabilities are built into the adder so the unit becomes a ALU (arithmetic and logic unit). A logic function is performed by the ALU under opcode control. In addition, the shift operation is also performed via the ALU. Specifically, a one-bit shifter can be placed in the final stage to skew the adder output to the left or right by one bit. That is to say, the ALU can perform one-bit shift operations under opcode control. It should be mentioned that an advanced shifter may be a separate unit that uses combinational logic of gates to shift a number of bits in one clock. To illustrate the shift concept, a one-bit shifter design that uses two-level AND – OR gates is described below.

3.6.3.1 Shift Logical Right

A shift operator is unary operator, and the operand may be in a register, or in memory. That is, when an operand is fed to one input bus of the adder and zeroes are fed to the other input bus, the adder output is same as the operand. For SHR (shift logical right) one-bit operation, there is a control signal that tells the shifter what to do. In consequence, the shifter skews the adder output to the right by one bit. Figure 3.7a shows a 16-bit operand before and after a one-bit SHR operation. As indicated by the arrows, the bits are skewed to their immediate right, a zero flows into the sign, and the LSB is shifted out into the carry bucket.

3.6.3.2 Shift Arithmetic Right

What does a one-bit SAR (shift arithmetic right) mean? For SHR, a zero flows into the sign bit, but for SAR the sign bit remains unchanged. Figure 3.7b shows a 16-bit operand before and after a one-bit SAR operation. As indicated by the arrows, the bits skew to their immediately right and the LSB is shifted out into C, but the sign bit retains its original value. If N is a signed integer using twos complement notation, the shift arithmetic right N by one bit is equivalent to taking the square bracket of N divided by two as shown below:

$$\text{SAR N by 1 bit} = [\, N / 2 \,]$$

Many one-bit SAR examples are given in Table 3.4 and their verifications at the bit level are left as an exercise.

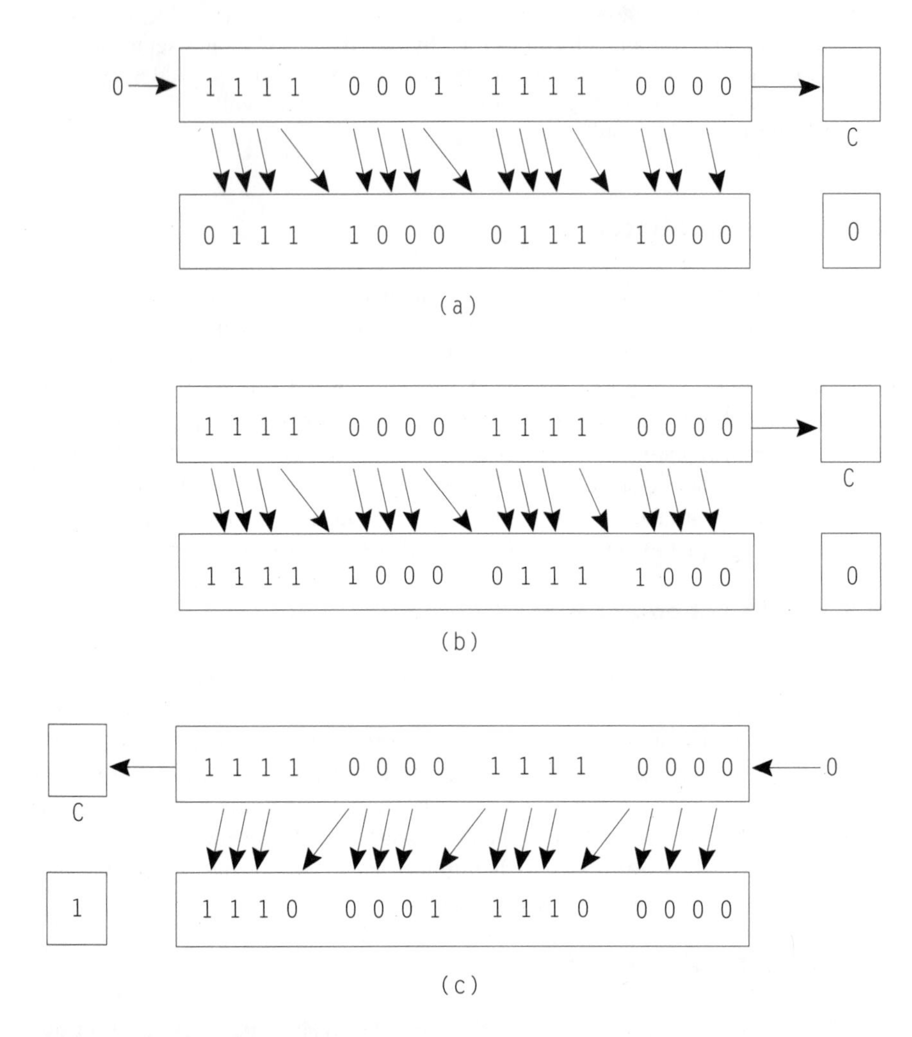

Figure 3.7 One-bit shift operations: (a) SHR, (b) SAR, and (c) SHL.

Table 3.4 One-Bit Shift Arithmetic Right Examples

Notation	Answer	
[5 / 2]	2	
[11 / 2]	5	
[1 / 2]	0	{cleared}
[-11 / 2]	-6	
[-1 / 2]	-1	{unchanged}

3.6.3.3 Shift Logical Left

Figure 3.7c shows the 16-bit operand before and after a one-bit SHL (shift logical left). As indicated by the arrows, the bits skew one bit to their immediate left, a zero is shifted into the LSB, and the sign bit is shifted out into C. After a one-bit left shift, the signed integer is multiplied by two. However, if the sign bit changes after the operation from zero to one, or vice versa, that indicates an overflow. In other words, the result is no longer correct due to a programming error, which is also the reason not to implement an opcode for SAL (shift arithmetic left).

The IU generates various control signals as described below:

NS: no shift
SHL: shift left
SHR: shift right, logical or arithmetic
ARI: arithmetic type

If the operand is 32 bits long, the logical equations for the shifter output are:

Sh0 = S0 NS + S1 SHR
Shi = Si NS + Si-1 SHL + Si+1 SHR for i = 1, 2, ... 30
Sh31 = S31 NS + S30 SHL + S31 ARI

where Shi is the shifter output at bit i, and Si is the adder sum bit i. The AND – OR gate implementation for generating Shi is shown in Figure 3.8. For non-shift operations, only no shift (NS) is one, so the adder sum bits are routed to the bus without skewing. For shift left operations, only SHL can be one, so the adder sum bits are skewed 1 bit to the left. For shift right operations, SHR is always one and the ARI bit tells whether to retain the sign bit: one means yes, zero means no. This is a clever design, and its explanation is left as an exercise. For generating Sh0 to Sh30, the sum bit is skewed to its immediate left if SHR is one, and to its immediate right if SHL is one. The sign bit Sh31 is S31 if NS is one, retains S31 if ARI is one, and is S30 if SHL is 1.

3.6.4 Status Register

In a third generation computer, a status register (SR) was developed to record the partial running environment of the current program in execution. The SR is also known as the flags (F) register, or the program status word (PSW) register. After an adder operation, the four condition bits constitute the CC (condition code) as described in the following:

3.6.4.1 Carry Bit

The C (carry) flag bit in the SR (status register) serves a dual function after an arithmetic operation. After an add operation it is a carry, but after a subtract operation it is a borrow. That is, after an add operation, the carry output generated from

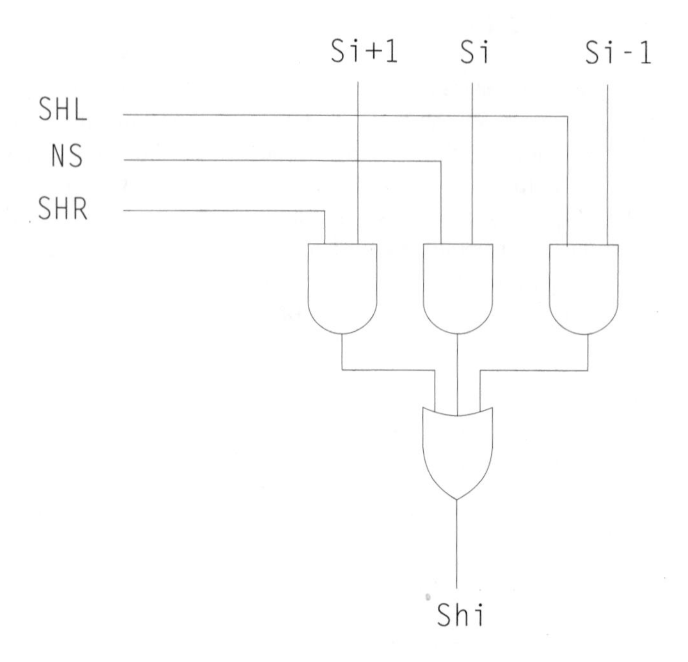

Figure 3.8 AND – OR gate implementation for generating the bit Shi.

the adder is stored in a flip-flop known as the carry bit or the carry bucket. If 32-bit arithmetic is used, the carry output (C32) at bit position 31 is stored. On the other hand, if eight-bit arithmetic is used, the carry output (C8) at bit position seven is stored. Let us try two eight-bit examples. The first example is to add 4 to 3. After the operation, the carry bit is set to zero meaning no carry output as shown below:

```
    0000 0011
+   0000 0100
  ___________
    0000 0111
```

The second example is to add zero to any source operand. Obviously, after the operation the carry output is zero, as is the C flag.

After a subtract operation, if no carry output is generated from the adder, the C flag in the status register is set to one. Thus, the complement of the carry output from the adder is stored in the C flag, and zero means no borrow. Again, we try two eight-bit examples. First, after subtracting 4 from 3, the carry flag is set to one meaning a borrow, because after adding the twos complement of the subtrahend to the minuend, no carry output is generated, as shown below:

```
             1        {Cin bit}
     0000 0011        {+3}
+    1111 1011        {ones complement of +4}
  ------------
     1111 1111        {with no carry output}
```

The sopd1 (+3) is one adder input, the ones complement of sopd2 (+4) is the other input, and the Cin bit is one. Perhaps the borrow bit makes more sense visually if we perform the subtract operation by hand as shown below:

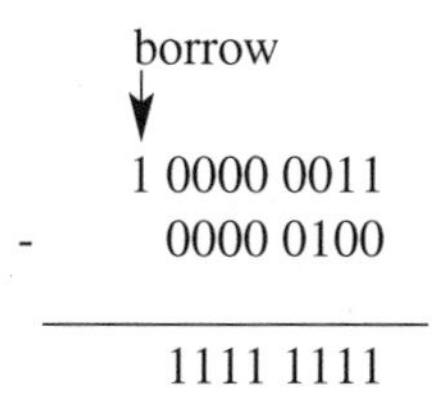

The second example is to subtract zero from any operand. Because the ones complement of zero means all ones and the Cin bit is one, a carry output is definitely generated by the adder. Consequently, the carry flag is set to zero, meaning no borrow.

Furthermore, the carry flag is set after a shift left or a shift right operation, so we summarize as follows:

```
CASE operation of
add:         IF a carry output is generated,
             THEN C  ←  1;      {carry}
             ELSE C  ←  0; ENDIF;
sub:         IF no carry is generated,
             THEN C  ←  1;      {borrow}
             ELSE C  ←  0; ENDIF;
shift left:           C  ←   MSB;
shift right:          C  ←   LSB;
ENDCASE;
```

3.6.4.2 Sign Bit

After an operation through the adder, the sign bit after the shifter (SH31) is clocked into this flip-flop.

3.6.4.3 Zero Bit

After performing an arithmetic or logical operation through the adder, a decoder is used to detect the output of the shifter. If all the bits are zeroes, the Z flag is set to one.

3.6.4.4 Overflow Bit

After an adder operation or a shift left operation, if the result is not correct, the overflow bit is set. For arithmetic operations, an overflow occurs when the two inputs to the adder have the same sign (+ or -), but the sign of the output changes. There are four possibile triggers of an overflow, as listed in Table 3.5.

Table 3.5 Overflow Conditions after an Adder Cycle

Input A	Operator	Input B	Output
(+)	+	(+)	(-)
(+)	-	(-)	(-)
(-)	+	(-)	(+)
(-)	-	(+)	(+)

The adder has two inputs, A and B, and the sign of a 32-bit input is enclosed in a pair of parentheses. When adding two positive integers, if the sign of output changes to negative, it is an overflow condition indicating that the result is too large for the machine to handle. In other words, in twos complement arithmetic the result can not exceed the largest positive integer (2^{31} - 1). This condition may also occur if we subtract a negative integer from a positive integer. When adding two negative integers, if the sign of output changes to positive, it is an overflow. That means the magnitude of the negative result is larger than the magnitude of the most negative integer (-2^{31}). This same condition may also occur if we subtract a positive integer from a negative integer.

Detecting an overflow condition remains a challenge. Intuitively, overflow occurs if both inputs to the adder have the same sign, but output has a different sign. The inputs and the output are specified in Table 3.6.

Table 3.6 Truth Table of Overflow after an Adder Cycle

A31	B31	S31	OA
0	0	0	0
0	0	1	1
0	1	0	0
0	1	1	0
1	0	0	0
1	0	1	0
1	1	0	1
1	1	1	0

We see that A31 is bit 31 of the first input, B31 is bit 31 of the second input, S31 is the sign bit of output, and OA means overflow after add. Note that it is impossible to generate an overflow if the two inputs have opposite signs. Therefore, we have the logic equation:

$$\begin{aligned} OA &= A31\ B31\ \backslash S31 + \backslash A31\ \backslash B31\ S31 \\ &= C32\ \backslash C31 + \backslash C32\ C31 \end{aligned}$$

The simplification from the first line to the second line is left as an exercise. An overflow condition occurs whenever the carry input to the MSB is different from its carry output.

A shift left (SHL) operation is equivalent to multiplying the integer by two. After the operation, if the sign bit changes, it is an overflow. If the logic variable OS is used here to stand for overflow after shift, we obtain its equation as shown below.

OS = \A31 S31 SHL + A31 \S31 SHL

Note that a shift right operation does not generate an overflow. After merging the two equations, the overflow (O) bit equation is shown below:

O = OA + OS
= C32 \C31 + \C32 C31 + \A31 S31 SHL + A31 \S31 SHL

3.7 ADDRESSES IN AN INSTRUCTION

As far as programming is concerned, all computers can be grouped into two classes: register machines or stack machines. In a register machine, one, two, or three addresses are coded in an add instruction, but in a stack machine, the add instruction has no addresses at all. For a binary operation we need two source operandi and after the operation we need a destination operand to store the result. If the number of addresses in an add instruction is less than three, one or more operandi are implied, that is, the number of addresses is reduced because the CPU knows exactly where to find the operand. How many addresses should be coded in an instruction? In early computers, the add instruction used only one memory address. In modern computers, the add instruction has two or three addresses. In general, the address in an instruction can be a memory address or a register address.

3.7.1 Three-Address Machine

A three-address machine means that there are three addresses coded in a binary arithmetic or logical instruction. For example, we show the assembly language statement of an add instruction below:

add dopd,sopd1,sopd2 ;M[dopd] ← M[sopd1]+M[sopd2]

where sopd1 is the memory address of the first source operand, sopd2 the memory address of the second source operand, and dopd the memory address of the destination operand. The RTL in the comment block says that the CPU fetches sopd1 and sopd2, performs the integer add operation, and stores the result in dopd. The order of addresses in an AL statement is logical because the physical order in an instruction may be different from machine to machine. For the add operation, the positions of the two source operandi are irrelevant, but for the sub (subtract), operation the minus sign in the middle means sopd2 is subtracted from sopd1. Such a design has the advantage that after the operation, the two source operandi remain unchanged. The format of a three-address instruction on VAX is shown in Figure 3.9a. The VAX machine supports either two addresses or three addresses in an arithmetic instruction. In a three-address instruction, the one-byte opcode is followed by three operand fields. Each operand field may consist of up to five bytes — an eight-bit addressing mode and a 32-bit long memory address, as shown in Figure 3.9b.

The addressing mode serves as an eop which tells the IU where and how to find

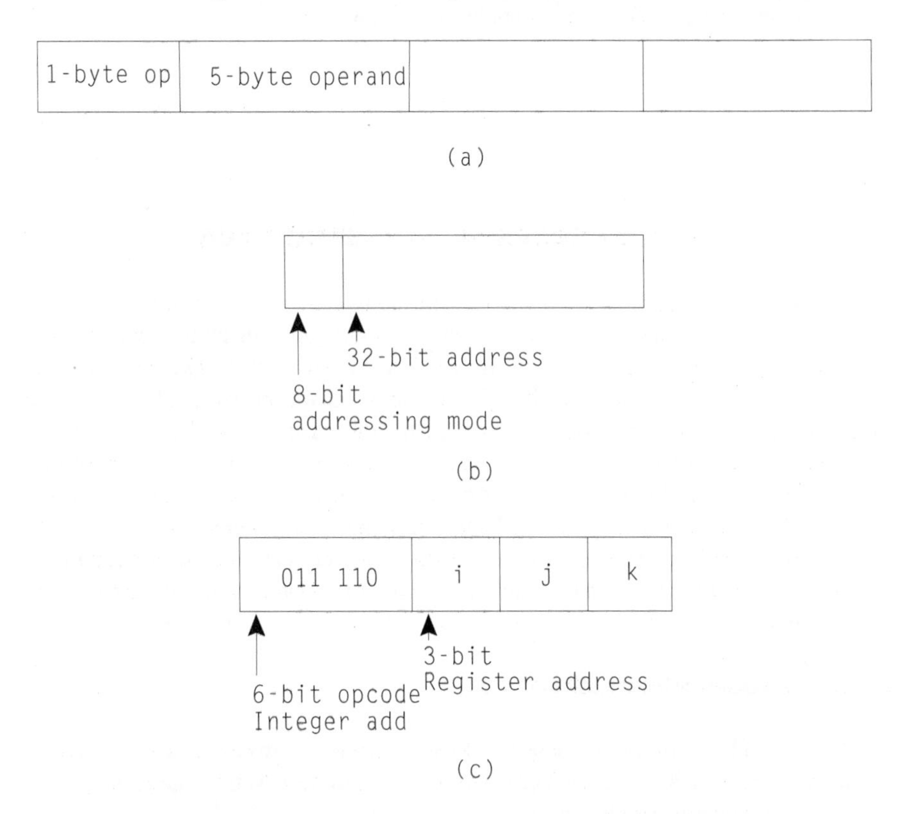

Figure 3.9 Three-address instructions: (a) the 21-byte instruction on VAX, (b) the 8-bit addressing mode with a 32-bit memory address, and (c) the 15-bit instruction on CDC6600.

the operand. The total instruction length is 16 bytes as computed below:

$$1 + 5 * 3 = 16$$

An instruction for extended divide specifies two destination addresses: one for the quotient and another for the remainder. Thus, this instruction is 21 (1 + 5 * 4) bytes long which has four long memory addresses contained in it. If an instruction has four memory addresses, its length is increased and so is the instruction fetch time. It also takes longer for the CPU to fetch many operandi in memory. It is reasonable to suggest that no more than two memory addresses should exist in an instruction. One modern proposal is to specify one memory address in a load or store instruction so that an operand can be moved to or from memory. After the operand is loaded into a register, the CPU can perform register-to-register operations.[16,48] After the operation, the result is a register that can be stored in memory by a store instruction. This idea was first pioneered on the CDC6600, a three-address machine. On such a machine, three three-bit register addresses are coded in the instruction, as shown in Figure 3.9c. The add instruction is 15 bits long as computed below:

$$6 + (3 * 3) = 15 \text{ bits}$$

where the opcode 36 in octal means integer add; i, j, and k are three-bit register addresses ranging from 0 to 7. The following assembly statement is creative in that i = 5, j = 2, and k = 3.

```
IX5    X2 + X3  ;X[5] ← X[2] + X[3]  {integer add}
```

where I specifies integer data type, X denotes the 60-bit operand register set, and the plus sign in the operand field means add. For any i, j, and k, we can generalize the integer add operations as follows:

$$X[i] \leftarrow X[j] + X[k] \quad \text{for } i = 0, 1, ..., 7; \; j = 0, 1, ..., 7; \; k = 0, 1, ..., 7$$

3.7.2 Two-Address Machine

A two-address machine means that only two addresses are specified in a binary operation instruction, arithmetic or logical. An add instruction in AL (assembly language) is shown below.

```
add    sopd1, sopd2  ;M[sopd1] ← M[sopd1] + M[sopd2]
```

The symbol sopd1 is the memory address of the first source operand and sopd2 is the memory address of second source operand. This logical ordering is based on an IBM mainframe or Pentium. The first source operand is implied as the destination operand so the original content in sopd1 will be destroyed after the operation. This design has an advantage that the instruction length is shorter, because the number of addresses is reduced from three to two. The mainframes usually allow two full memory addresses coded in an instruction.[26] The Pentium processor, however, allows one full memory address at most in an instruction in order to reduce the instruction length.[29]

3.7.3 One-Address Machine

All early computers were one-address machines so as to reduce cost. On such a machine, the acc (accumulator) is implied as the first source operand as well as the destination operand. Since two addresses are implied, only one address is coded in an add instruction. It should be mentioned that the address is a memory address. The AL statement for such an add instruction with one address is shown below:

```
add    sopd2      ;acc ← acc + M[sopd2]
```

In the statement, acc is the accumulator and sopd2 represents the address of the second source operand in memory. After the operation, the acc contains the result.

3.7.4 Zero-Address Machine

A stack machine is a zero-address machine because there are no addresses coded in an arithmetic or logical instruction. It uses a stack in lieu of working registers. The stack resembles one used in a restaurant, as depicted in Figure 3.10a. The plate on top is the top of the stack (TOS) that changes after each push or pop. Push stack means that a new plate is placed on top. The push instruction has an address field, so the new plate can be brought on top. Pop stack means that a plate is taken off the stack. The second plate below the previous top emerges as the TOS after a pop. In practice, a stack is simulated in memory and each memory word is a stack element that represents a plate. The stack is a last-in-first-out (LIFO) device. Some stack machine instructions are shown below:

```
push   #17      ;TOS is 0011 in hex
push   #18      ;TOS is 0012 in hex
add             ;pop stack into sopd2
                ;pop stack into sopd1
                ;perform (sopd1 + sopd2)
                ;push result onto stack
pop    result   ;pop TOS into result
```

where sopd1 means the first source operand and sopd2 means the second source operand. Assume that the stack element is a memory word of 16 bits. After executing the first push, the TOS contains 0011 bits in hex or 17 in decimal, and the # sign means that the number itself is pushed onto the stack, as shown in Figure 3.10b. After executing the second push, the TOS contains 0012 bits in hex (18 in decimal), as shown in Figure 3.10c. The execution of the add instruction is done in four steps.

First, pop 18 off the stack as sopd2. Second, pop 17 off the stack as sopd1. Third, perform the add operation (sopd1 + sopd2) via the adder. Finally, push the result onto the stack, as shown in Figure 3.10d. Note that the TOS now contains 0023 bits in hex, which is 35 in decimal. The last pop instruction is executed in two steps. First, it pops 0023 off the stack in one memory cycle as the destination operand. Second, it stores the operand in a memory location named result in another memory cycle. After executing the four instructions, the stack goes back to the previous state as shown in Figure 3.10e. Because there is no address needed in an arithmetic or logical instruction, the stack machine code is short. Its design principles are discussed in the final chapter.

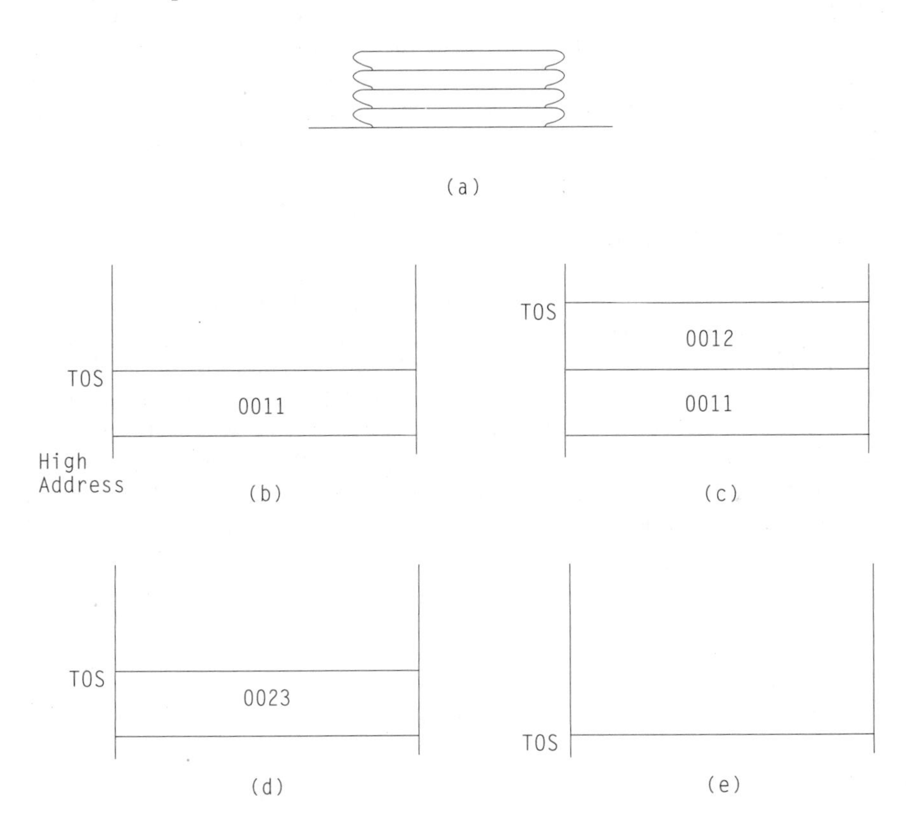

Figure 3.10 Stack operations: (a) stack in a restaurant, (b) push first source operand, (c) push second source operand, (d) add operation, and (e) pop stack into destination.

3.8 HARDWARE EXECUTION

A computer program can generate computation results using one of two methods: hardware execution, or software interpretation. In regard to hardware execution we have two approaches: hardwired logic and microprogrammed logic. It is also possible to combine the two approaches into a hybrid approach. To differentiate between hardware execution and software interpretation, we have the following definitions.

1. Hardware execution means that the instruction is loaded into the IR at a fixed location in the hardware that is not subject to change.
2. Software interpretation means that the instruction is loaded into a memory word named IR whose address can be changed per each run.

Hardware execution is much faster than software interpretation. Furthermore, hardware execution by hardwired logic is faster than execution by microprogrammed logic. The former uses logic gates and flip-flops to interpret an instruction in the IR, and the latter uses a microcode routine in ROM (read only memory) to interpret the instruction in the IR. Commonly, we use the symbol μ for micro, which is 10^{-6}, so μprogram means microprogram or microcode.

3.8.1 Hardwired Logic

Hardwired logic implies gates and flip-flops not including ROM. The instruction opcode in the IR is decoded by AND gates to generate a sequence of timing control signals. In Figure 3.11a, the memory data bus is 16 bits wide so it takes one memory cycle to obtain a 16-bit instruction in the IR. The slash (/) indicates a bus, 16 specifies the number of wires or bus width, and the Enable IR In (EII) signal may be a clock input to IR. If the lower 12-bit field in the IR is a memory address, it is placed on the address bus whenever the Enable IR Out (EIO) signal is on. Figure 3.11b depicts a simplified block diagram without showing the enable signals and bus width.

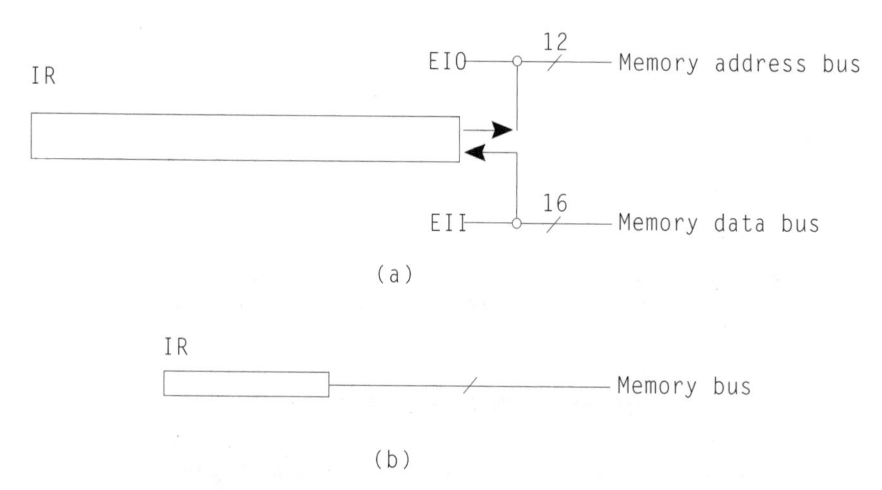

Figure 3.11 Instruction register and memory bus: (a) to address bus and from data bus, and (b) simplified diagram.

The hardware steps to execute the one-address add instruction are as follows. The control logic in the IU needs to generate three timing signals for controlling the operations. The t1 signal in Figure 3.12a enables the lower 12 bits in IR on the address bus, and at the same time a memory read request is issued. The t2 signal in Figure 3.12b fetches the operand from the data bus into a temp register. The t3 signal in Figure 3.12c enables the acc to the input bus of the adder, the temp register to the other input bus, and its trailing edge clocks the adder output into the acc.

The hardwired logic approach is fast, but it lacks flexibility. If there is a design error, we need to make hardware or wiring changes on the chip. As an alternative, we can place a group of microinstructions (microcode) in ROM along with a minimal set of hardwired logic to support the execution of microcode. It is still necessary to debug a smaller set of hardwired logic that supports the microcode execution. Any errors found in the ROM code can be fixed without changing wiring or circuit.

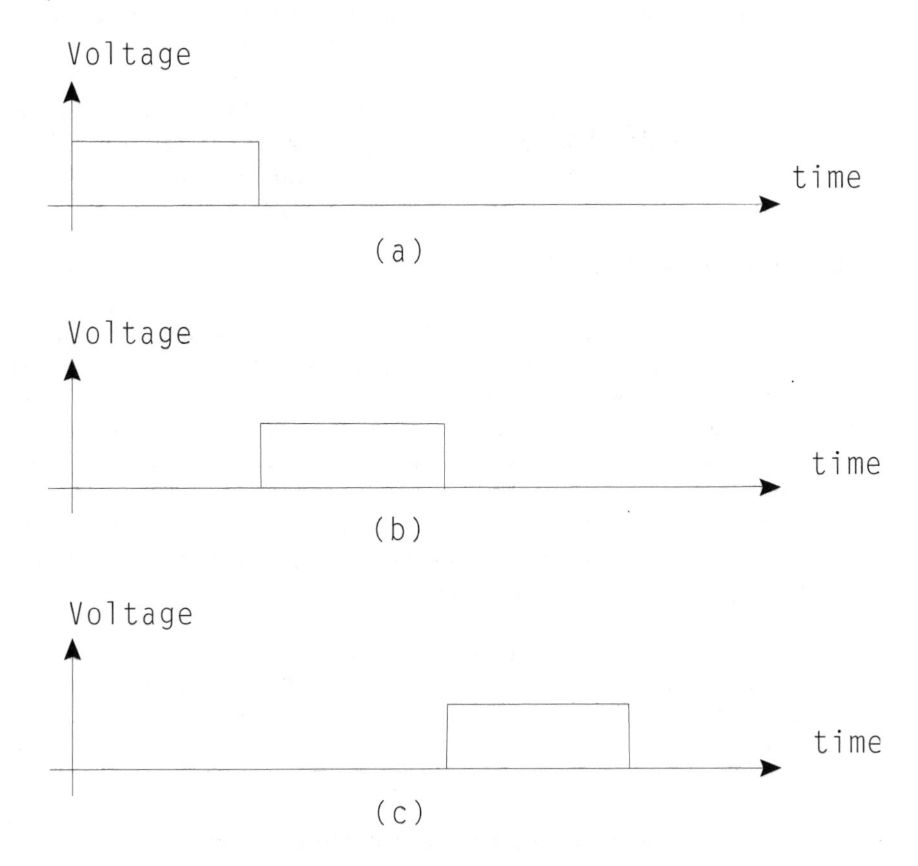

Figure 3.12 Three timing control signals: (a) t1, (b) t2, and (c) t3.

3.8.2 Microprogrammed Logic

Microprogrammed logic execution means that a sequence of microinstructions in ROM is executed to interpret a target instruction in the IR. The interpreter is a microcode program. In comparison, the speed of microcode execution is slower than hardwired logic, but it provides more flexibility. The ROM, also called control store, resides in the instruction unit (IU). The execution of microcode in ROM provides control timing signals to interpret and execute a target instruction. The interpreter in ROM along with its supporting hardwired logic is referred to as the microcode engine. Such a design needs a different set of instruction register and program counter for executing the microcode.

It is easy to interpret the more complex target instructions using microprogrammed logic. As a simple example, the multiply instruction may be decomposed into many shift and add operations. The algorithm for a sequential microcode engine to interpret such an instruction is described later in Chapter 6. At the present time, we introduce the microcode concept that each target instruction is fetched, interpreted, and executed by many microinstructions stored in ROM. Remember that the CPU still needs hardwired logic to obtain, interpret, and execute the microcode in ROM. The speed using microcode logic is proportional to the amount of hardwired logic employed in the engine: the more hardware, the higher the speed.

3.8.3 Hybrid Logic

The hybrid logic approach is derived from microprogrammed logic mainly because of its flexibility.[66.] The instruction fetch stage is hardwired. The decode unit or stage uses the opcode of a target instruction to map it into one or many microinstructions fetched from a ROM. For example, an integer add or subtract instruction between two CPU registers is mapped into one microinstruction. A complex instruction, e.g., divide, sine, or cosine, is mapped into several microinstructions defined as a microorder. The decode unit stores the mapped micro-order in a microcode pool waiting for its turn to execute. When all the resources required by the execution unit are available, the micro order is dispatched from the microcode pool to the execution unit. The target instruction fetch job can be done by the memory bus control unit, so a stream of instructions flows into an instruction queue or pipe on a continuous basis. The key idea is that a high-speed execution unit consists of many functional units that can operate at the same time. The hybrid logic approach is attractive in that it gains both speed and flexibility.

3.9 SOFTWARE INTERPRETATION

Software interpretation is another option for a user to submit an executable program to be interpreted by another program running on a host computer. If we hide the host machine in a dark room, computational results are obtained such that the user can not differentiate whether the program is interpreted or executed. The submitted program is composed of target machine code, but it is treated as data to the interpreter on the host. The interpreter instructions execute on the host in two phases. First, the target machine instruction is placed in a memory location called the IR. Next, the opcode is interpreted in the IR and results are generated.

The interpreter is often written in assembly code as a computer simulator that reads each target instruction from memory in the IR, interprets, and executes. In assembly code, the IR is a symbolic name placed in column one before a define storage statement to mean an address definition. Because the IR is a memory variable, its address may change for each run. In fact, all the working registers of the target machine are simulated as memory variables whose addresses may change for each run.

3.9.1 Host Machine

The computer hidden in the black box is the host machine on which the software interpretation program executes. Computational results are generated just like those from hardware executions. The raw speed of the host machine determines how much time the interpreter requires to complete its job. If the host machine runs a multiprogramming OS, multiple users can execute on the host concurrently. The word concurrently has a different connotation than the word simultaneously. The former means that multiple users can take turns executing on one CPU, so that they can all finish within a specific time interval. The latter means that multiple users can execute at the same time, which is only possible if we have multiple CPUs in the system. That is, each user has his own processor to fetch an instruction at the same instant, interpret, and execute. On a single CPU, the multiprogramming OS acts like a traffic policeman who decides which program will execute next and which program will take a break. Consequently, multiple user programs can execute during the same interval phenomenon known as multiprogramming.

Running under a multiprogramming OS, the interpreter is loaded into a memory partition allocated by the OS. In other words, the loading address of the interpreter is determined by the OS. Since the IR is a memory variable in the partition, its address may also change for each run. The host machine executes the interpreter that further interprets the target machine code.

3.9.2 Target Machine

The target machine, after being built can execute the users' program directly by hardwired logic or microprogramming logic. However, if the target machine is not available, we can instruct the interpreter to interpret its instruction set as long as we understand the target machine spec thoroughly. The interpreter program is a tool with which we can develop an OS for the target machine. First we need other programming tools, such as an editor, an assembler, or a compiler. If the host computer is different from the target computer, the assembler is a cross assembler that runs on the host but generates code for a different target machine. Likewise, the compiler is a cross compiler.

Note that a computer simulator can also be used as a teaching tool because students are not allowed to modify an existing OS running on a university computer. Thus, under a simulator on the host that may be the target machine, the operating system can be developed by using the system software tools. As interpreted by the simulator, the new OS can be fully debugged.

3.10 SUMMARY POINTS

1. A target machine instruction must be stored in memory before its execution can begin.
2. Instructions and data are alike in a computer.
3. A simple CPU has two major hardware units: the IU and the EU.
4. The IU also called the CU, contains the logic to obtain an instruction from memory into IR, interpret the opcode, and generate various control timing signals for executing the instruction.
5. The EU has a set of working registers, and an ALU is where the operand execution takes place.
6. A working register provides the temporary storage designed as an ordered set of flip-flops. The adder in the EU is a crucial hardware component, so the faster the better.
7. After executing an instruction, the IU increases IAR by the length of the instruction in bytes.
8. If the instruction length is fixed, the IAR is a counter.
9. The number of wires on an address bus determines the size of memory or physical addressing space.
10. A wider data bus can transfer many bits in parallel so the future trend is to have 64 bits or more on the data bus.
11. The internal computer operations can be described by RTL.
12. The memory contains two registers: MAR and MDR. Both are buffers that interface with the internal circuitry in the memory.
13. The execution of an instruction is divided into two phases: instruction fetch and operand execution.
14. If no memory operand is specified, operand execution may begin right after instruction fetch.
15. The instruction is executed by an infinite hardware loop that can halt by executing a specially designed instruction.
16. Instructions are executed in sequence unless there is a branch instruction.
17. A branch instruction changes the IAR so the next instruction is fetched from a new address.

18. When the CPU executes a branch instruction, the new branch address specified in the code is loaded into the IAR.
19. A conditional branch instruction tests a condition first, and if the result is true, it branches.
20. After a program terminates, the OS resumes control to select and execute the next program.
21. In a register machine, the add instruction may contain three addresses, two addresses, or one address. In a stack machine, the add instruction contains no address.
22. Hardware execution means that the IR is fixed in hardware design while software interpretation means that the IR is in memory. Regardless of the approach, each instruction is obtained, interpreted, and executed.
23. There are three hardware execution approaches: hardwired logic, microprogrammed logic, and hybrid logic.
24. In computer simulations, the host machine and the target machine may be the same.

PROBLEMS

1. Explain the functions of an instruction register and an instruction address register.
2. If all the instructions are four bytes long, how can we design the IAR as a binary counter? Justify your answer.
3. Explain the functions of MAR and MDR in memory. Are they buffers?
4. Given a one-address add instruction, use RTL to describe:
    a. the instruction fetch phase
    b. the operand execution phase
5. Given a 20-bit address bus, what is the physical addressing space of memory?
6. Design a four-bit look-ahead binary adder and specify the logical equations for its carry output.
7. By expanding the carry look-ahead concept to a 16-bit adder ideal for a 16-bit PC, what should the equations be for C15 and C16?
8. By expanding the carry look-ahead concept to a 64-bit adder, what should the equations be for C63 and C64?
9. Explain the carry input (Cin) signal at the LSB of an adder?
10. A microinstruction may contain a shift control field that has four enable bits defined below:

    NS = No Shift
    SHL = Shift Left
    SHR = Shift Right
    ARI = Arithmetic Type

For the NS and SHL signals, if any of them is one, the other three bits in the field must all be zeroes. For shift logical right, the SHR bit is on but the ARI bit is off. For shift arithmetic right, both SHR and ARI are on. We may use two separate enable signals, one for shift logical right enable (SLRE) and one for shift arithmetic right enable (SARE). Interestingly, they can be encoded into ARI and SHR, and the logic is simpler. The truth table is shown below:

Input		Output	
ARI	**SHR**	**SARE**	**SLRE**
0	0	0	0
0	1	1	0
1	0	—	—
1	1	0	1

The dash means a don't care because both ARI and SHR must be on at the same time. If the operand has 32 bits, we obtain the logical equations for the shifter output as follows:

Sh0 = S0 NS + S1 (SLRE + SARE)
Shi = Si NS + Si-1 SHL + Si+1 (SLRE + SARE) for i = 1, 2, ... 30
Sh31 = S31 NS + S30 SHL + S31 SARE

where Shi is the shifter output at bit i, and Si is the adder sum bit i. Use the don't cares and prove that the following equations are correct.

a. SARE = ARI
b. (SARE + SLRE) = SHR

Therefore, the final logic equations are chosen as follows:

Sh0 = S0 NS + S1 SHR
Shi = Si NS + Si-1 SHL + Si+1 SHR for i = 1, 2, ... 30
Sh31 = S31 NS + S30 SHL + S31 ARI

11. What are the logical equations of the one-bit shifter output after a 32-bit binary adder?
12. The overflow condition after add (OA) is true as shown below:

OA = \A31 \B31 S31 + A31 B31 \S31
= \C32 C31 + C32 \C31

a. From the condensed five-variable truth table, we see that either answer is correct by intuitive inspection. Derive the second line from the first by means of the dashes (does not matter).

b. Using 4-bit arithmetic, verify that
   i. 0 + (-8) = -8
   ii. 0 - (-8) generates overflow. {Remember that Cin = 1.}

13. Explain the functions of the carry (C), sign (S), and zero (Z) flags in SR.
14. Why is an overflow condition not generated after a shift right operation, arithmetic or logical?
15. What is the key difference between hardware execution and software interpretation of a target instruction?
16. Name the three design approaches to execute a target instruction by hardware. Which approach do you like best?
17. Describe the execution sequence of an add instruction in a stack machine.
18. Which machine architecture do you prefer, a two-address machine or a three-address machine? Justify your answer.
19. Why can the adder width decide the performance of a computer?
20. What is the reason to write a computer simulator for a target machine that is also the host machine?

CHAPTER 4

Assembly Language Principles

4.1 INTRODUCTION

An assembler translates a source program from AL (assembly language) to object code. Because a mop (machine opcode) statement is always translated into one instruction, AL is referred to as machine language. The AL is used as a tool to understand instruction executions at the functional level. In practice, AL is used to write real-time applications for its speed and efficiency. In some cases, an application program consists of two parts: one in high-level language and one in assembly code. After translation, the two object modules are linked into one executable module. In a PC, the BIOS (basic I/O system) and the boot routine are all written in assembly code, and their executable routines are placed in ROM. It is also quite common for the OS to be written in a high-level language with assembly code inserted, which is also called in-line assembly. For debugging convenience, a C compiler often translates the source statements into assembly code, then into object code. This chapter introduces the Pentium processor, its instruction set, and the basic addressing modes. One objective of this chapter is to develop some basic AL coding skills.

4.2 PENTIUM PROCESSOR

The Pentium processor, also called the IA-32 (Intel architecture-32 bits) supports two-address instructions in general. As the processor is backward compatible with the previous generations of processors, the basic ISA (instruction set architecture) remains the same for the 8086, 386, and 486 families.[28] The 8086 is a 16-bit machine, while the rest are 32-bit machines as determined by adder size. The 16-bit applications cover editors, word processors, stock market quotations, WWW (World Wide Web) browsing, e-mail, and games. For such applications, a 16-bit adder in the CPU would be adequate. For 32-bit applications, we need a 32-bit adder that can perform computations at a higher speed.

Regarding operand address, each instruction can specify at most one address directly pointing to the operand in memory. A 16-bit address can access a 64 KB memory segment but a 32-bit address can access a 4 GB memory segment. The memory address in an instruction is a relative offset with respect to the segment. Cleverly, a 16-bit segment base register is used to map a program address, i.e., offset into a linear memory address of 20 bits or 32 bits. Granted that a large memory can be used to store large arrays. However, as developing software skills is concerned, a memory of 1 MB is more than adequate.

4.2.1 Register Set

The IR resides inside an instruction queue and is left justified, as shown in Figure 4.1a. The IU fetches the instruction stream continuously from memory into the queue. The Pentium processor supports four accumulators, known as data registers. The register size is 16 bits for 8086 but 32 bits for Pentium. On Pentium, the 32-bit data registers are named EAX, EBX, ECX and EDX, where E denotes extended. As the low order 16-bit word of a 32-bit data register is called AX, BX, CX, or DX, the Pentium can execute 8086 instructions as well. One unique feature is that the low order or high order byte of each low order 16-bit data register can be accessed by an instruction. For example, the 16-bit AX consists of two bytes: AH (AX high) and AL (AX low). Either byte, AH or AL, can be accessed by an instruction.

In programming, some of the registers are used to store addresses and they are referred to as address registers. The number of bits in an address register determines the program addressing space. On an 8086, four 16-bit address registers are named SI (source index), DI (destination index), SP (stack pointer), and BP (base pointer) as shown in Figure 4.1c. On Pentium, the address registers are extended to 32 bits, namely ESI, EDI, ESP, and EBP. They are mainly used as address registers, or index registers. The indexing scheme is innovative in that a hardware register is added to the program address to make an EA (effective address). Note that one of the four accumulators, BX or EBX, can be specified as an index register. The mnemonic B stands for base, a variation of indexing.

On an 8086, there are four segment registers named CS (code segment), DS (data segment), SS (stack segment), and ES (extra segment). The segment register concept is intricate in that the segment register is used to map a program address into a physical memory address as explained later. Normally, CS points to the code segment, DS points to the data segment, SS points to the stack segment, and ES also points to a data segment. The Pentium processor adds FS (F segment) and GS (G segment) for data segments so that the total is six as shown in Figure 4.1d.

Figure 4.1e shows an IP (instruction pointer), EIP (extended IP), F (flags), and EF (extended F). Again, the extension names are for Pentium processors. The IP is a program counter that contains a 16-bit offset pointing to the instruction in the code segment. The flags register contains status information, i.e., a running environment, for the current program.

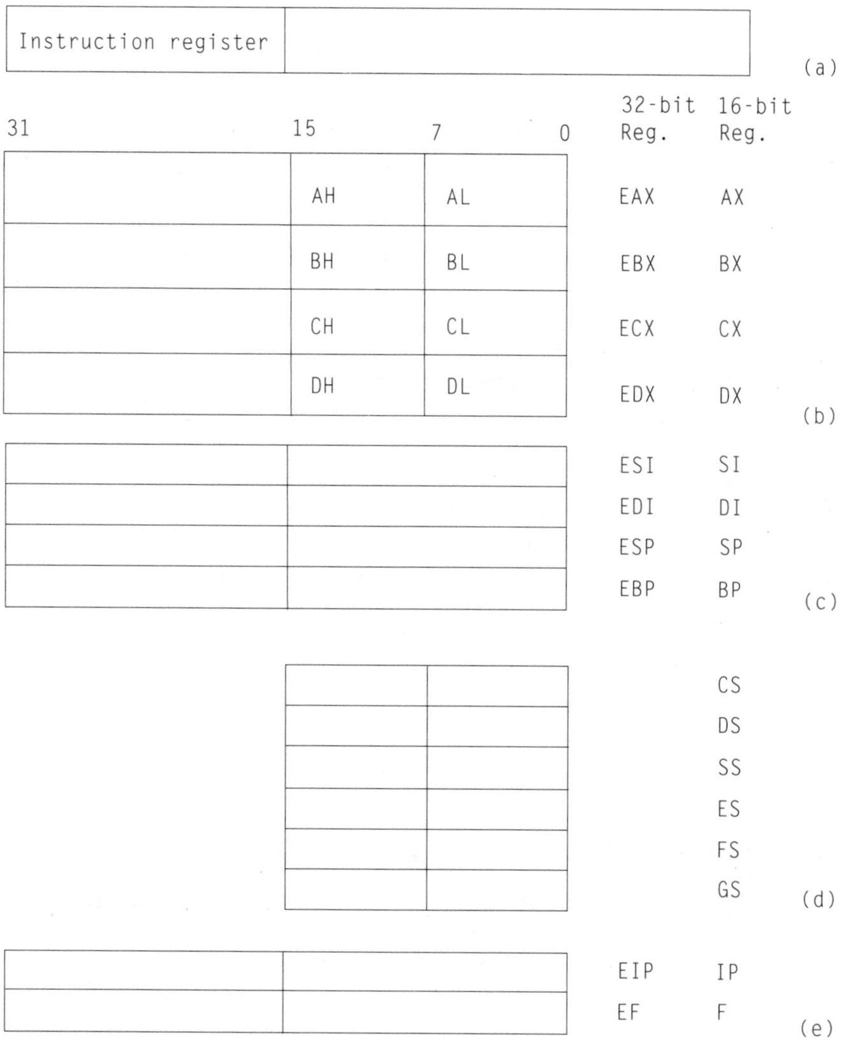

```
Instruction register                                        (a)

                                          32-bit  16-bit
31              15        7          0    Reg.    Reg.
                AH        AL              EAX     AX
                BH        BL              EBX     BX
                CH        CL              ECX     CX
                DH        DL              EDX     DX      (b)

                                          ESI     SI
                                          EDI     DI
                                          ESP     SP
                                          EBP     BP      (c)

                                                  CS
                                                  DS
                                                  SS
                                                  ES
                                                  FS
                                                  GS      (d)

                                          EIP     IP
                                          EF      F       (e)
```

Figure 4.1 Register set: (a) instruction queue, (b) data registers, (c) address registers, (d) segment registers, and (e) instruction pointer and flags register.

4.2.2 Segment Concept

The Pentium is a segment based machine. A segment is a contiguous block of memory that is logically intact. A segment may contain instructions, data, or stack, but a composite segment may contain a combination of those. The Pentium processor can operate in any of the modes listed below:

1. Virtual 8086 mode
2. IA-32 real mode
3. IA-32 protected mode

The Virtual 8086 mode really means IA-16 real mode because the data register is 16 bits and the offset, i.e., program address, is 16 bits. Thus, each segment size is limited to 64 KB. The IA-32 real mode means that the data register is 32 bits but the offset is still 16 bits so the segment size is 64 KB. The IA-32 protected mode means three things. First, the data register is extended to 32 bits. Second, the offset is extended to 32 bits to access a segment of 4 GB. Third, the CPU uses tables in memory to map a 32-bit program address into a memory address.

4.2.2.1 Real Address Computation

When the processor operates in IA-16 or IA-32 real mode, the segment register contains a base address pointing to the segment. The base address in a segment register is a 16-bit paragraph number, and each paragraph is a block of 16 bytes. Therefore, the base address can be visualized as a 20-bit linear byte address with four trailing zeroes implied. Before accessing an operand in memory, its real memory address, i.e. physical memory address, is computed as the sum of the offset and the base address after being shifted to the left by four bits. Mathematically, the offset is added to the paragraph number multiplied by 16. In other words, the lower four bits in the offset remain unchanged in the final 20-bit linear address. This extra add cycle is performed at run time so a total of 1 MB of physical memory can be accessed by an instruction. Which segment register should be used to generate the 20-bit linear address? The default rule is specified in Table 4.1.

Table 4.1 Default Rule to Select a Segment Register

Type of Reference	Segment Register	Default Selection Rules
Instruction	CS	All instruction fetches.
Stack	SS	All stack pushes and pops. All memory references using BP as the base register.
Local Data	DS	All operand references except stack and string destination.
Destination (Strings)	ES	Destination reference of all string operations.

When fetching an instruction, the CS and IP pair is used to compute the physical memory address. The symbol CS:IP represents a 20-bit linear address. The colon (:) means that the IP is added to (CS * 16) at run time, so the adder output is a 20-bit linear memory address. For instruction retrieval, the default base is in the CS and the offset is in the IP.

In Figure 4.2a, the CS contains 13C6 and the IP contains 0100, so the 20-bit linear address is 13D60. Note that the paragraph number in the CS is appended with four implied zeroes. As shown in Figure4.2b, the lower four bits in the offset do not go through the adder, therefore, the adder is 16 bits. We show the arithmetic in hex on the left and its binary equivalent on the right. We conclude that the offset in an instruction is a logical or virtual address as seen by the programmer. At run time, this logical address is mapped into a physical address.

For data operand retrieval, the instruction contains the 16-bit offset and the DS contains a base address in paragraphs. The 20-bit linear memory address is computed as the sum of the offset and DS after being shifted four bits to the left. That is, the DS contains a 20-bit base address in bytes with four trailing zeroes implied. In other words, the DS points to the first byte in a segment. The 16-bit offset is a relative displacement with respect to to the beginning of a segment. It is possible to have multiple data segments in a program because the PAS (physical addressing space) is 1 MB.

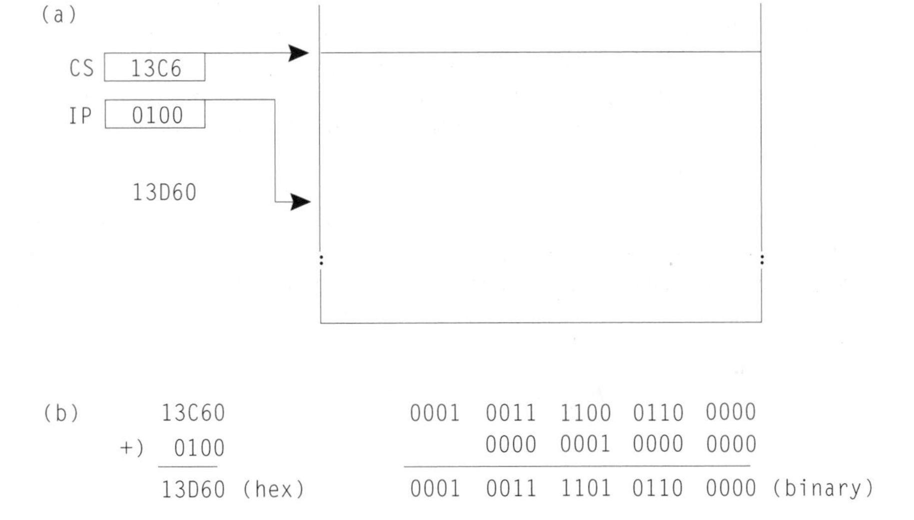

Figure 4.2 Instruction retrieval: (a) code segment register and instruction pointer and (b) 20-bit linear address computation.

A segment register and offset pair must be used to access an operand in memory. The operand can be instruction, data, or stack. An instruction may contain a field known as an addressing mode that tells the CPU how to find the operand. Indexing is an addressing mode that tells the CPU to compute the EA (effective address) as the sum of an offset and the content of a hardware register. After indexing, the CPU still needs to add the base in a segment register to the EA to form the 20-bit linear memory address. For data operand retrieval, the default base is DS as long as SI, DI, or BX is specified as the index register. However, remember this exception. If BP (base pointer) is specified as the index register, the default segment base is SS. This is a clever design so that an operand in the middle of the stack can be accessed.

Just like other computers, the stack in memory grows from high address to low address during a push operation. Whenever a push or pop instruction is executed, SS serves as the segment register and SP (stack pointer) contains an offset pointing to the TOS (top of the stack). As we view the memory, the stack is upside down and the address of the TOS is represented by SS:SP, which represents a 20-bit linear address of (SS * 16 + SP).

In a string operation, the source operand reference is just like any other operand, so an offset is added to the base in DS. However, the destination reference coded in an instruction uses the ES segment register as the default for the sake of convenience. The key concepts are summarized below:

1. To fetch an instruction, CS is always the base.
2. To access the TOS, SS is always the base for stack instructions. To access an operand in memory using BP as the index register, SS is the default base.
3. To access an operand in memory with SI, DI, or BX as the index register, DS is the base.
4. For string operations, the source operand uses DS as the base, but the destination operand reference uses ES as the base.
5. To access an operand, source, or destination, if a one byte prefix is placed in the front as a segment override, the prefix tells the CPU which segment register is the base.

In real mode, the PAS is 1 MB but the operand size can be 16 or 32 bits. Note that one or more segments may be overlapped in memory. As a special case, Figure 4.3 shows that all four segment registers contain the same base address.

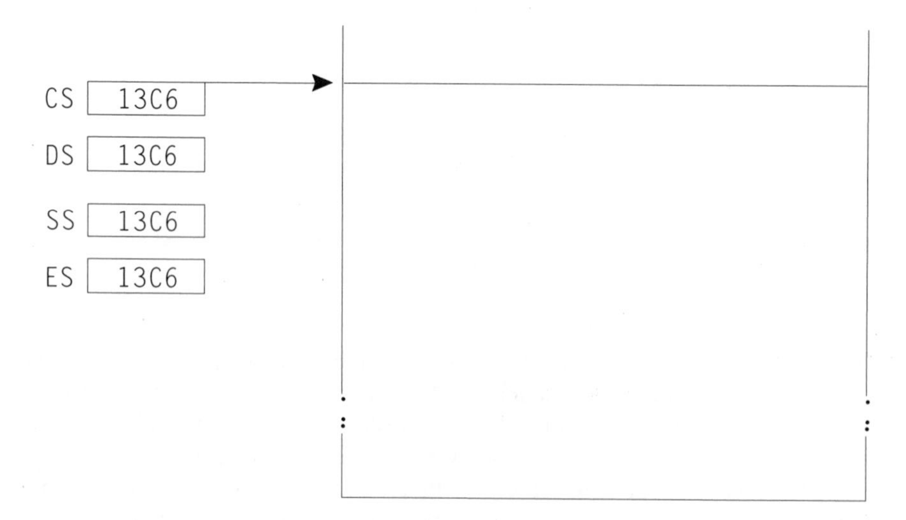

Figure 4.3 All the segment registers contain the same base address.

4.3 ASSEMBLER LANGUAGE STATEMENT FORMAT

An assembly language program consists of many statements, each of which is a line or string of ASCII characters divided into four fields as shown in Figure 4.4. The language uses a free format so one or more spaces can be used to separate two fields and there is no fixed boundary between two fields. A space line may be inserted between statements for the purpose of clarification. The four fields, address, opcode, operand, and comment, constitute a message that tells the assembler what to do. The AL statement is case insensitive.

Figure 4.4 Assembly language statement format.

The first field may contain a symbol as an address definition. The symbol, if existent, represents the name of a memory address (location) where code or data bits are stored. Thus, the terms symbol, symbolic name, and symbolic address are all synonymous. Next is the op field: mop (machine op), pop (pseudo op), or macro call. A mop is translated into an instruction, a pseudo op is not, and a macro call is usually expanded into many statements defined as a macro body. A pseudo op is referred to as an assembler directive because it passes information to the assembler. A pseudo op does not generate code except define constant or define storage. A define constant code generates data in memory with a specific bit pattern, while a define storage code merely reserves a memory block with whatever bits in it. The term constant is similar to a mathematical constant except in assembly programming the defined constant can be changed by instructions in a program at run time.

In a mop statement, the operand field specifies how to locate the operand. For define constant, the operand field specifies the bit pattern. There may be zero, one, or more operandi separated by commas as the delimiter. The fourth field is comment which is optional. Each comment begins with a semicolon (;), and may start in column one as a separate line.

4.3.1 Symbolic Address

There are two symbolic address types: relocatable and absolute. An absolute symbol is translated into a binary number in the address field of an instruction by the assembler. This is an absolute address that requires no further modifications. In contrast, a relocatable symbol is translated into the code as an address that is subject to further changes during linking and loading. There is a big difference between a symbol definition and a symbol reference. The former means that a symbol regardless of its address type, must be defined before the opcode field in a statement. The latter means that the defined symbol may be referenced in the operand field of a statement before or after its definition.

4.3.1.1 Symbol Definition

A symbol placed before the opcode is an address definition. Because a symbol is the name of a memory address as translated by the assembler, each symbol can only be defined once in a program. In the case that the same symbol is defined more than once, the assembler flags this error and generates no object code. It should be stressed that a symbol referenced in a high-level language statement represents the content of a location, but a symbol referenced in an AL statement represents the address of a location. That is to say, a symbol coded in the operand field is an address reference and the assembler translates it into a binary address in the instruction. The MASM (Microsoft assembler), editor, and debugger are available on the web or from some text books.[31] The language is intuitive but some semantic rules need to be clarified. For example, a symbol for a data location has no colon suffix, but a symbol for an instruction location must have a colon suffix to mimic the label used in a high-level language. As a result of this restriction, a program symbol may be placed after column one in the line as long as it is preceded by spaces. It should be mentioned that for most assemblers, a symbol must be defined in column one. Some MASM symbol definitions are shown below:[81]

```
HUNDRED     equ  100       ;equate the absolute symbol HUNDRED to
                           ;100 in decimal

opd1        dw   100       ;define a 16-bit word containing the
                           ;integer 100 in decimal whose address
                           ;is named opd1.

opd2        dw   255       ;define a 16-bit word containing 255
                           ;in decimal whose symbolic address opd2
                           ;is 2 bytes greater than opd1.

b0010:      sub  ax, ax    ;ax ← ax - ax
                           ;b0010 is the symbolic address or label
                           ;of the sub instruction.
```

The equate (equ) statement defines the symbol HUNDRED as absolute with a value of 100 in decimal. An absolute symbol is coded in upper case letters because it is treated as a numeric absolute address. The next two statements are also pseudo ops because they generate data instead of instructions. The first dw (define word) asks the assembler to prepare a 16-bit memory word in the segment whose address is opd1 and whose value, i.e., content, is 100. The next dw defines a 16-bit word containing 255 in decimal whose symbolic address opd2 is 2 bytes greater than opd1. The symbolic address of the sub instruction is defined as b0010 (backward 0010). In practice, a relocatable symbol usually starts with a lower case letter followed by letters or digits. The subsequent English term in such a symbol may start with an uppercase letter so as to improve readability. For example, we may use intExit to stand for interrupt exit. In the sub instruction, because the two source operandi are identical, the result is always zero in AX after execution. If a symbol exists before an empty line, it represents the address of the next instruction in sequence.

4.3.1.2 Symbol Reference

If a symbol appears in the operand field of an instruction, it is a symbolic address reference. A symbol can be defined only once but can be referenced many times in a program. In contrast to a high-level language, a symbol reference is translated into the code as a binary address, i.e., an unsigned integer. The symbol may be referenced in the operand field of a statement before or after the position where the symbol is defined. The assembler knows how to translate each symbol reference in the code as an offset with respect to the beginning of a segment. A coding example is shown below:

```
            mov    ax, opd1 ;ax ← M[ds:opd1]
            jmp    skip010  ;ip ← skip010
            ...
skip010:    jmp    skip010  ;ip ← skip010 {infinite loop}
```

The mov (move) instruction really means copy so the second source operand in memory named opd1 is copied into AX, which is the destination. After execution, the content at opd1 remains unchanged. In the transfer language, the 20-bit linear address is symbolically represented by ds:opd1, a segment base and offset pair. The M[ds:opd1] notation can be abbreviated as M[opd1] as long as we understand that ds is the default segment base. A jmp (jump) instruction passes control to another instruction whose address is specified in the operand field. The first jmp instruction passes control to the instruction at skip010. At this address, a second jump instruction passes control to itself, that is, after executing the jmp instruction, the 16-bit offset in the instruction is copied into an ip pointing to itself. Therefore, the second jmp skip010 instruction is executed repeatedly in an infinite loop; this is purposefully designed to facilitate debugging. The jmp instruction is the same as an unconditional goto statement in a high-level programming language.

4.3.2 Address Expression

In the operand field, we may specify an address expression as shown below:

```
mov  cx, opd1+2  ;cx ← M[ds:opd1+2]
                 ;cx contains 255 in decimal or
                 ;ff in hexadecimal.
```

The address expression opd1+2 tells the assembler that two is added to opd1, and the sum is placed in the code as an offset. As a matter of fact, the assembler has the intelligence to interpret the address expression, so the addition is done at assembly time. After the instruction is executed, the memory content at an offset that is two bytes greater than opd1 is copied into cx. Because the symbol opd1 represents a relocatable address, the numeric two is an absolute address, so after adding the sum is treated as a relocatable address. After execution, cx contains 255 in decimal. Again, M[ds:opd1+2] can be written as M[opd1+2] with the understanding that ds is the base. The basic rules to determine the address type of an address expression are listed below:

1. If an absolute address is added to or subtracted from a relocatable address, the result is relocatable.
2. If an absolute address is added to or subtracted from an absolute address, the result is absolute.
3. If a relocatable address is subtracted from another relocatable address, the result becomes absolute.
4. A relocatable address can not be added to another relocatable address because doing so makes no sense.

Most assemblers treat all addresses equally regardless of the length of an operand. Nevertheless, MASM performs type checking on an operand based on its length, e.g., byte, word, or double word. Assume that we use a mov (move) instruction to store a byte in opd1 that is defined as a word. Because the length attribute is not compatible, the keyword byte ptr (pointer) is placed in front of the symbolic address so as to override the type checking rule. Some examples are shown below:

```
mov  byte ptr opd1, cl      ;store low order byte
mov  byte ptr opd1+1, ch    ;store high order byte
mov  opd1, cx               ;same as above
```

In the first instruction, the cl is copied into M[ds:opd1], the low order byte. The second instruction copies ch into M[ds:opd1+1], the high order byte. Therefore, as far as function is concerned, the two mov byte instructions combined are equivalent to the third mov word instruction.

4.4 PSEUDO OPS

A pop (pseudo op) is used to pass information to the assembler and it does not generate a machine instruction. Except for define constant and define storage, a pop does not even generate code. A pop may signal the beginning or the end of a procedure, segment, or program. A pseudo op is also called an assembler directive so to differentiate from a mop.

4.4.1 Define Data Operand

There is a subtle difference between define constant code and define storage. A define constant generates a specific bit pattern in memory while a define storage code reserves memory block with any bit pattern.

4.4.1.1 Data Operand in Memory

The four generic data types are byte, word, double word, and quad word. Let us examine the length of each operand and not the bits contained within it. As shown in Figure 4.5a, a byte has eight ordered bits located at address N. Because Pentium is a little endian computer, in a memory word its low order byte is first stored at address N, and its high order byte is stored at address N+1 as shown in Figure 4.5b. The bit number ranges from 0 to 15 on top, and b0 is the LSB. As in a double word, the low order word is followed by the high order word as shown in Figure 4.5c. In each word, the low order byte is stored first. A quad word has 8 bytes as shown in Figure 4.5d with the low order double word stored before the high order double word. In each double word, as usual, the low order word is stored first. It should be mentioned that an IBM mainframe is a big endian computer. Not only does its word have 32 bits but the byte ordering in the word is also reversed.

MASM uses the mnemonic DB (define byte) to define a sequence of bytes. If the operand field has a specific bit pattern, it is define constant. If the operand field is specified as a question mark, it means define storage or reserve byte. We can verify the following examples in an AL program. In the listing file, a defined constant statement has its code image shown on the LHS, but for define storage, only a question mark is displayed.

Example 1: Define byte.

Code image	Statement		
30 31 32 33	db	'0', 31h, 62q, 51	;Define 4 bytes
?? ?? ?? ??	db	?, ?, ?, ?	;Define 4B storage
04 [??]	db	4 dup (?)	;Same as above

On the LHS (left hand side), the code is shown in hex. The byte sequence is from left to right, and the leftmost byte has the lower address. The first DB op generates four bytes as specified by its operand field. The first byte is an ASCII zero that is a zero enclosed in single quotes. The second byte is 31 in hex as denoted by the suffix h (hex). The third is 62 in octal as denoted by the suffix q (octal). The fourth byte 51 has no suffix, so it is in base 10 as the default. As we can see from the code image, the byte sequence is 30 31 32 33. Note that there is a space between every two bytes for clarification purposes and the space character is not part of the code. The second statement contains four question marks in its operand field, so any random bit pattern is acceptable, that is, we use the question mark symbol in the operand field to

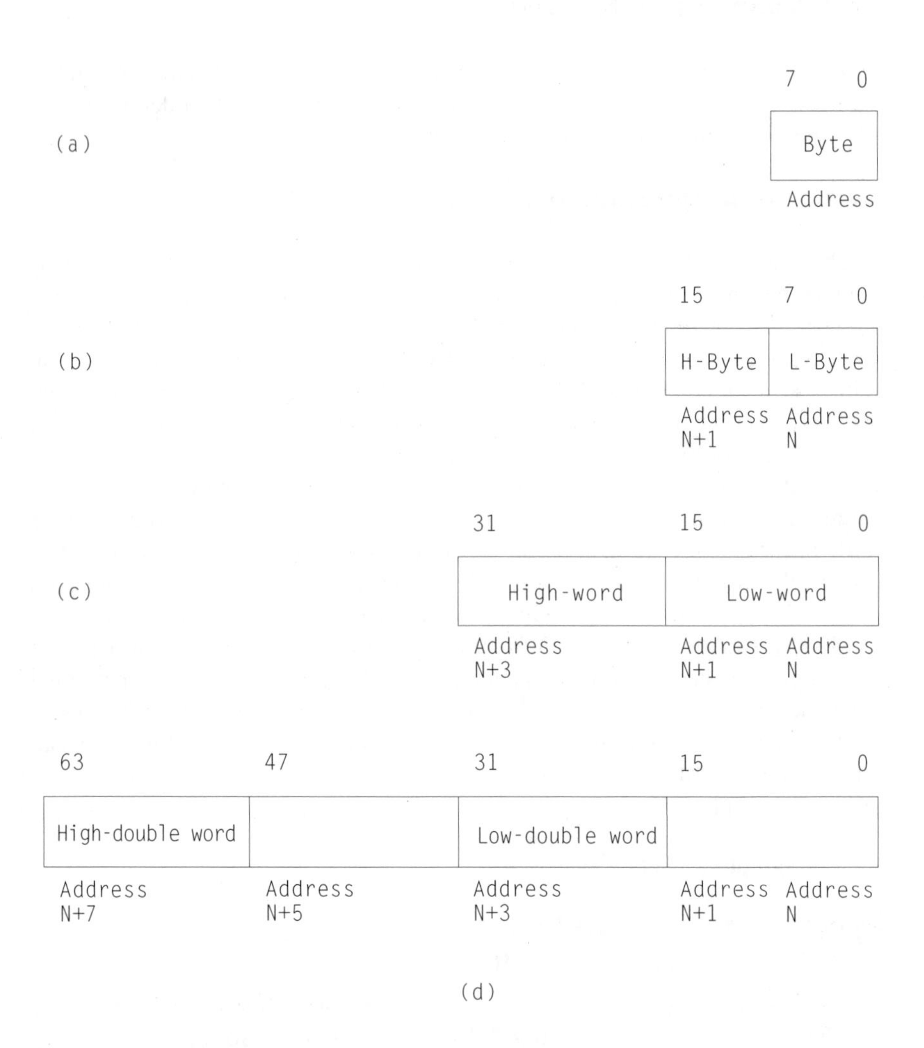

Figure 4.5 Basic data operand lengths: (a) byte, (b) word, (c) double word, and (d) quadruple word.

define storage. The third statement has an integer four followed by the dup (duplicate) keyword, followed by a question mark enclosed in parentheses. The 4 dup keyword means that whatever is specified in the parentheses is duplicated four times. In the code image, we show symbolically that 04 is followed by [??], which means four bytes with any bit pattern. If the symbol ? is changed to zero, then 4 dup specifies that a byte zero is to be duplicated four times.

Example 2: Define word.

Code image	Statement		
0123 FFFF	dw	0123h, -1	;Define 2 words
FFF0	dw	0fff0h	;Define 1 word

In the 16-bit code image on the left-hand side, there is no space between two bytes, which symbolizes a logical representation. In memory, the word 0123 in hex has a byte ordering. in which 23 is stored first at address N followed by 01 as shown in Figure 4.6a. The second operand -1 is 16-bit ones in twos complement notation.

If a number has the suffix h (hex), it is a hexadecimal constant. Note that if the number starts with a non-digit (a to f), it is necessary to add a prefix zero to differentiate it from a symbol. For example, in the second statement, we have 0fff0h in the operand field which means 12 ones followed by 4 zeroes and the prefix zero is not part of the number. In fact, the define constant statement merely tells the assembler to prepare the specific bit pattern in memory. Therefore, the constant is not different from a memory variable that can be modified as the result of instruction execution. If a numeric constant is specified in the operand field of an instruction, it is an absolute address, not subject to change by assembler, linker, or loader.

Example 3: Define double word and quad word.

Code image	Statement		
000000FF	dd	255	;Define decimal 255 in double word
??	dq	?	;Define a quad word
3FB999999999999A	dq	0.1	;Define a double precision float ;whose value is decimal 0.1

The first dd (define double word) statement defines four memory bytes containing decimal 255, as shown in Figure 4.6b. The byte ordering is FF at a lower address followed by three bytes of zeroes. The next dq (define quad word) statement has the ? keyword in the operand field, so it means define storage of one quad word. The third statement defines a quad word that contains 0.1 in double precision and its 32-bit logical representation is shown on the LHS. To request the 80-bit extended precision, simply change dq to dt (define ten bytes).

4.4.1.2 Address Constant

A memory address constant can be stored in memory at assembly time. That is to say, an address may be stored in a memory location that has a symbolic name. In a high-level language, this symbol represents a pointer variable. On a Pentium, a word may contain an offset while a double word may contain an offset, followed by a segment base address. A pointer variable in a high-level language contains an address, as shown by the next two examples.

(a)

Address	
N	23
N+1	01
N+2	FF
N+3	FF
N+4	F0
N+5	FF

(b)

Address	
N	FF
N+1	00
N+2	00
N+3	00
N+4	??
N+5	??
N+6	??
N+7	??
N+8	??
N+9	??
N+10	??
N+11	??

Figure 4.6 Memory representations in hex: (a) three words and (b) one double word followed by one quad word.

Example 4:

```
ptr1       dw    opd1        ;16-bit offset opd1
farptr1    dd    DSEG:opd1   ;16-bit offset opd1
                             ;followed by DSEG
```

The symbol ptr1 (pointer 1) is the name for a 16-bit memory location that contains the offset opd1. Note that ptr1 is an address definition, but opd1 is an address reference that is, ptr1 contains an address symbolically represented by opd1 which is defined somewhere else. In the second line, farptr1 (far pointer 1) is the symbolic name for an offset of a 32-bit double word. The first word contains an offset named opd1 and the second word contains a base address named DSEG. Thus, the offset opd1 is stored at a memory address named farptr1 and the base address DSEG is stored at an address farptr1+2. The offset and the segment base represent a far address. Again, farptr1 is an address definition and its operand field has two references specified as DSEG:opd1. The assembler does not know the base address of DSEG at assembly time. However, after linking and loading the correct 16-bit paragraph number is prepared at farptr1+2 before program execution.

Example 5: Address constant and literal symbol.

```
acon2    dw    =100
=100     dw    100
```

In the example, the first statement defines a pointer variable whose symbolic address is acon2. In the operand field, we see that =100 is specified as a literal symbol. The assembler knows to create the bit pattern for decimal 100 in a memory location whose address is further stored in another location named acon2.

4.4.2 Other Pseudo Ops

Other popular pseudo ops are shown in Table 4.2.

Table 4.2 Other Pseudo Ops

Name	Description
assume	Assume that at run time a segment register contains a base address as specified
end	End of program
endm	End of a macro definition
endp	End of procedure
ends	End of segment

Name	Description
equ	Equate an absolute symbol to a numeric value
exitm	Exit macro expansion
extern	External reference symbols
macro	Beginning of a macro definition
org	Origin sets a LC (Location Count) variable in the assembler to a specified value
proc	Beginning of a procedure
public	External definition symbols
segment	Beginning of a segment
title	Title of the program

4.4.2.1 Segment

Recall that a segment is a block of bits in memory. A segment pseudo op is used to declare a segment that has a name in the front. A segment name is just like any other symbolic address except for two things. First, when it is used in a mov instruction as the source operand address, the 16-bit paragraph number coded in the instruction is the source operand. Second, in the last statement of a segment, we must use ends to signal the end of a segment. However, the same segment name is placed in front of ends as a delimiter. Thus, the same segment name shows up twice in column one. For this reason, a segment name is usually coded in upper case letters so it looks different in appearance. A data segment is declared below:

```
DSEG    segment
msg     db      'Hello world.$'
DSEG    ends
```

The name DSEG (data segment) before the segment op is the base address definition. The same symbol placed again before the ends op is just a delimiter. The segment contains a ASCII string whose symbolic address is msg (message). The last character ($) serves as a terminator to the OS if the string needs to be displayed on screen.

A program stored in one or more segments may be comprised of one or more procedures. A procedure is a block of instructions grouped in a code segment. A procedure starts with the proc pop (procedure pseudo op), and its last statement is endp (end procedure) pseudo op. The very last statement in a program is the end pop. In the end statement, an address may be specified in the operand field as its entry point (i.e., starting address) when execution begins.

4.4.2.2 Assume

The assume pop statement informs the assembler about the setting of segment registers at run time. For example, consider the following:

assume cs:CSEG, ds:CSEG, ss:CSEG

where the colon between the segment register and its base address really means that the two terms are equal. The statement is not executable, so it disappears after assembly. It tells the assembler that the segment name CSEG is the base address stored in all three segment registers at run time. This assumption allows the assembler to prepare the offset of a memory operand correctly in the code according to the specified base. Since the OS has its own rules to set segment registers before a program executes, a user program usually needs to set the data segment register correctly before accessing the operand in it.

4.5 THREE BASIC ADDRESSING MODES

During instruction execution, an operand may reside in memory or in the CPU. If the operand is in memory, a memory cycle is issued to fetch the operand from memory to the CPU. If an operand is 16 bits long and the processor has a memory data bus of eight bits, it takes two physical memory cycles for each time only a portion of the operand is retrieving. The three basic memory addressing modes are specified in Table 4.3.

Table 4.3 Three Basic Memory Addressing Modes

Name	Memory Cycles Required to Fetch the Operand
Memory Direct	One
Memory Immediate	Zero
Memory Indirect	Two

By definition, memory direct addressing means that it takes one memory cycle to fetch the operand. In contrast, memory immediate addressing means that it takes zero memory cycles to obtain the operand. The operand is in the CPU, so memory immediate really means no memory reference. If the operand is coded as part of an instruction, then it is called an immediate operand in regard to memory reference. If it takes two or more memory cycles to fetch an operand, it is memory indirect. Memory direct is the number one addressing mode. It is more powerful than memory immediate because the former can simulate the latter but not vice versa.

4.5.1 Memory Immediate

In general, memory immediate means that the operand is in the CPU. In practice, an operand can be coded in the address field of the instruction as shown in Figure 4.7a. The instruction under execution resides in the IR but it still takes one clock cycle to fetch the operand bits in the IR. That is to say, an immediate operand means that no memory reference required. Note that an immediate operand in the instruction can be the source, but it makes absolutely no sense for it to be the destination. Most assemblers require that in the operand field, a pound sign (#) be placed in front of the symbolic address or numeric absolute address to mean memory immediate. For example, #opd1, #100, etc. are commonly used to indicate memory immediate addressing.

4.5.2 Memory Direct

Memory direct addressing mode means that it takes one memory cycle to fetch the operand. In concept, the lower 16-bit field in the IR contains an address where the operand is stored in memory, as shown in Figure 4.7b. Specifically, the instruction contains an address directly pointing to the operand in memory.

A hardware register can be used to modify the memory address in an instruction before accessing the operand. This is called indexing and is commonly recognized as the second major programming concept in computing. More extensive discussions on indexing are provided in the next chapter. Register direct means that the operand is stored in a CPU register. Register indirect means that the CPU register contains a direct address pointing to the operand in memory. If the operand is register direct, it takes zero memory cycle to fetch the operand. Thus, based on our basic definitions, we can classify register direct as a derivative of memory immediate. As regards the CPU, it takes equal time to access an operand in IR or in any other working register. Similarly, register indirect is a derivative of memory direct because it takes only one memory cycle to fetch the operand. The term indirect has caused some confusion because we have two accesses in total, the register and the memory. It must be mentioned that first generation computers only supported memory direct. Later, more addressing modes, such as memory immediate and memory indirect were added to enhance programming convenience. For the sake of satisfying curiosity, some of the MASM examples are explained below:

1. An absolute symbol or numeric address without indexing specifies a memory immediate operand.

Code image	Statement		
B8 0064	mov	ax, HUNDRED	;ax ← 0064 {HUNDRED equ 100}
B8 0064	mov	ax, 64h	;ax ← 0064
B8 0064	mov	ax, 50*2	;same as above
B8 0064	mov	ax, opd2-opd1+98	;same if opd2=opd1+2

Note that the instruction has three bytes. The opcode B8 specifies move AX memory immediate, where AX is implied as destination. The address field contains 0064, but physically, the byte 64 in hex is stored before 00.

2. A relocatable symbol with the keyword offset in the front specifies memory immediate as shown below:

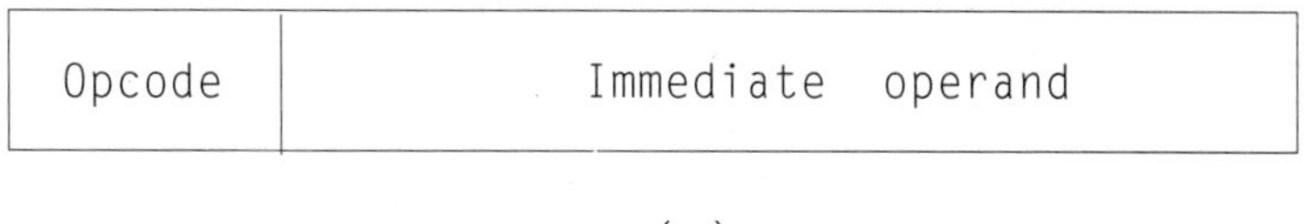

(a)

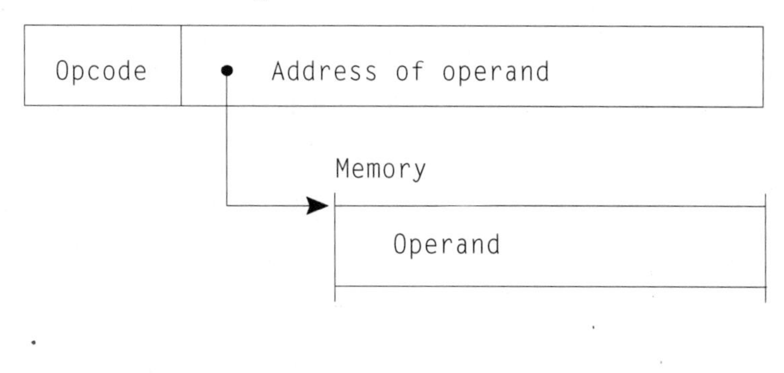

(b)

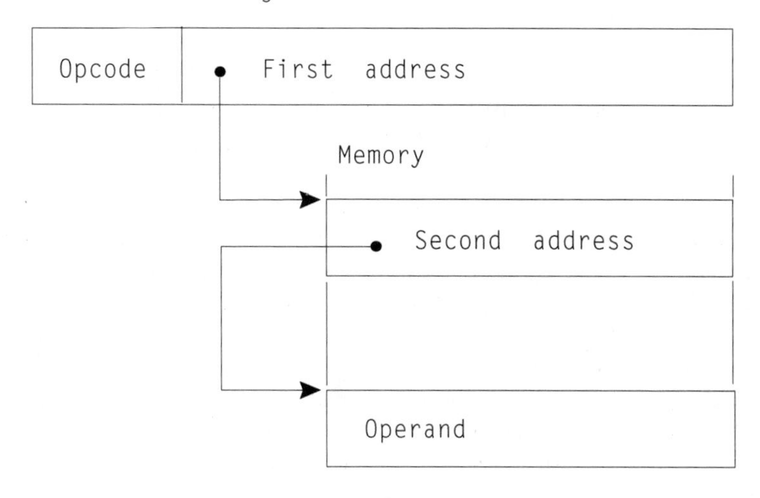

(c)

Figure 4.7 Three basic memory addressing modes: (a) memory immediate, (b) memory direct, and (c) memory indirect.

```
mov  ax, offset opd1      ;ax ← opd1
mov  di, offset opd1+2    ;di ← opd1+2
```

3. A segment name means an immediate operand as follows.

```
mov  ax, CSEG             ;ax ← CSEG
```

The segment name CSEG, often written in uppercase, is treated as an absolute symbol so the 16-bit base address, a paragraph number, is copied from the IR <15:0> into the AX.

4. A numeric address or absolute symbol with the keyword ds: in front means memory direct.

```
mov  ax, ds:80h           ;ax ← M[ds:80]  {memory direct}
```

5. A relocatable symbol in the instruction, with or without indexing, means memory direct as shown below:

```
mov  ax, opd1             ;ax ← M[ds:opd1]
mov  ax, opd1[si]         ;ax ← M[ds:opd1+si]
```

6. An absolute symbol or numeric address with indexing means memory direct as shown below:

```
mov  ax, [si]             ;ax ← M[ds:si]
mov  dx, 100[si]          ;dx ← M[ds:64+si]
mov  dx, HUNDRED[si]      ; same as above
```

The source operand of the first instruction is register indirect, i.e., the SI (source index) register contains the address of an operand. The source operand of the second instruction is memory direct with indexing. The offset 100 in decimal form is added to the index register named SI at run time. That is to say, the EA (effective address) is computed as the sum of the content in SI and the offset 0064h. After that, the result is added to DS to form the 20-bit linear address.

If a debugger is used to disassemble an instruction, we can examine the opcode to verify the addressing mode specified. By doing single step execution, we can easily figure out the rules on addressing mode and on operand type. That is to say, the debug program prints out the syntax based on the bit pattern in the instruction.

4.5.3 Memory Indirect

Memory indirect addressing mode means that it takes two memory cycles to fetch the operand. The instruction contains a first address of a memory word where a second address is stored as shown in Figure 4.7c. After the first memory read cycle, the second address is fetched in the CPU, and this second address directly points to the operand. Since pointer means address, the instruction contains a pointer to a pointer that points to the operand in memory. Immediately after the first memory read cycle, a second memory read cycle is issued to fetch the operand. Therefore, it takes a total of two memory read cycles to fetch the operand. Because memory is relatively slower than CPU, memory indirect is not supported by most of the mainframes.[5,26] Pentium does not support memory indirect either, so we study memory indirect as a programming concept. It should be mentioned that memory indirect was supported by other computers.[40,51] On such a machine, an indirect address symbol has the @ symbol placed in front to indicate memory indirect, e.g., @opdPtr1, @100, etc.

4.6 MACHINE OPS

Each mop (machine op) is translated into an instruction by the assembler. All the machine ops can be grouped into six categories: load and store, arithmetic, logical and shift, compare and branch, I/O, and system control. Even though the opcode mnemonic is the same, the syntax of an operand specified in a move, arithmetic, logical, or shift instruction may further reveal the operand length, which is either eight bits, 16 bits, or 32 bits.

In computing, the four basic instructions are load, store, subtract, and conditional branch. Any application program can be written in these instructions, so other instructions are designed only to make programming faster and easier.

4.6.1 Load and Store

A load instruction is to fetch an operand from memory to the CPU, and a store instruction does the reverse by storing an operand from the CPU to memory. As influenced by PDP-11, the Pentium processor can use a mov (move) instruction to do both load and store because the only difference is in direction. Note that move does not means physically move, instead it means copy, because the source operand remains intact. The instruction needs two operandi. If the destination operand is register and the source is memory, it is load; otherwise, it is store. The mnemonic mov may be translated into a different opcode depending on the direction of the move and the size of the operand.

4.6.2 Push Stack vs. Pop Stack

The stack is a dynamic data structure that can be simulated in memory. In other words, a memory read cycle can be issued to access an operand in a stack. The segment statement is a pseudo op to declare a stack as follows:

```
SSEG    segment     stack
        dw          80 dup (?)
SSEG    ends
        ...
        push    cs  ;SP ← SP - 2
                    ;M[SS:SP] ← CS
        pop     ds  ;DS ← M[SS:SP]
                    ;SP ← SP + 2
        push    cx  ;save CX
        ...
        pop     cx  ;restore CX
```

The symbol SSEG, the segment pop, and the keyword declare a stack segment named SSEG (stack segment) that is defined as the base address. As shown in Figure 4.8a, the base address happens to be 19DF in hex, which is a paragraph number. The dw statement reserves a block of 80 words equivalent to 160 bytes. The OS sets SP to this initial value (00A0 in hex) before the program executes. Because an 8086 would not allow a mov instruction to transfer the CS directly to the DS, the same goal is accomplished by pushing the CS first, then popping it into the DS. At the machine level, the push instruction decreases SP by two first and stores the CS in memory as

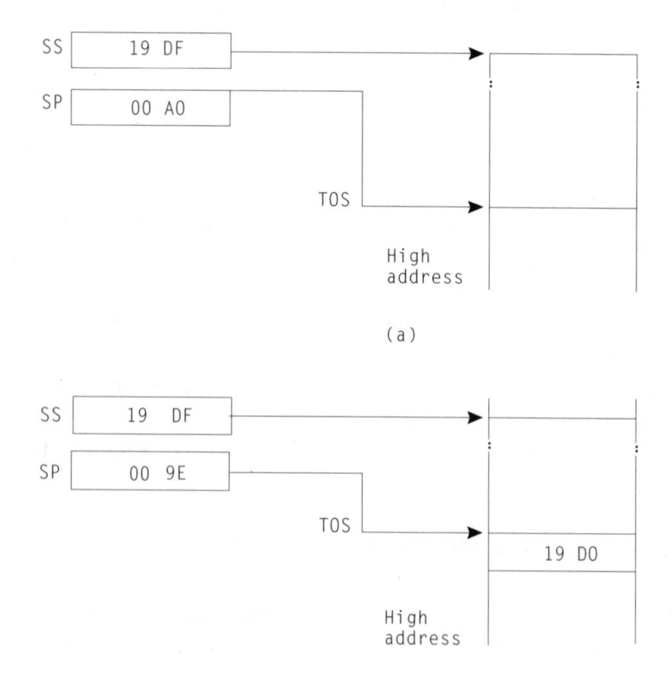

Figure 4.8 Stack segment: (a) initial state and (b) after pushing CS.

the TOS that has 19D0, as shown in Figure 4.8b. The next pop instruction loads the TOS into the DS, then increases the SP by two so the stack goes back to its original state before the push. In AL programming, we can save a register on a stack and fetch it later. The third instruction pushes CX on the stack to save. After the computation block, the last instruction pops the CX off the stack to restore.

4.6.3 Minimal Assembly Language Program

We are ready to write a minimal AL program in a file named mini.asm as listed below:

```
        title    minimal AL program
        assume   cs:CSEG, ds:CSEG
CSEG    segment
main:   push     cs                 ;OS sets cs to CSEG
        pop      ds                 ;ds points to CSEG
        mov      dx, offset msg     ;pointing to string
        mov      ah, 09h            ;display string
        int      21h                ;system call
        mov      ah, 4ch            ;system termination
        int      21h
msg     db       'Hello world.$'    ;$ is the terminator
CSEG    ends
SSEG    segment  stack
        dw       80 dup (?)
SSEG    ends
        end      main
```

The two segments in this program are named as CSEG and SSEG. Perform the following job steps:

1. masm /l mini.asm
2. link mini.obj
3. mini.exe

In the first job step, the /l option asks the assembler to generate a listing file named mini.lst. Upon request, the assembler translates the source into an object file named mini.obj and generates the mini.lst file containing pure ASCII. Some assemblers may prompt questions. If so, simply answer the prompt. The second step links mini.obj into an executable file named mini.exe. The third step loads and executes mini.exe. As a result, the 12-character "Hello world." is displayed without the single quotes because it is not part of the string.

The title statement asks the assembler to place the message in the operand field in the listing file. The assume statement is a pop (pseudo op) to inform the assembler that the program offsets are relative to the segment named CSEG, and the data offsets are also relative to CSEG. Before a .exe file executes, MSDOS prepares a PSP (program segment prefix) of 256 bytes in front of the machine code, as shown in Figure 4.9. The PSP contains the addresses of some error routines, the two built-in file control blocks for the purpose of accessing files, any arguments passed on the command line, etc. Before passing control to an .exe file, the OS sets the DS, and ES pointing to PSP, the CS pointing to the code segment named CSEG, and the SS pointing to the stack segment named SSEG. If an entry point is specified in the end statement, then the IP is set to the entry point, otherwise it is set to 0000. For this

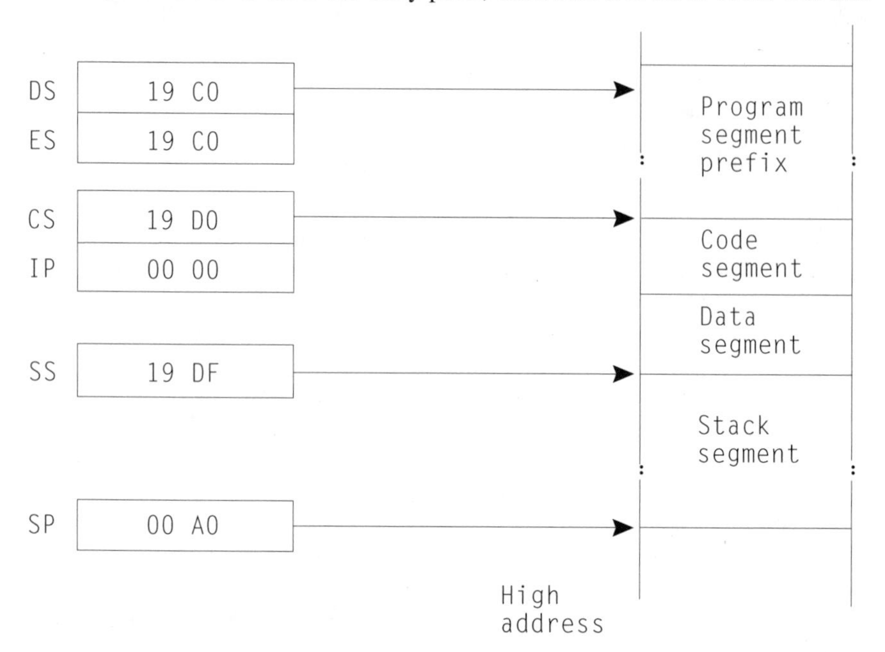

Figure 4.9 Initial register settings for an .exe file.

program, the OS sets the IP to the offset named main that happens to be 0000. In addition, the OS sets the SP to 00A0 as the high address of a stack segment declared to be 80 words (160 bytes).

Because the DS points to the PSP before instruction executes, it is necessary for the programmer to set the DS correctly to the code segment before any data can be referenced. Because the OS sets the CS to the code segment, we simply push the CS and pop it into the DS. The next two instructions prepare two arguments for an OS call. Therefore, the first mov instruction uses the memory immediate addressing mode to copy the offset named msg into the DX. That is, DS:DX points to the string in memory. The second mov instruction copies the one byte immediate operand 09 into AH as the display function code. The fifth instruction is int (interrupt), and the number 21 in hex serves as an identifier. This instruction is a system call for requesting a service from the OS. That is to say, after executing this instruction, the CPU passes control to the OS in low memory. In other words, the current program relinquishes control to the OS after the call. Subsequently, the OS examines the function code in AH and locates the address of the string in DS:DX. Then, the OS prepares the string in screen memory for the display processor to interpret. Because the $ sign is a terminator to the OS, after seeing the $ sign, the OS quits display and passes control back to the user program. The sixth instruction sets the function code 4C in AH to ask for termination. After another system call, the OS resumes control. Thus, we can place the string "Hello world.$" below the last instruction because it is not executed.

4.6.4 Arithmetic

The two basic arithmetic operations are add and subtract (sub). If sign–magnitude notation is used to represent a negative integer, an add operation can be simulated by sub, but not vice versa.

$$\text{sopd1} + \text{sopd2} = \text{sopd1} -(-(\text{sopd2}))$$

The instruction below has sopd1 in AX and sopd2 in BX.

```
add  ax, bx          ;ax ← ax + bx
```

This instruction can be simulated by instructions as follows,

```
mov  temp1, ax       ;M[temp1] ← ax
mov  ax, 0           ;ax ← 0
sub  ax, bx          ;ax ← -bx
mov  temp2, ax       ;M[temp2] ← -bx
mov  ax, temp1
sub  ax, temp2       ;ax ← ax + bx
```

The temporary variables are named temp1 and temp2 in memory to store intermediate results. It should be stressed that if twos complement notation is used, then both add and sub instructions are equally capable. Consequently, one adder is adequate for both types of operations in a simple CPU.

In programming, if the second source operand is one, we have two special instructions: inc (increment) and dec (decrement). In either instruction, only sopd1 is specified because sopd2 is implied as one. However, the opcode for an eight-bit operand is different from a 16-bit operand. If sopd1 specifies the length of an operand, the assembler prepares the opcode accordingly. In the case that no information about the size of the operand is provided in the AL statement, it is necessary to use a special keyword before the addressing mode to specify the operand size. Some of the coding examples are shown below:

```
inc     ax                  ;ax ◄— ax + 1
inc     opd1                ;M[opd1] ◄— M[opd1] + 1
                            ;Opcode is based on opd1: byte, or word.
inc     word ptr 4[si]      ;16-bit operand in memory
inc     byte ptr 4[si]      ;8-bit operand in memory
dec     si                  ;si ◄— si - 1
```

The neg (negate) instruction requires only one operand, so it is a unary operation. The CPU negates an operand by feeding zero as one adder input, the ones complement of the operand as the other adder input, and meanwhile turning on the Cin (carry in) bit. Then, the output of the adder is clocked into the operand if it is a register. For example, the operand in AX is replaced by its twos complement as shown below.

```
neg     ax                  ;ax ◄— 0 - ax
```

4.6.4.1 Multiple Precision Arithmetic

A program can be written to handle multiple precision arithmetic. Recall that after executing an add instruction, the CF (C) bit in the flags register is set if there is a carry output generated from the adder. However for a sub instruction, the C bit is set if no carry is generated. That is to say, the C bit is a carry for an add instruction but a borrow for a subtract instruction. The two examples below illustrate this important concept.

Example 1:

```
mov     al, 3               ;al <— 0000 0011 (binary)
add     al, 4               ;al = 0000 0111
                            ;C <— 0
```

Example 2:

```
mov     al, 3
sub     al, 4               ;al = 1111 1111
                            ;C ◄— 1
```

Two special instructions, adc (add with carry) and sbb (subtract with borrow), for multiple precision arithmetic operations are introduced. Intuitively, the adc instruction signifies that the carry flag in F is also added to the LSB. By the same token, the sbb instruction signifies that the C flag is also subtracted from the minuend. A subtract operation means that the twos complement of the subtrahend is added to the minuend. Because of a subtract operation, a no carry output condition sets the C flag to one. Thus, subtracting an extra one from the subtrahend puts it in ones complement (i.e. twos complement minus one). That is to say, if there is a borrow, the ones complement of the subtrahend is added to the minuend. The next example shows how to perform 32-bit add and subtract operations using a 16-bit adder.

Example 3:

```
data1      dd     00008001h
data2      dd     00008002h
result1    dd     ?
result2    dd     ?
           ...
;32-bit result1 = 00010003 after add.
           mov    ax, word ptr data1
           add    ax, word ptr data2      ;add
           mov    word ptr result1, ax    ;C bit remains unchanged
           mov    ax, word ptr data1+2
           adc    ax, word ptr data2+2  ;add with carry
           mov    word ptr result1+2,  ax
;32-bit result2 = ffffffff (signed -1) after subtract.
           mov    ax, word ptr data1
           sub    ax, word ptr data2      ;subtract
           mov    word ptr result1, ax    ;C bit remains unchanged
           mov    ax, word ptr data1+2
           sbb    ax, word ptr data2+2    ;subtract with borrow
           mov    word ptr result1+2, ax
```

Note that the operand field has the keyword "word ptr" to override the defined double word data type. After executing the add routine, the physical byte sequence is 03, 00, 01, 00 in a memory location named result1. The logical representation of the bits in result1 is 00010003 in hex. After executing the subtract routine, the byte sequence is ff, ff, ff, ff in a memory location named result2. Both routines work correctly because the C bit does not change after executing a mov instruction.

In a simple processor or controller, an adder is often used to implement integer multiply or divide operations. The micro steps are described in Chapter 6 which deals with microprogrammed CPU design. The mul (unsigned multiply) instruction is different from imul (integer multiply). The div (unsigned divide) instruction is different from idiv (integer divide). The mnemonic i really represents an integer that is signed. The mul instruction multiplies two unsigned numbers and the product is unsigned. The imul instruction multiplies two signed numbers and the product is signed. For example, if the signs of two binary numbers differ, the product is negative in twos complement notation.

Note that the multiplier is implied so only the multiplicand is specified in the instruction as an operand in a data register or memory. For eight-bit multiply instruction, the eight-bit multiplier is pre-loaded in AL and the eight-bit multiplicand can be in an 8-bit data register or in memory. Before the operation starts, the CPU clears AH. After the operation, the 16-bit product is in AX. For 16-bit multiply instructions, the multiplier is pre-loaded in AX, and the 16-bit multiplicand can be in a 16-bit data register or in memory. Before the operation starts, the CPU clears DX. After the operation, the 32-bit product is in DX^AX that means DX is concatenated with AX.

Example 4:

```
mov     al, 101         ;ah is cleared by hardware
mov     cl, 50
mul     cl              ;ax ◄— al * cl {8-bit mul}
```

After the eight-bit unsigned mul, ax contains 13ba in hex or 5050 in decimal.

Example 5:

```
mov     ax, -101
mov     bx, -50
imul    bx  ;dx^ax ◄— ax * bx  {16-bit imul}
```

After 16-bit imul, the register pair dx^ax contains 000013ba in hex. If the divisor bx is changed to opd1, a word defined in memory, then the product is represented by ax * M[ds:opd1].

Example 6:

```
mov     eax, 1010
mov     ebx, 500
imul    ebx ;edx^eax ◄— eax * ebx  {32-bit imul}
```

In the above example, after multiplying two 32-bit signed numbers, the 64-bit signed product is in edx^eax. For a divide operation, the dividend is implied. For an eight-bit signed divide, the AX must be pre-loaded with the 16-bit dividend. After the divide, AL contains the quotient and AH contains the remainder that has the same sign as the dividend, as shown below.

Example 7:

```
mov     ax, 255         ;ax ◄— 00ff  {255 in decimal}
mov     bl, 10          ;bl ◄— 10    {base 10}
idiv    bl
```

After the eight-bit idiv, al contains 19 in hex as the quotient, and ah contains five as the remainder whose sign obeys the dividend.

For 16-bit divide, the 32-bit dividend is implied in DX^AX. For unsigned divide, DX should be cleared if the dividend has 16 bits. For signed divide, DX should be padded with the sign bit of AX all the way from bit zero to bit 15 by executing the cwd (convert word to double) instruction. This instruction has one-byte opcode 99 in hex but no operand because they are implied, that is, the sign bit of AX is extended all over into DX. In other words, if the sign of AX is zero, DX contains zeroes. If the sign bit of AX is a one, DX contains all ones. The divisor must be specified in a data register or memory. After a 16-bit idiv, the quotient is in AX, and the remainder, in DX, has the same sign as the dividend. To remember that after dividing, AX contains the quotient and DX contains the remainder, think logically that if the remainder is zero, a subsequent signed multiply should restore the dividend into DX^AX. A 16-bit signed divide example is shown below:

Example 8:

```
mov     ax, 5050      ;ax ◄— 13ba  {5050 in decimal}
cwd                   ;convert from word to double
mov     bx, 10        ;bx ◄— 10   {base 10}
idiv    bx
```

The register pair dx^ax has the dividend and bx is the divisor. After idiv, ax contains 01f9 in hex (505 in decimal), and dx contains a perfect zero as the remainder.

In the case that the quotient is too big for the hardware to handle, an overflow condition occurs so the CPU sets the O (overflow) bit in the F register.

4.6.4.2 Default Operand Size: 16-bit vs. 32-bit

The Pentium processor can support a default operand size of 16 bits or 32 bits, which also means real or protected mode. After loading an executable file, the OS sets a bit in a control register: one for a 32-bit operand and zero for a 16-bit operand. Interestingly, the processor uses the same opcode in an instruction for either a 16-bit or 32-bit operation. For example, the one-byte opcode for cwd (convert word to double) or cdq (convert from double to quad) is 99 in hex. If the default size is 16 bits, the one-byte opcode extends the sign bit in AX into DX. If the default operand size is 32 bits, the same 1-byte opcode extends the sign bit in EAX into EDX.

However, with a default size of 16 bits the processor supports two sets of instructions operating on either 16 bits or 32 bits. This is done by placing a one-byte prefix 66 in hex in front of the opcode to change the default operand size. Since either operand size can be the default, this prefix actually selects the non-default size. That is to say, if the default size is 16 bits, the prefix changes the size to 32 bits. If the default size is 32 bits, then the prefix changes the size to 16 bits. The prefix can be thought of as an eop (extended opcode).

When the default operand size is set at 16 bits, instructions with no prefix operate on 16-bit registers but instructions with a prefix operate on 32-bit registers. However, it is necessary to inform MASM that the program executes in real address mode with data registers to be either 16 bits or 32 bits. For this purpose, two pseudo ops are placed in front of the program and their ordering can not be reversed as follows.

```
.model medium
.586
....{MASM code}
```

That is to say, MASM places the prefix 66 in front of the opcode for a 32-bit operation. Consequently, we may issue instructions to handle either operand size as shown below.

Code image	Statement		
B8 FFFF	mov	ax, -1	;AX <— FFFF
66 B8 FFFFFFFF	mov	eax, -1	;EAX <— FFFFFFFF
99	cwd		;extend sign in AX into DX
66 99	cdq		;extend sign in EAX into EDX

In the first two AL statements, the address expression -1 is absolute so the addressing mode is memory immediate. The first instruction has a 16-bit immediate operand. The second instruction has the prefix, and its immediate operand is 32 bits. The third instruction extends the sign of the 16-bit AX into DX. The fourth instruction has the prefix to extend the sign of the 32-bit EAX into EDX.

4.6.5 Logical

The logical operations are performed on a bit-wise basis. Therefore, each bit in the operand is treated as a logical variable. Some logical instructions are given below.

Example 1:

```
mov     ax, 5005h       ;ax ← 0101 0000 0000 0101
not     ax              ;ax ← .NOT. ax
                        ;ax = 1010 1111 1111 1010
```

The not instruction performs the ones complement operation on a unary operand. Thus, every bit in AX is flipped from one to zero, or from zero to one.

Example 2:

```
mov     ax, 0affah      ;ax <— 1010 1111 1111 1010
or      ax, 0f000h      ;ax <— ax .OR. 1111 0000 0000 0000
                        ;ax = 1111 1111 1111 1010
```

The or instruction is used to pad a particular field in an operand with a specific bit pattern. The padding field is shown as the leading four ones followed by 12 zeroes in the immediate operand. As a result of execution, the upper four bits in the first source operand in AX are padded with ones but its lower 12 bits remain unchanged.

Example 3:

```
mov     ax, 0fffah      ;ax ← 1111 1111 1111 1010
mov     cx, 000fh       ;mask
and     cx, ax          ;ax =  0000 0000 0000 1010
```

The and instruction is used to mask out a field in an operand. In this example, the operand in AX is "anded" with the CX register that contains a mask of four ones preceded by 12 zeroes. At each bit position, the bit in the destination operand is one if the bit in the first source operand is one and the bit in the second source operand is also one. The lower four bits of AX are masked out in CX.

Example 4:

```
mov     ax, 1           ;ax ← 0000 0000 0000 0001 binary
xor     opd1, ax        ;M[ds:opd1] ← M[ds:opd1] .XOR. ax
xor     opd1, ax        ;LSB is toggled back.
```

The xor instruction has many applications. For example, it can be used to toggle a bit in an operand. In doing so, we set the LSB (least significant bit) of the second source operand to one with 15 leading zeroes and opd1 may contain any bit pattern. Based on the definition, the resulting bit is one if the two source bits differ. Therefore, the leading 15 bits in opd1 are "exclusive ored" xor with zeroes so they remain unchanged, but the LSB is flipped each time the instruction is executed. After executing the second time, the LSB in the first source operand is toggled back with the same upper 15 bits.

4.6.6 Shift

Recall that the two major types of shifts are arithmetic and logical. The purpose of a shift instruction is to move bits around in a register or in memory. The operand may have a different size: 8 bits, 16 bits, or 32 bits, and the shift count may be an immediate operand coded in the instruction or placed in a register. The Pentium chip also supports rotate shifts in either direction, some without C (carry) in the rotate operation and others with the C bit.

4.6.6.1 SHR (Shift Logical Right)

The shift logical right operation specifies that a bit zero flows into the MSB of the operand, and its LSB flows into the C flag.

4.6.6.2 SAR (Shift Arithmetic Right)

For a shift arithmetic right instruction, the sign bit in the operand remains intact and its LSB flows into C.

Example 1:

```
mov     ax, 1        ;ax ← 0000 0000 0000 0001 binary
sar     ax, 1        ;ax = 0, i.e. [ax / 2]
                     ;C ← 1 {Carry}
```

Example 2:

```
mov     ax, -1       ;ax <— 1111 1111 1111 1111 binary
sar     ax, 1        ;ax = -1, i.e. [ax / 2]
                     ;C <— 1 {Carry}
```

4.6.6.3 SHL (Shift Logical Left)

For any logical shift, left or right, the bit flowing into the register is always a 0 and the bit flowing out is the carry bit in the F register. To shift logical left one bit, a bit zero flows into the LSB of the operand, and its MSB flows out into C. The SAL (shift arithmetic left) instruction is identical to an SHL (shift logical left) instruction with no difference in hardware design. To shift arithmetic an operand one bit to the left means to double its arithmetic value provided that its sign bit remains unchanged. A change in the value of the sign bit from zero to one or from one to zero after shifting indicates an overflow condition, and the result is no longer correct. Based on this, SHL will handle both operations, and some coding examples are shown below:

Example 3:

```
mov     si, 7        ;si ← 0000 0000 0000 0111 binary
sal     si, 1        ;si = 0000 0000 0000 1110
                     ;O ← 0 {No overflow}
                     ;C ← 0 {No carry}
```

Example 4:

```
mov     ax, 8000h     ;ax ◄— 1000 0000 0000 0000 binary
sal     ax, 1         ;ax = 0000 0000 0000 0000
                      ;O ◄— 1 {Overflow}
                      ;C ◄— 1 {Carry}
```

4.6.6.4 ROL (Rotate Left) vs. ROR (Rotate Right)

The rotate shift is like a circular shift. The operand is viewed as a shift register because the LSB and MSB are considered adjacent. For a 16-bit operand, ROL or ROR means a 16-bit rotate shift, and the difference lies only in the direction. If the shift count is one, the ROL instruction rotates the operand to the left as shown in Figure 4.10a. The top figure depicts the bit pattern in an operand before the shift, and its MSB flows back into the LSB as shown in the bottom figure after the shift. The ROR instruction rotates the operand to the right, and the LSB flows into the MSB as shown in Figure 4.10b.

4.6.6.5 RCL (Rotate with Carry Left) vs. RCR (Rotate with Carry Right)

The rotate with carry operation includes the C bit in the middle between the LSB and the MSB. That is, for a 16-bit operand, we have a 17-bit rotate shift. If the shift count is one, the rotate with carry left is shown in Figure 4.10c and the rotate with carry right is shown in Figure 4.10d. This feature is for multiple precision arithmetic: add, subtract, multiply, and divide. To shift arithmetic right 32-bit DX^AX by one bit, we have the following code:

```
sar     dx, 1         ;sign bit remains the same
                      ;C ◄— lsb in DX
rcr     ax, 1         ;msb in AX <— C
                      ;C ◄— lsb in AX
```

All the instructions are normally executed in sequence and the ordering of instruction execution can be altered by executing a branch instruction. This is a very important programming concept, as described in the next section.

4.7 PROGRAM SEQUENCE CONTROL

Both unconditional branch instructions and conditional branch instructions are used for sequence control. The unconditional branch instruction has an address in the code, and this address is placed in the IAR regardless of the CC. Thus, the instruction at this new address is to be fetched, interpreted, and executed in the next cycle. We say that control is passed to a different location in the program. The conditional

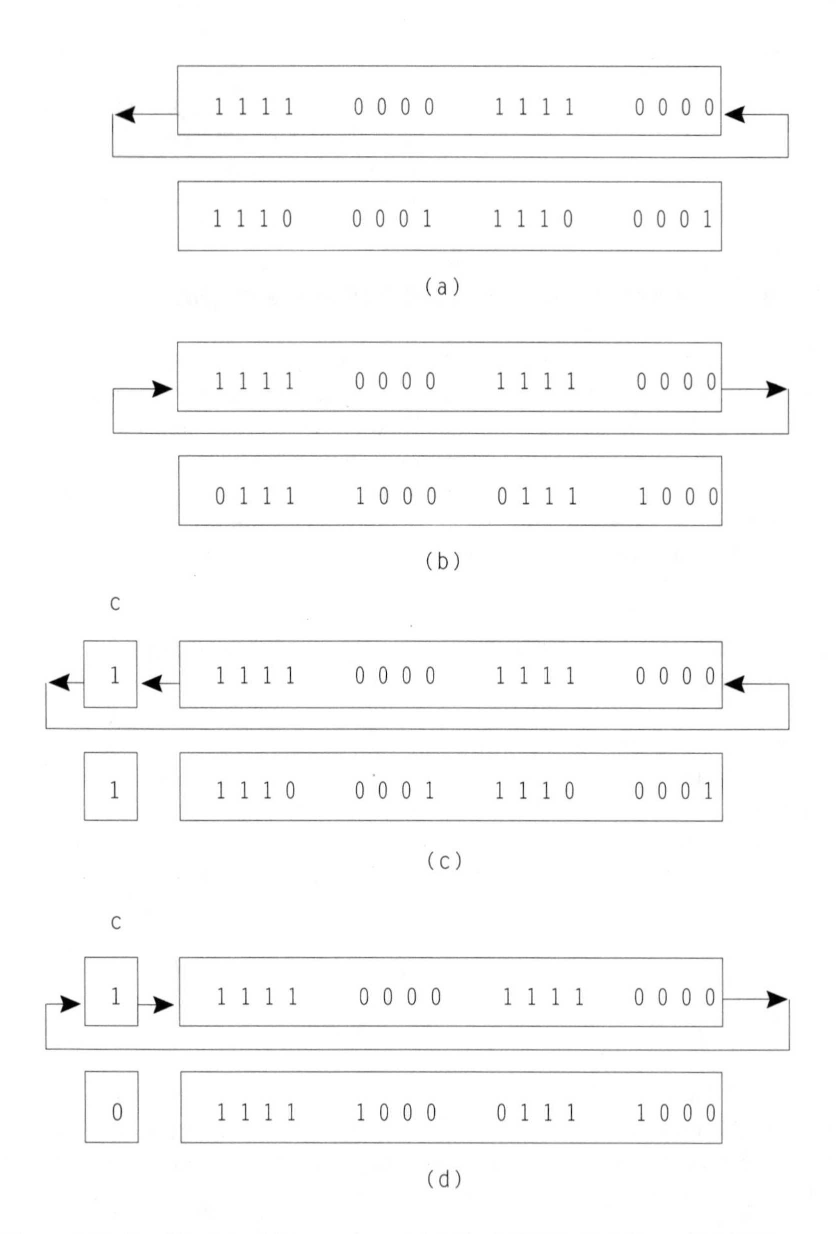

Figure 4.10 One-bit rotate shift operations: (a) ROL, (b) ROR, (c) RCL, and (d) RCR.

branch also contains an address, and its opcode specifies a condition to be tested first. If the condition is true, this new address is placed in the IAR. Otherwise, the next instruction in sequence is executed and this branch instruction is a nop (no operation).

4.7.1 Flags Register

The F register contains the partial CPU status information of the current program in execution. Figure 4.11 shows the F register supported by the 8086, and some of the bits are not used.

				11	10	9	8	7	6		4		2		0
				O	D	I	T	S	Z		A		P		C

Figure 4.11 Flags register.

Note that the Pentium uses the EF register that has 16 bits added in the front for system control. If the mnemonic F is dropped in the naming convention, then CF (carry flag) becomes C (carry), etc. The flag bits in the F register are described below:

- CF (b0): The carry bit is set after executing an add, subtract, shift left, or shift right instruction.
- PF (b2): The odd parity bit for the destination operand is computed below:

  Exclusive Or all the bits in the destination operand into result;
  IF (result .EQ. 0),
  THEN P ← 1;
  ELSE P ← 0; ENDIF;

Mathematically, by adding all the bits in the destination operand including the parity bit, we obtain an odd number.

- AF (b4): The auxiliary carry is the carry output generated from b3 (i.e., fourth bit) of the adder. The A bit can be tested in a routine performing BCD arithmetic: add or subtract. If the A bit is one, the four-bit BCD result must be adjusted to make it correct.
- ZF (b6): The zero flag indicates that the destination operand contains all zeroes.
- SF (b7): The sign flag is copied from the sign bit of the destination operand.
- TF (b8): The trace flag, if it is set to one, triggers a trace interrupt after executing each instruction. This design feature allows the debugger to single-step through a program.
- IF (b9): The interrupt enable flag tells the CPU to accept any pending interrupt request from inside or outside. If this bit is zero, the CPU puts all the interrupt pending signals on hold. That means the interrupt is temporarily disabled by the CPU.

- DF (b10): The direction flag is used for string operations, i.e., to move or search a block in memory. A D bit one means up, so the addresses in SI and DI will be decremented by one automatically after each byte operation. A D bit zero means down so the addresses in SI and DI will be incremented by one after each byte operation.
- OF (b11): The overflow flag indicates a fault condition after an add, subtract, or shift left operation. That is, the destination operand is no longer correct.

4.7.2 Compare

The four bits (C, Z, S, and O) in the F register constitute the CC (condition code). The CPU sets the CC based on the result of an add, subtract, shift, or logical instruction. A cmp (compare) instruction is designed to set the CC according to the result of a comparison. During its execution, the CPU does an internal subtract, which means the result is not stored into the destination, so both sopd1 and sopd2 remain unchanged after compare. As the CPU sees no difference between signed and unsigned numbers, the CC is set strictly based on adder output. We show the format of a compare instruction below:

```
cmp     sopd1, sopd2
```

In concept, sopd1 denotes the first source operand and sopd2 denotes the second source operand. Recall that no more than one memory direct address can be specified in the code.

```
cmp     cx, 10          ;compare cx with 10.
cmp     bx, cx          ;compare bx with cx.
cmp     cx, kount       ;compare cx with M[ds:kount].
```

The first instruction compares cx with an immediate operand 10. The second instruction compares two registers: bx and cx. The third instruction compares the count register cx with a memory direct operand.

4.7.3 Unconditional Branches

The unconditional branch instruction tells the CPU to branch regardless. The terms unconditional branch and unconditional jump are synonymous. The next instruction is fetched from an address coded in the branch instruction. Because the Pentium processor is a segment based machine, it supports near jump and far jump as discussed below.

4.7.3.1 Jump: Near vs. Far

A near jump (jmp) instruction is three bytes long and has an opcode followed by a two-byte relative offset with respect to IP after advancing. That is to say, the IP is first incremented by three, then the offset is added to it. Therefore, the offset is relative to the next instruction right below the branch. The instruction is a near jump that passes control within the segment. In contrast, the far jump instruction can pass control out of the segment. The keyword far ptr is placed in front of the address to ask the assembler for a different opcode. Note that the far jump instruction is five bytes long — an opcode, an two-byte offset with respect to segment, and a two-byte segment base address. Upon its execution, the offset is copied into the IP and the base address is copied into the CS. Some jump instructions are shown below:

Loc		**Statement**		
000C		jmp	L2	;near jump
...		dw	170 dup (0)	
0163	L2:	jmp	far ptr L3	;far jump
0168	L3:	. . .		

The left-hand side shows the Loc (location) count in bytes. Between the two jmp instructions a 340B memory block is defined. The first jmp instruction passes control to L2 and the second jmp instruction passes control to L3 as a far address for the purpose of illustration. The run time code for both instructions are shown below, and the transfer language describes the execution.

Code image	**Loc**		**Statement**	
E9 0154	000C	jmp	L2	;ip ← 0154 + 000F
				; {0163}
00 00 . . .	000F	dw	170 (0)	
EA 0168 0CB1	0163	jmp	L3	;ip ← 0168
				;cs ← 0CB1
	0168	. . .		

The LHS shows the run time code image. After the opcode, the 16-bit logical representation of the offset follows. As the jmp L2 instruction at location 000C in the code segment is executed, the IP is increased by three — the length of instruction. Consequently, the IP becomes 000F (000C + 3) pointing to the next instruction. The block size is decimal 340 (0154 in hex) in bytes as coded in the instruction. When executing the jmp instruction, the CPU adds this relative offset 0154 to the IP after advancing, so the final answer is 0163 as the new IP. At this address, we see the far jmp instruction which is five bytes long. Upon its execution, the IP becomes 0168 and the CS becomes 0CB1. Because the CS remains unchanged, we do a far jump within the same segment.

4.7.4 Conditional Branches

A conditional branch instruction tells the CPU to branch if the specified condition is true. That is, the execution of a conditional branch provides two possibilities. First, if the condition is true, the CPU fetches the next instruction at the address specified in the branch. Otherwise, the CPU increases the IP by the length of the branch instruction so that the next instruction in sequence is fetched. The opcode specifies the condition and the test outcome decides whether or not to branch. There are two types of conditional branch instructions, based on either signed compare or unsigned compare. Signed compare means that both operandi are treated as signed integers using twos complement notation, and unsigned compare means that both operandi are treated as absolute unsigned integers. That is, if a negative integer in twos complement is treated as an unsigned integer, it is greater than a positive integer because its sign bit one is treated as the MSB. Testing the condition, the CPU knows to branch or not, as explained in Chapter 7 which discusses microcode.

Some popular conditional jumps are jle (jump if less or equal) and jg (jump if greater) for signed compare. The counterparts for an unsigned compare are jbe (jump if below or equal) and ja (jump if above). Other instructions include je (jump if equal), jne (jump if not equal), js (jump on sign), etc. The js instruction also means jump on negative because the negative sign bit is one.

4.7.4.1 Addressability Problem

Like most of the microcomputers, a conditional branch instruction is two bytes long, an opcode followed by an eight-bit address. This eight-bit field contains a signed integer relative to the IP so the range is from -128 to +127. The addressability problem exists if the branch address is out of range, as often flagged by the assembler in a program. Fortunately, the unconditional jump instruction supports a longer address, so we can use it to solve the problem as shown below:

Example 1:

```
jle     b0010
```

Example 2:

```
jg      skip0020
jmp     b0010
skip0020:
```

The jle instruction passes control to b0010 based on the less or equal condition. If b0010 is out of range, we must use two instructions and add a new label named skip0020 pointing to the instruction below jmp. The first instruction is jg (jump on greater than) to skip0020 and the next instruction is jmp (jump) to b0010, which really means less or equal.

4.8 LOOPING CONCEPT

There are four basic programming concepts: looping, indexing, subroutine, and interrupt. Our goal is to study each one of them carefully in detail. Looping enables a sequence of similar operations to be repeated under control as implemented in both hardware and software design. A loop can be decomposed into four components as shown below:

1. Initialization
2. Operation
3. Increment counter
4. Test and branch

The initialization is done once to set the initial value of all memory variables. One of the variables is a counter that keeps track of the number of times the loop has been through. The operation block is the body of the loop to be executed so many times as determined by the initial counter. Usually, after each iteration, the counter is increased by one. The last step is to test the counter and determine whether or not to loop back. The above sequence is intuitive with the understanding that the counter counts up.

A specially designed loop instruction counts down after each iteration. The implied operand in the CX is decreased by one by hardware each time the loop instruction is executed. If the CX is not zero, then control is passed to the address coded in the instruction, otherwise the next instruction below the loop is executed. Therefore, we initialize a loop count in the CX before the loop starts.

```
        mov     cx, 16    ;loop count
...
b0010:  int     21h
        loop    b0010     ;cx ◄— cx - 1
                          ;IF (cx .NE. 0),
                          ;THEN ip ◄— b0010
```

After executing the loop instruction 16 times, the CX reaches zero so the next instruction below the loop is executed. The loop instruction decreases the counter, tests, and branches.

In a high-level programming language, a loop may have the test and branch step right after initialization. Hence, if the condition is fulfilled, we execute the loop, otherwise we exit. After the body, there is an unconditional jump to pass control back to step two so the condition is tested again. The parts are the same but the sequence is slightly different as follows.

1. Initialization
2. Test and branch
3. Computation
4. Increment counter
5. Goto step two

4.8.1 Looping Examples

Let us use the loop instruction to construct a program that displays the string "Hello world." a total of 16 times. We declare absolute symbols for the control characters. The OS knows to interpret the character and prepares the screen memory for the display processor to interpret and execute. The modified program is listed as follows:

```
        title     h16 program
DSEG    segment
CR      equ       0dh
LF      equ       0ah
BEL     equ       07
msg     db        'Hello world.', CR, LF, BEL, '$'
                  ;$ is the terminator
DSEG    ends
SSEG    segment   stack
        dw        80 dup (?)
SSEG    ends
CSEG    segment
        assume    cs:CSEG,ds:DSEG
main:   mov       ax,DSEG         ;memory immediate
        mov       ds,ax
        mov       cx,16           ;loop count
        mov       dx,offset msg   ;memory immediate
        mov       ah,09h          ;display function code
b0010:  int       21h             ;system call

        loop      b0010
        mov       ah,4ch         ;terminate function code
        int       21h
CSEG    ends
        end       main
```

Because we declare a data segment named DSEG, the assume statement ds:DSEG means that the data offsets are relative to DSEG. The mov instruction can not move an immediate operand into the DS, so we perform two steps: move the base address named DSEG to AX, and move AX to DS. Because the system call does not change CX, AX, and DX, we can use a loop instruction as controlled by the CX. The system call executes a total of 16 times, and each time the OS interprets the function code in AH and the address of string in DS^DX. The two characters <cr><lf> mean carriage return and line feed, so the OS knows to skip a new line when placing the ASCII text in screen memory for the display processor to interpret. The character <bel> means bell, so the OS rings the bell, i.e., makes a sound through the speaker. The "$" serves as an end delimiter that informs the OS to quit the display. As a consequence, the string "Hello world." is on screen in a single column 16 times. During the 16th time, when the loop instruction is executed, the CX is decremented to 0, i.e. no more branch. Then, the next two instructions execute as a system call to ask the OS for termination, so control is not returned.

The second routine computes the summation of a series 1 + 2 + ... + 100. Algebraically, the answer is (1 + 100) * 100 / 2, which is 5050 in decimal or 13ba in hex. We can write a programming loop to compute the answer in a straightforward way. In the first approach, the test and branch is placed as the last step. In the second approach, the test and branch is moved right after initialization. We change the code somewhat so a new address label is introduced as the exit address.

```
Approach 1:
          sub     ax,ax       ;clear ax
          mov     cx,1        ;initialize counter
again:    add     ax,cx       ;ax contains the partial sum
          inc     cx          ;increment counter
          cmp     cx,100      ;check upper limit
          jle     again       ;test and branch

Approach 2:
          sub     ax,ax       ;clear ax
          mov     cx,1        ;initialize counter
again:    cmp     cx,100      ;check upper limit
          je      exitLoop    ;test and branch
          add     ax,cx       ;body of the loop
                              ;and ax has the partial sum
          inc     cx          ;increment counter
          jmp     again
exitLoop: . . .
```

4.9 MACROS

Recall that the opcode field in an AL statement may specify a macro call. A macro is defined as a block of ASCII text, so macro processing really means ASCII in, ASCII out. Since a macro call may generate many AL statements in the source, macro is an open subroutine in system design. This novel concept is used in low-level language, high-level language, and word processing applications. A macro processor can be designed as a pre-assembly routine that performs two functions. First, it processes the macro definition in the source. Second, it expands the macro upon a call. A macro is defined only once, but it may be called many times. In general, a macro should be short and is often written as a system call interface.

4.9.1 Macro Definition

A macro definition consists of three parts, header, body, and trailer, and it is usually placed in front of the program. For example, we may have a macro defined as a system call to write string as shown below:

```
wrstr    macro msg      ;macro wrstr
         mov   dx, offset msg
         mov   ah, 09h
         int   21h
         endm
```

The first line is the header. In column one, we specify a macro name that is unique. In this case, the name is chosen as wrstr (write string). In the opcode field, the keyword macro is a pop (pseudo op). In the operand field, a list may specify zero, one, or more dummy arguments separated by a comma or space. After the header, the next three lines constitute the macro body. A macro body is usually comprised of many AL statements. A dummy argument may appear anywhere in the body as an ASCII string. In the above example, the second line has a dummy argument msg that is to be replaced by a passing argument during a call. The trailer contains the pop endm (end of macro).

The macro processor scans the opcode field in each AL statement for the keyword macro. If it is found, the macro processor builds a list data structure to store the definition.

4.9.2 Macro Expansion

The macro processor scans the opcode field for a macro name. After finding it, the macro processor expands the macro definition block into the source. That is to say, each macro call may generate many extra lines of AL code. During macro expansion, each instance of a dummy argument in the macro definition block is replaced by the argument respectively passed in the list. As the macro processor copies the macro definition block from the data structure statement by statement, all the ASCII characters in a line remain unchanged, except for the dummy arguments which are replaced by the passing arguments. In the following, we see that the passing argument msg001 replaces the dummy argument in the macro body after the call.

```
        wrstr     msg001        ;macro wrstr
+       mov       dx, msg001
+       mov       ah, 09h
+       int       21h
```

As the macro name wrstr is specified in the opcode field to indicate a call, three lines are expanded into the source including any comment in the line. However, the first line indicates a macro call that is only placed in the listing file but not in the source file. In the listing file, each expanded line has a plus sign in column minus two to indicate that the line is inserted due to macro expansion.

4.9.3 I/O System Calls

4.9.3.1 String I/O

We have discussed a basic system call to write string. In order to read a string from the keyboard, the OS requires a buffer format as explained below.

```
buf   db          81   ;Maximum number of input characters
                         allowed
                       ;including <cr>
      db          ?    ;The OS stores the actual number of input
                       ;characters excluding <cr>
str1  db          81   dup (?) ;buffer to store input characters
                                ;including <cr>
      . . .
      mov  dx, offset buf
      mov  ah, 0ah          ;read string ended with <cr> and echo
      int  21h
      mov  al, buf+1        ;fetch the actual character count
      mov  ah, 0
      mov  si, ax           ;si contains the character count
      mov  str1[si], '$'    ;replace <cr> with the $ sign as
                            ;terminator
      mov  dx, offset str1
      mov  ah, 09h          ;write string
      int  21h
```

Assume that the user enters the string BOB<cr>. After hitting the enter key, the OS stores the string BOB<cr> in the buffer named str1 and stores the actual character count 03 in the second byte. If a null string is entered, the character count is 00 with only <cr> stored in str1. In order to write the same string out, the character <cr> is changed into a $ sign as a terminator.

4.9.3.2 Character I/O

System calls are provided to read one character from the keyboard or write one character. This design feature allows an application program to take immediate action as soon as the user hits the keyboard. The three basic character I/O system calls are listed below.

```
1. mov    ah, 01h  ;read one character in AL and echo.
   int    21h
2. mov    ah, 08h  ;read one character in AL,
                   ;but don't echo.
   int    21h
3. mov    dl, 'A'
   mov    ah, 02h  ;write one character in DL that
                   ;includes the $ sign.
   int  21h
```

4.9.3.3 .com File

As an important concept, a .com (compact) file has no header. That is to say, every bit in the file is pure code. When the file is loaded into memory, the code is not modified at all. Before executing a .com file, MSDOS also prepares a PSP in front of the code and sets the CS, DS, SS, and ES all pointing to the PSP as shown in Figure 4.12. In addition, the OS sets the IP to 100 in hex and the SP to FFFE pointing to the last word in the segment.

There are three requirements to prepare a .com file. First, the size of a .com file code can not exceed (64K–256) bytes because the PSP occupies 256B in the segment before the code. Second, the code must not contain any segment base addresses. That is to say, all segment registers point to PSP with no further change. Relative offsets in the code are fine because they require no modifications after loading. Third, the entry point must be 100 in hex. Because the assembler can not generate a .com file directly, we rely upon a utility program named exe2bin (exe to binary). This program converts a .exe file into an .com file and after conversion the .exe file remains unchanged. Note that the .exe file header contains information about the relocatable addresses and this file has such information. Thus, the exe2bin program simply chops off the header as well as the 256B garbage code in the front, and writes the pure code section into the .com file.

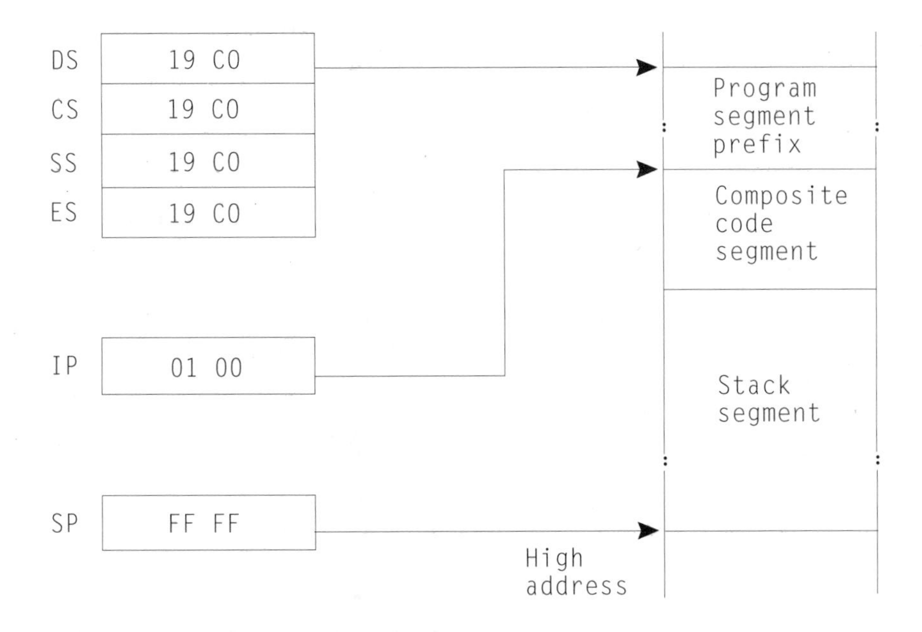

Figure 4.12 Initial register settings for a .com file.

The following program computes the summation of the series 1+2+...+100 and displays the result in decimal integer. In order to generate a .com file, the org statement informs the assembler to set the LC (location count) to 100 in hex, which is the size of the PSP as well as the offset of the first instruction. Therefore, the .exe file has a 256B garbage code in the front. The structure below shows how to generate a .com file:

```
;1. edit sumpgm.asm
;2. masm /l sumpgm.asm  {the option /l requests listing.}
;3. link sumpgm.obj
;4. exe2bin sumpgm.exe sumpgm.com  {conversion}
;5. sumpgm >>sumpgm.asm
;
```

The symbol >> is the redirect and appended operator so the OS appends the output to the source file instead. After execution, the output reads as:

```
;The summation of the series 1+2+..+100 is 5050 in decimal.
        title    sum program
CSEG    segment
        assume   cs:CSEG,ds:CSEG
        org      100h
main:   sub      ax,ax     ;clear ax
        mov      cx,1
```

```
again:   add    ax,cx        ;ax contains partial sum in a loop
         inc    cx           ;increment counter
         cmp    cx,100       ;compare with upper limit
         jle    again        ;loop back if not 100 times
         mov    di,offset answer+3  ;points to the last char
         mov    bx,10        ;base 10
b0010:                       ;convert binary to decimal ascii
         cwd                 ;convert word to double
         div    bx
         add    dl,30h       ;add bias to make ASCII
         mov    [di],dl      ;store decimal ASCII in the msg
         dec    di
         cmp    ax,0
         jne    b0010        ;loop if quotient .NE. 0.
         mov    dx,offset msg       ;ds:dx points to string
         mov    ah,09h       ;display function
         int    21h          ;system call
         mov    ah,4ch       ;terminate function
         int    21h          ;system call
msg      db     ';The summation of the series 1+2+..+100 is'
answer   db     4 dup (?)
         db     ' in decimal.$'
CSEG     ends
         end    main
```

4.10 SUMMARY POINTS

1. An assembly language statement may represent a machine instruction. Because of this one-to-one relationship, assembly language is referred to as the machine language.
2. If a program has two parts, one may be written in high-level language and the other in assembly code. The linker can combine the two object modules into one executable module.
3. Assembly code may also be inserted in a high-level language program, a phenomenon known as in-line assembly.
4. An assembly statement may be a machine op, pseudo op, or macro call.
5. A machine op is always translated into an executable instruction, a pseudo op is not, and a macro call is usually expanded into many assembly statements grouped as a macro.
6. An address expression may be specified in the operand field and the assembler interprets and evaluates the address expression into a single numeric address in binary after assembly.

7. An address constant in a program means that the assembler prepares an unsigned integer in memory as an address or pointer to another memory location containing an operand.
8. The three basic addressing modes are memory immediate, memory direct, and memory indirect.
9. Any application program can be decomposed into four basic instructions: load, store, subtract, and conditional branch.
10. There are two special instructions, adc (add with carry) and sbb (subtract with borrow), that are used to perform multiple precision addition, subtraction, multiplication, and division.
11. A logical instruction performs a bit-wise logical operation in that each bit in the operand is treated as an independent logical variable.
12. The shift instructions are used to shift an operand in a register or memory. The operand size may be eight bits, 16 bits, or 32 bits, and the shift count may be an immediate operand coded in the instruction or a register direct operand.
13. The four flag bits C, Z, S, and O in the F register are set after executing an instruction, e.g., arithmetic, shift, or logical, and they collectively constitute the CC (condition code).
14. In design, the CPU tests the CC based on the opcode and decides whether or not to branch. The two types of conditional branch instructions are based on either signed compare or unsigned compare.
15. Any programming loop can be decomposed into four parts: initialization, computation, increment counter, and test and branch.
16. A macro is an open subroutine.
17. A .com (compact) file contains nothing but pure code.

PROBLEMS

1. Name eight 32-bit working registers of a Pentium processor.
2. What are the default rules to use CS, DS, or SS to form the 20-bit memory byte address before retrieving the operand in memory?
3. What default segment register should be used if BP is specified as the base pointer in the instruction?
4. What are the three possible assembly language statements?
5. What is an address expression coded in the instruction? How does the assembler handle it?
6. What is an address constant in memory?
7. What are the three basic addressing modes?
8. What is the difference between register direct and register indirect?
9. What are the four basic ones machine instructions?

10. Explain why the C (carry) bit is really a borrow after executing a subtract instruction.
11. If we need to toggle the LSB in an operand each time in a loop, what logical instruction can be used to accomplish this goal?
12. What is the difference between ROL (rotate left) and RCL (rotate with carry left)?
13. What are the four components in a programming loop?
14. What is the A (auxiliary carry) bit in the F register for?
15. Explain why the je (jump equal) and jz (jump zero) instructions are exactly the same.
16. Because a machine op also passes information to the assembler, do you agree that pseudo op is a more meaningful term than assembler directive?
17. What are the three requirements to write a .com program?
18. On an IBM PC, click the icon sequence Start-Program-command Prompt to see the window. Modify the mini.asm program to conform with .com format. First, the org 100h statement is added. The first two instructions are no longer needed so the mini.com file has 24 bytes. After masm, link, exe2bin, we enter dir mini.com to see the file size. A new linker may produce a .com file directly if option /t (tiny) is specified as follows:

    link /t mini.obj

```
        title           Mini.com has 24 bytes
CSEG    segment
        assume  cs:CSEG, ds:CSEG
        org     100h
start:  mov     dx, offset msg  ;memory immediate
        mov     ah, 09h         ;display function
        int     21h             ;system call
        mov     ah, 4ch         ;system termination
        int     21h
msg      db     'Hello world.$'
CSEG    ends
        end     start           ;entry point
```

19. In the command window, type help or ? followed by a command name for help about edit, debug, etc. Type debug mini.com to see the debug prompt. Enter a debug command, e.g., r (register), u (unassemble), t (trace), etc. Single-step the execution of each instruction.
20. Because the loop instruction does not change the condition code in the F register, we can use it to control a loop that performs multiple precision arithmetic. Write an AL loop in virtual 8086 mode to subtract the 64-bit integer data2 from the integer data1 and store the result in a quad word named answer. Before the loop starts, issue clc to clear the carry flag. Inside the loop, use sbb instead of sub. What is the logical representation of the result, namely M[ds: data1] - M[ds: data2] <63:0>?

```
data1     dd        00008001h
dd                  0000ffffh
data2     dd        00008002h
dd                  0ffff0000h
answer    dq        ?
```

21. Write an AL program to read a string from the keyboard. Use indexing to scan the string and write a message to say whether or not it is a palindrome. Some palindromes are shown below:

    ABLE WAS I ERE I SAW ELBA
    MADAMIAMADAM
    UHU UHU ... UHU UHU

CHAPTER 5

Computer Architecture — General Features

5.1 ADDRESSING MODES

The instruction set repertoire depicts the architecture of a computer. Besides opcode, there are two major issues in designing an instruction: first, how many addresses are placed in an instruction, and second, where to find each operand. Thus, an instruction contains an opcode along with an addressing mode and/or addresses. That is to say, the opcode tells the CPU what to do and the addressing mode tells how to find the operand. An instruction may require no operand, e.g., clc (clear carry), stc (set carry), nop (no operation), etc. An instruction may perform a unary operation that requires only one operand, or an instruction may perform a binary operation that requires two operandi. Depending on the type, instructions may have different length. Take a unary operation as an example. The format of such an instruction has an opcode and one addressing mode, as shown in Figure 5.1a. The instruction becomes longer if it also contains a memory address as shown in Figure 5.1b. The addressing mode can be thought of as an eop to pass additional information to the CPU. In practice, the addressing mode can be explicit or embedded as part of the

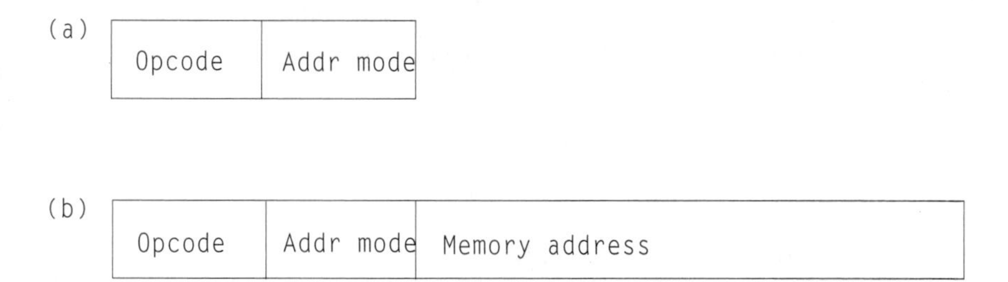

Figure 5.1 Unary op instruction format: (a) opcode and addressing mode and (b) opcode, addressing mode, and memory address.

main op. Depending on its addressing mode, a shorter instruction may have its operand in memory, while a longer instruction may have its operand elsewhere. As the addressing mode specifies where the operand is found, we may have the basic memory immediate, memory direct, and memory indirect.

Each one of the three basic addressing modes may be coupled with indexing, which was the second major programming concept developed in the second generation computers. That is, a hardware register is used to modify the memory address in an instruction. In other words, the hardware register is added to the address in the instruction to form the final address. Because the hardware register may contain a relative index, it is called the index register. Conceptually, more addressing modes can be derived by coupling indexing with memory immediate, memory direct, or memory indirect.

5.2 INDEXING

The indexing scheme was first introduced to work with memory direct, so EA (effective address) is computed as the sum of a register and the address coded in an instruction as follows:

Effective address = address coded in the instruction
+ content of an index register, if specified

The operation is done by the adder at run time, and the sum bits are placed on the address bus to fetch the operand from memory. Concerning software, the instruction contains a base address and the hardware register contains an index or offset. The effective address points to the operand in the middle of a memory block as shown in Figure 5.2a.

The base address points to the memory block, and the index register contains an offset pointing to an entry below the base. Since a memory block is also called an array in high-level programming languages, an instruction using memory direct with indexing can access the entry in an array. Each time the index register is changed in a loop, the effective address is also changed. Therefore, the same instruction can access a different entry in an array each time it is executed. In other words, the memory direct addressing mode with indexing allows an instruction to modify the memory address of an operand. Henceforth, a loop can be written to process all the entries in an array as shown below:

```
;A loop to store 1, 2, 3,.., up to 100 in a memory block named
;intArray to mean integer array.
...
intArray  dw  100 dup (?)
...
mov       di, 0           ;initial index
mov       ax, 1           ;counter
```

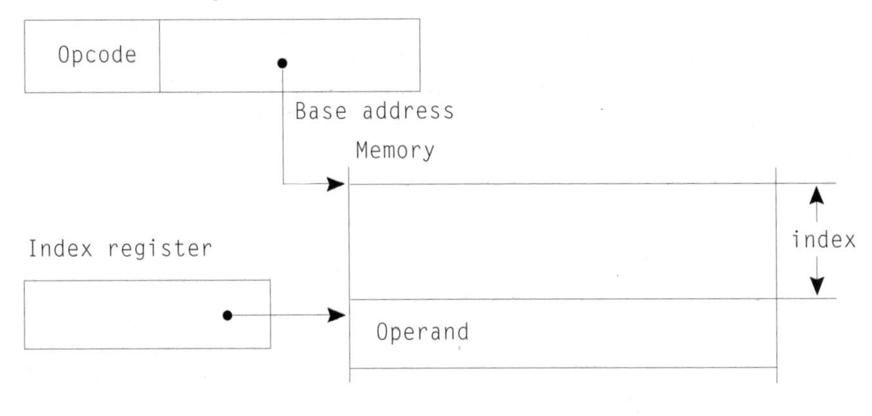

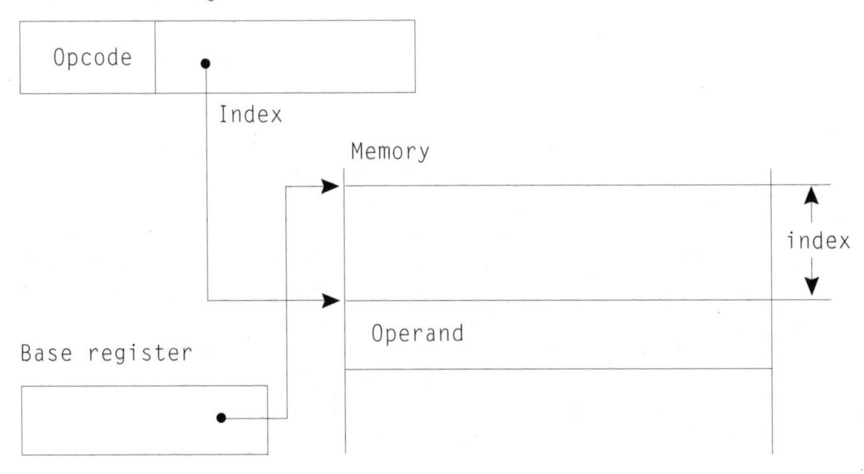

Figure 5.2 Hardware index registers: (a) the register contains an index and (b) the register contains a base address.

```
b0010:  mov     intArray[di], ax   ;store counter in array
inc     ax                         ;increment counter by 1
add     di, 2                      ;increment index by 2
cmp     ax, 100
jle     b0010
. . .
```

This loop stores integers from 1 to 100 in an array named intArray from top to bottom in ascending order. The mov instruction uses the DI index register to change the effective address from time to time as the loop is executed. After the store operation, the count in AX is increased by one and DI is increased by two because each word is two bytes long. As the loop is stepped through, the count in AX is stored into a different word in the array. The following structure shows the use of a loop instruction to cut the code short:

```
        mov   di, offset intArray + 198   ;di points to the last
                                          ;word {Note 1}
        mov   cx, 100                     ;down counter {Note 2}
b0010:  mov   [di], cx                    ;store count from bottom to top
                                          ;{Note 3}
        sub   di, 2                       ;decrement di by 2 {Note 4}
        loop  b0010                       ;{Note 5}
```

- **Note 1:** The source operand is memory immediate and the destination operand is register direct. After execution, DI contains an offset (intArray + 198) pointing to the last word in the array named intArray. The + operator is interpreted by the assembler at assembly time, and the sum result is placed in the code.
- **Note 2:** The source operand is memory immediate and the destination operand is register direct. After execution, the CX count register contains 100 as the initial integer to be stored at the bottom of the array.
- **Note 3:** The mov instruction stores CX in the memory location whose offset is in the destination index register DI.
- **Note 4:** The sub instruction decreases DI by two so the next time the same instruction is executed, the effective offset will be two bytes less.
- **Note 5:** The loop instruction decreases CX by one, and then performs the test and branch function. If CX is not zero, control is passed to the beginning of the loop. As the same mov instruction executes the second time, it will store 99 into the second entry from the bottom. That is to say, the loop stores all the integers 100, 99, 98, ..., and zero in reverse order. Upon exit, the integers are correctly stored in the array in ascending order.

If memory direct is used with indexing, many variations can be derived in regard to the usage of a register. An index stored in the register is called indexing. A base address is stored in the register is called based addressing. An index register specified as PC that contains the address of the instruction currently under execution is called relative addressing. Nonetheless, from a hardware viewpoint, a register is a bunch of flip-flops. From the software viewpoint, a register can be used to store data, an index, or a base address. If a register is used to store data or an address, it is a GPR (general purpose register).[26] In coding, the three address registers DI, SI, and BX are all interchangeable. It is also possible to use BP as an index register to access a data segment provided that a DS segment prefix override is specified (DS:BP). There are many variations derived from indexing as explained in the following sections.

5.2.1 Based Addressing

A hardware register containing a base address, as shown in Figure 5.2b, is known as based addressing. Therefore, the register is referred to as the base register. In hardware design, there is no difference between an index register and a base register. However, there is a subtle difference between indexing and based addressing in software design. If our goal is to store the integer 100 into the last word of a 100W block, we can use either approach. The first approach is to use indexing as shown below:

```
mov     di, 198             ;di contains 198
mov     intArray[di], 100
```

The first instruction places 198 in the DI, and the second instruction stores 100 in a word that is 198 bytes below the base address of intArray. The second approach is to use based addressing as follows:

```
mov     di, offset intArray
mov     198[di], 100
```

Because the goal is to add the base and the offset, it does not matter where they are positioned. The first approach is to place the base address in the instruction and the index in a register. The second approach is to place the index in the instruction and the base address in a register. As far as the hardware register DI is concerned, there is no difference between an index and a base.

5.2.2 Relative Addressing

When the PC is used as an index register, relative addressing is employed, as the EA (effective address) is computed as the sum of an offset in the instruction and the PC. If the relative addressing mode is used to fetch an instruction or an operand, the code is relocatable. That is, the instruction can be positioned anywhere in memory and the address field in the instruction does not need be modified. Microcomputers usually support a relative address in a conditional branch instruction. For example, the conditional branch instruction on an 8086 uses one-byte opcode followed by one-byte relative offset. During adding, this relative address is sign extended so that the offset must range from 127 to -128 with respect to the IP. If the condition is true, the relative offset is added to IP, else IP is increases by two. The assembler will flag an addressability error if the address is out of range.

5.2.3 Register Direct vs. Register Indirect

The term direct refers to a group of bits accessed only once directly with respect to a given reference point. Therefore, in addressing mode, register direct and jump direct are similar, but memory direct is different. The memory direct addressing mode means that the address in an instruction directly points to the operand in memory. On the other hand, the register direct addressing mode means that the register address coded in the instruction directly points to the operand in a register. In other words, register direct means an operand in the register. Memory access is memory immediate with indexing and the address field is zero, which may be implied. In concept, an offset zero is added to an index register to make the effective address. Nonetheless, if addressing mode bits are used to specify register direct, the offset zero is implied, so it can be omitted from the instruction. Consequently, the add operation is also bypassed because it is not necessary.

Indirect means that the bits, fetched the first time with respect to a given reference point, constitute an address pointing to the operand. Using memory indirect as an example, it takes more than one memory cycle to fetch the operand. In contrast, register indirect means that a hardware register contains the address of an operand in memory. Treating the register as a reference point, two accesses are performed to obtain the operand. The first time, the register is accessed to fetch the operand address, and the second time, the memory is accessed to fetch the operand. Nevertheless, only one memory cycle is involved, so register indirect really means memory direct with indexing, and the address is zero. Again, if addressing mode bits are used to specify register indirect, the address zero does not need to be part of the instruction. As a result, the add operation is also bypassed, as it is deemed unnecessary. Needless to say, register indirect is a variation of indexing that is used to access individual elements in an array. For example, a mov instruction using register indirect stores the integer 100 into the last word of a 100W array named intArray, as shown below:

```
mov  di, offset intArray + 198

mov  [di], 100              ;same as mov  0[di], 100
```

The first mov instruction has the source operand specified as memory immediate (intArray + 198) and the destination operand as register direct. After execution, the DI contains an offset pointing to the last word in intArray because the + operator is interpreted by the assembler at assembly time. The second mov instruction has the source operand specified as memory immediate and the destination operand as register indirect. As a result of its execution, the integer 100 is stored into the memory location identified by the DI. By placing an offset zero in front of the index register DI, we obtain memory direct with indexing.

5.2.3.1 Jump Direct vs. Jump Indirect

The jump direct instruction causes confusion because the following instruction is treated as a memory direct operand. In terms of execution, the address field in the IR is copied into the PC. That is, the operand is immediately available as part of the IR. Therefore, its addressing mode is really memory immediate. Similarly, jump indirect means that the instruction contains an address in memory where the address of the next instruction is stored. In terms of execution, the addressing mode is really memory direct because it takes one memory cycle to obtain the branch address in the PC. Because of that, PCs and other mainframes all support register indirect and jump indirect, but not memory indirect.

5.2.4 Double Indexing

When memory direct addressing mode is specified, it is sometimes desirable to specify two index registers, a process known as double indexing. That is, the EA is computed as the sum of three bit strings as shown below:

```
Effective address  = address in the instruction
                   + content of the first index register if specified
                   + content of the second index register if specified
```

Add the first register to the address, and then add the second register to the result to obtain the final effective address. Two run-time adder cycles are needed to compute the EA. Double indexing is attractive to process an array of records (i.e., structures). If one register is called the base and the other register is called the index, we have what is known as based-indexing, which is just a different name for double indexing.

5.2.4.1 Based-Indexing

Based-indexing becomes important when we need to process an array of many heterogeneous records and each record contains many different data types. For example, a personal record may consist of name, age, salary, hiring date, etc. Given a name, the person's age and salary may be displayed by a search program using based-indexing. As regards hardware, there is no difference between the two index registers. As far as software is concerned, we can program the two registers differently as long as the effective address remains the same. We code two examples to illustrate the concept. Assume that each record has a name field of four characters and an age field of 16 bits (short integer), and we intend to fetch the age of the third record into AX as shown below:

```
              ...
ageRecord     db     'Dave'
              dw     17
              db     'Mary'
              dw     21
              db     'Max '
              dw     91
              ...
; ————————————Approach 1: Array base address in the instruction.
```

```
        mov   bx, 6*2                 ;bx ←— 000C {record base}
        mov   si, 4                   ;si ←— 0004 {record index}
        mov   ax, ageRecord[bx + si] ;ax = 005b
;
; ——————————Approach 2: Array base address in bx.
        mov   bx, offset ageRecord
        mov   si, 6*2
        mov   ax, 4[bx + si]          ;ax = 005b
```

Because each record has two fields, a four-character name followed by a two-byte age, the record length is six bytes. In approach 1, the base register contains an offset (6*2) pointing to the record, and the index register contains an offset pointing to the age field in the record. After interpreting the multiply (*) operator, the assembler places the result 000C in the address field as memory immediate because it is a numeric number. The array base is coded in the third mov instruction. As shown by the pointers in Figure 5.3a, the instruction contains the array base, the base register contains a record base, and the index register contains a record index. In approach 2, we rearrange the pointers differently as shown in Figure 5.3b. That is, the instruction contains a record index, the base register contains the array base, and the index register contains a record base.

5.2.5 Auto-Indexing

Since indexing is used to modify the memory address, we can add the capability to modify the index register as part of the execution. One attractive feature is pre-decreasing, i.e., to decrease the index register by the length of the operand before accessing an operand in memory. The other is post-increasing, i.e., to first access an operand in memory and then increase the index register. Both addressing modes are classified as auto-indexing because the index register is automatically modified. The value to be decreased or increased is usually determined by the length of the operand.

In the case that we implement a stack in memory, the push and pop operations uses the auto-indexing concept in that SP (stack pointer) is the index register. That is to say, during a push operation the index register is first decreased before writing an operand into memory. During a pop operation, SP is increased after reading an operand from memory.

5.2.6 Pre-Indexing vs. Post-Indexing

Assume that a computer supports memory indirect addressing coupled with indexing. That is to say, the CPU performs two extra operations before accessing an operand: one is memory indirect and the other one is indexing. However, there are two design options: indexing before memory indirect or indexing after memory indirect. The former is known as pre-indexing and the latter is known as post-indexing. Either addressing mode requires two memory cycles because the instruction contains

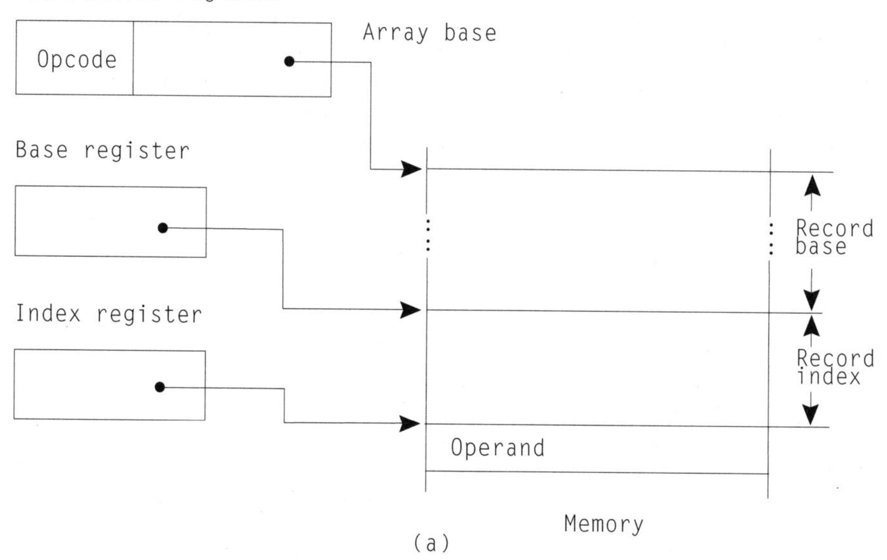

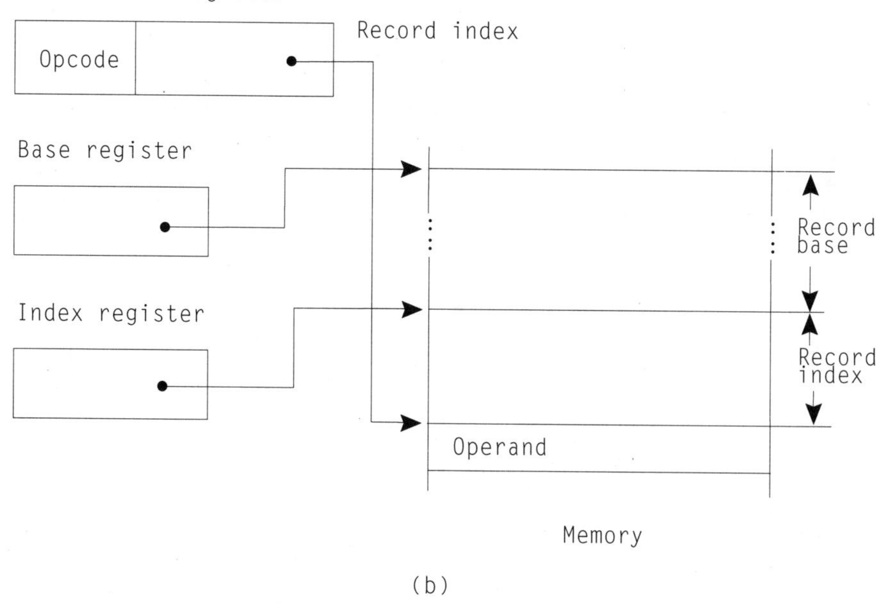

Figure 5.3 Based-indexing: (a) the array base address is in the instruction and (b) the array base is in the base register.

an indirect address. Let us read an operand in memory using pre-indexing. That is, indexing is done before retrieving the direct address in memory as shown in Figure 5.4a. After adding the indirect address coded in an instruction to the specified index register, the EA (effective address) is placed on the address bus to fetch a second direct address in memory. Thus, after the first read cycle, the direct address of an operand is fetched into the CPU. By placing the direct address on the address bus during another read cycle, the CPU fetches the operand in memory.

Instruction register
Opcode
First address
Memory
Second address
Index reg
Operand
(a)

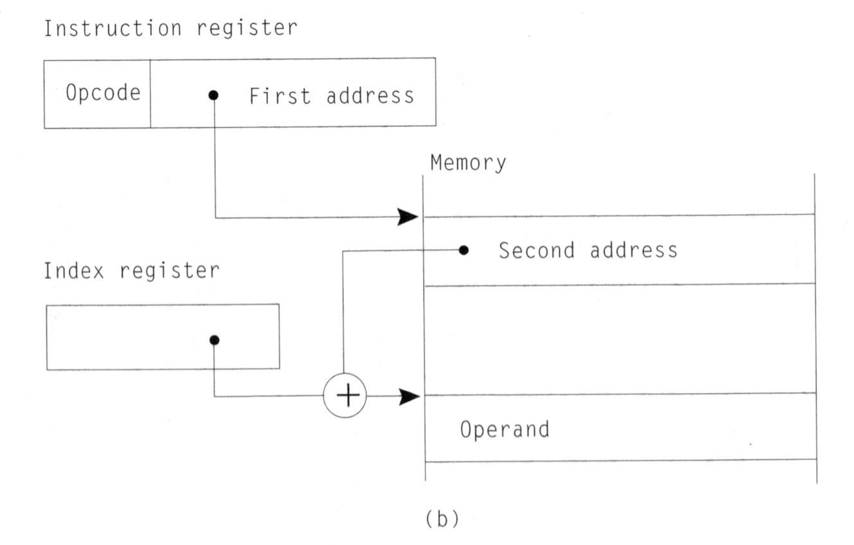

Figure 5.4 Memory indirect with indexing: (a) pre-indexing and (b) post-indexing.

Post-indexing means that indexing is performed after fetching the direct address in memory as shown in Figure 5.4b. Let us read an operand in memory using post-indexing. The instruction contains an indirect address pointing a second direct address in memory. After the first memory read cycle, the second direct address in memory is fetched into the CPU. Now, the CPU performs indexing by adding the index register to it to form an EA. By placing the EA on the address bus during another memory read cycle, the CPU fetches the operand in memory.

In conclusion, either pre-indexing or post-indexing can be specified as the addressing mode. For example, the 6502 supports pre-indexing, the Sigma 7 computer supports post-indexing, but the VAX machine supports both pre-indexing and post-indexing.[51,72] As indexing is as powerful as memory indirect, the future trend is not to support memory indirect because the extra central memory cycle is slow.

5.3 ADDRESSING MODES OF AN 8086

An 8086 processor supports a variable instruction length ranging from 1 byte to 7 bytes while a Pentium supports a superset so its instruction length varies from 1 to 11 bytes. In general, an instruction has one byte opcode followed by an eight-bit addressing mode, e.g. memory direct, memory immediate, register direct, register indirect, indexing, and double indexing. The processor does not support memory indirect. There are exceptions that the addressing mode is embedded into the opcode. As an example, the opcode to push stack or pop stack implies auto-indexing. Often times, one or more operandi can be implied in an instruction, so the code is short. However, they are some restrictions in coding as explained below.

1. An instruction can not contain more than one memory direct address so that at least one source operand must be in a register or as immediate.
2. A mov segment register instruction can not take an immediate source operand.
3. A multiply or divide instruction can not take a source operand that is memory immediate.

Note that the same set of code may support 16-bit operations for 8086 or 32-bit operations for Pentium. When a C compiler is used to do program development work, debugging tools can be used to disassemble the run-time code in memory. Knowing the target instructions is helpful, therefore, we study the layout of 8086 instruction set.

5.3.1 Instruction Format of an 8086

As hardware design is concerned, the 8086 instructions are divided into 4 groups with few exceptions. A unary operation instruction requires one operand but a binary operation instruction requires two operandi. However, if an operand is implied, it has no address specified in the instruction. As shown in Figure 5.5a, each instruction in the first group has only one-byte opcode because one or two operandi may be implied. The different fields in the opcode are defined in Figure 5.5e. Single opcode instructions include nop, clc (clear carry), stc (set carry), cbw (convert byte to word), cwd (convert word to double), in (input), out (output), etc.

5.3.1.1 Hardware First Operand vs. Second Operand

In a binary operation, the instruction needs two operandi. The location of the operand; is determined by the addressing mode or by the opcode if the operand is implied. Specifically, the opcode specifies the second operand and the addressing mode primarily specifies the first operand. Therefore, the ordering, first or second, is defined by physical hardware. Recall that the source operand one (sopd1) in the instruction is logical, and it also serves as the destination. But, in the instruction spec, sopd1 may or may not be the hardware first operand. In fact, either the hardware first operand or the hardware second operand can be specified as the destination.

Each instruction in the second group has one operand specified as shown in Figure 5b. This group includes all the unary op instructions plus the multiply and divide instructions. Even though such an operation is binary, only one source operand is specified because the other source operand is accumulator as implied.

The one-operand instruction has a length of up to five bytes, that is jointly decided by the opcode and the addressing mode. The first byte is a segment override prefix that is optional. If specified, the two-bit seg (segment) field specifies the override segment base register as defined in Figure 5e. The second byte contains the seven-bit opcode followed by the W (width) bit. If W is one, it is a word; otherwise it is a byte. The third byte has the two-bit mod (mode), three-bit eop (extended op), and three-bit R/M (register/memory). The two-bit mod determines whether the operand is in a register or in memory. If the mod is 11, the operand is in a register, so no displacement field is needed. In addition, the three-bit R/M along with the W bit specifies a register address, 16-bit or 8-bit. If Mod is not 11, then the operand is in memory. Consequently, the fourth and fifth bytes are needed to store its displacement, Disp-Lo (displacement low-order byte) and Disp-Hi (displacement high-order byte). If the mod is 00, the operand is in memory with no indexing. Finally, if Mod is 01 or 10, it is memory direct with indexing and the three-bit R/M further specifies single indexing or double indexing. Because only one operand is specified in the instruction, there are three spare bits left as the eop. That is, the seven-bit opcode and the three-bit eop jointly define an unique operation, such as not, inc, dec, neg, mul, imul, div, idiv, etc.

Figure 5c shows the format of a two-operand instruction whose second operand is register direct as determined by the opcode. As usual, the first byte is the segment override prefix which is optional. The second byte has a six-bit opcode followed by two bits: D (destination) and W. The D bit specifies which of the two operandi is the destination. If D is one, the hardware second operand is the destination; otherwise the first operand is the destination. The third byte has the two-bit mod, three-bit reg,

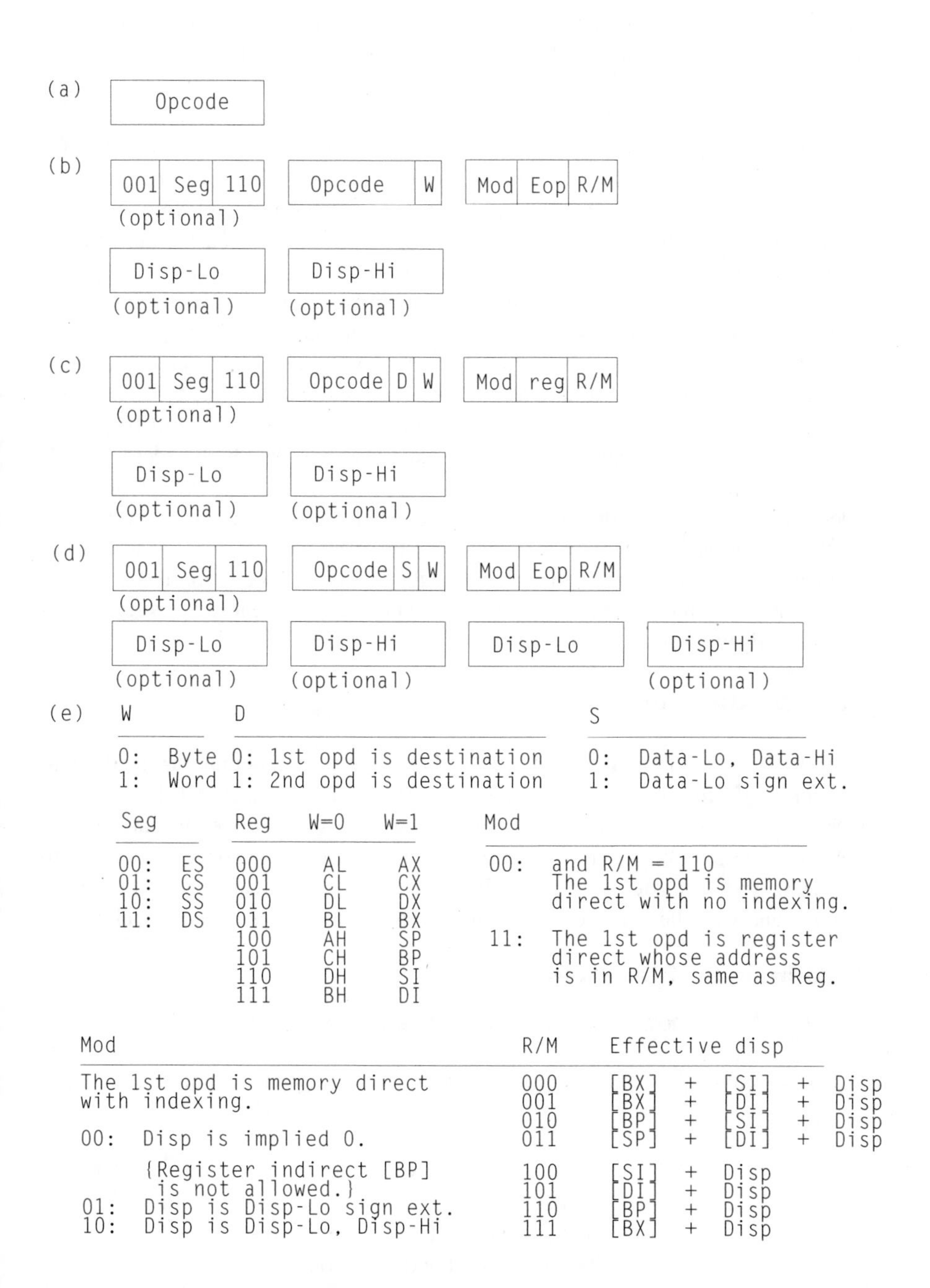

Figure 5.5 Instruction format of the 8086: (a) opcode only, (b) 1-operand, (c) 2-operand and the second operand is register, (d) 2-operand and the second operand is constant, and (e) field definitions.

and three-bit R/M. The two-bit mod decides whether the first operand is in a register or in memory. If the mod is 11, the first operand is register direct so no displacement is needed. Consequently, the three-bit R/M along with the W bit specifies a register address, 16-bit or eight-bit. If Mod is not 11, the hardware first operand is in memory so the fourth and fifth bytes are needed to store its displacement, Disp-Lo and Disp-Hi. If the mod is 00, the first operand is memory direct with no indexing. If the mod is 01 or 10, the three-bit R/M further specifies single indexing or double indexing. Because the second operand is a register, the three-bit Reg along with the W bit specifies a register address, 16-bit or eight-bit.

Figure 5d shows the format of a two-operand instruction whose hardware second operand is a constant (i.e., memory immediate), so its hardware first operand is the destination. The first byte is the optional segment override prefix. The second byte contains a six-bit opcode followed by two bits: S (short) and W. Note that S is meaningful only if W is set to one. If S is zero, the instruction contains the full 16-bit second operand, Data-Lo (data low-order byte) and Data-Hi (data high-order byte). If S is one, the instruction contains only Data-Lo and its sign is extended into its high-order byte during a 16-bit operation. The third byte is the two-bit mod, three-bit eop, and three-bit R/M. The two-bit mod and three-bit R/M jointly define the addressing mode of the hardware first operand just like the format in Figure 5b. If the two-bit mod is not 11, the hardware first operand is in memory, so two more bytes are needed to store its displacement, Disp-Lo and Disp-Hi.

5.3.2 Destination Bit

If the D bit in an instruction is one, the hardware second operand is the destination; otherwise the first operand is the destination. This means that the hardware operand merely provides a reference point in the spec — either operand can be specified as the destination. To further study the hardware operand concept, let us try to encode a register transfer instruction. Because either register can be specified as the first operand with the D bit set, we obtain two possible images as shown below:

Code image	Statement
8B F0	mov si, ax ;si ← ax {same as 89 C6}

The 16-bit code image (8B F0 in hex) can be further broken down into bits as shown below:

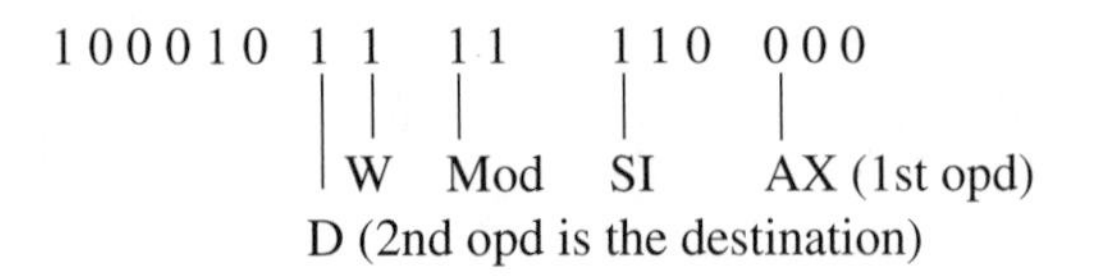

If the D bit in the instruction is one, the second operand is the destination. The six leading bits specify that the second operand is a register whose address indicates SI.

The mod field 11 tells that the first operand is in a register whose address indicates AX. In the transfer language, AX is the alias for R[0] and SI is the alias for R[6]. The same function, however, can have a different code image (89 C6) as follows:

```
100010 0 1   1 1    000  110
       | |   |       |    |
       | W   Mod     AX   SI (1st opd)
       D (1st opd is the destination)
```

Note that the D bit is zero in the opcode byte, so the first operand is the destination. As a consequence, the two register addresses are interchanged in the second byte to achieve the same function.

5.3.2.1 32-bit Encoding Extensions

When a Pentium processor operates in 32-bit operand mode, some of the encoding extensions are described below:

1. If the W bit is set, the operand length is 32 bits
2. If the S bit is set, sign extend the immediate data to 32 bits.
3. For disp and immediate data, each field can go up to 32 bits.
4. For memory operand addressing, if the R/M field is three-bit 100, another byte named SIB (scaled index base) is inserted as an extended addressing mode. As a consequence, not only can each of the 32-bit accumulators be specified as an index register, but double indexing is allowed as well.

5.4 SUBROUTINE LINKING

A subroutine is a group of instructions that performs a particular function, e.g., sorting, searching, etc. The calling routine issues a call instruction to pass control to the subroutine. After the subroutine completes its execution, it issues a return instruction to pass control back to the calling routine as shown in Figure 5.6. The terms subroutine, routine, procedure, and subprogram are all synonymous. Concerning logic flow, control first goes to the subroutine and returns to the calling routine in a close loop. This is the closed subroutine, the third major programming concept.

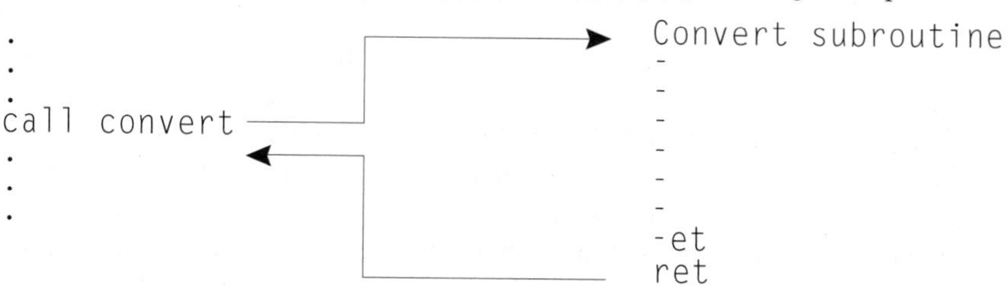

Figure 5.6 Subroutine linking.

The call and return instructions are special hardware design features. A call instruction is similar to an unconditional branch except the CPU saves a return address somewhere and knows how to fetch it upon executing a return instruction. That is, while executing a call, the CPU saves the return address and passes control to the subroutine. When the subroutine is completed, a return instruction is executed to pass control back to the calling program. In sum, several key concepts are listed in the following:

1. A subroutine may be called many times from different locations in the main program.
2. A subroutine may have multiple return instructions and each return means that the processing of a specific case is completed.
3. A subroutine is usually placed after an unconditional branch instruction, a return instruction, or a system termination call.
4. If working registers are modified by the subroutine, the subroutine may save the registers before modification and restore the registers before return.

In practice, the return address is stored in one of two locations as listed below:

- Stack
- Register

Commonly during a call, the return address is pushed on the stack. Thus, at the end of the subroutine a return instruction can pop the return address off the stack into program counter. As an alternative, the return address is placed in a register and the return instruction becomes a branch register indirect.[26,40]

Near Call and Near Return

There are two types of call on a PC: near and far. A near call is an intrasegement call to pass control within a segment. A far call is an intersegement call to pass control to a different segment. That is, the CS register remains the same in a near call, but changes in a far call. Let us study the run-time code of a near call on an 8086 as given below:

Code image	**Loc**	**Statement**
E8 32 00	003A	call convert ;call near ptr convert

where convert is the symbolic address or entry point of the subroutine declared in the same segment. The near call instruction has a one-byte opcode followed by a two-byte relative offset with respect to the IP. Before executing the call, assume that the IP contains an offset 003A pointing to the first byte E8 in the instruction. After interpreting the three-byte instruction, the IP is increased by three to become 003D, the return address pointing to the next instruction after the call. The second byte of the code image is 32 which is the low-order byte of the offset and the third byte is 00, the high-order byte. The execution of the call instruction involves save return address and branch. First, push the return address on the stack. Second, add the relative offset to the IP to form the branch address. The hardware steps are described by the transfer language below:

```
;sp <— sp - 2
;M[ss:sp] <— ip         {TOS is 003D}
;ip <— ip + 0032        {convert is 006F}
```

Obviously, the address or entry point of convert is 006F (003D + 0032). There are many variations of the ret (return) instruction, and each one has a different opcode even though the mnemonic is the same. For example, a pseudo statement is used to declare a procedure, near or far. For a near procedure, the ret statement is translated by the assembler into a near ret instruction as shown below:

Code image	**Statement**	
C3	ret	;near return ;ip <— M[ss:sp] ;sp <— sp + 2

The near return instruction has one-byte opcode C3, and upon its execution the return address is popped into the IP.

Far Call and Far Return

A far call instruction is longer in that the opcode byte 9A is followed by a far branch address as shown below:

Code image	**Statement**	
9A 6F 00 C6 13	call convert	;call far ptr convert

The far call pushes the far return address first and then branches. Because a far address is represented by four bytes: a two-byte offset followed by a two-byte segment base, more hardware steps are executed, as shown below:

```
;sp <— sp -2
;M[ss:sp] <— cs
;sp <— sp -2
;M[ss:sp] <— ip
;ip <— 006F
;cs <— 13C6
```

The offset in a far call instruction is an offset with respect to the segment base with the low-order byte positioned before the high-order byte. Consequently, a far ret instruction pops the far return address off the stack, as shown below:

Code image	Statement	
CB	ret	;far return ;ip <— M[ss:sp] ;sp <— sp + 2 ;cs <— M[ss:sp] ;sp <— sp + 2

5.4.1 Argument Passing

A subroutine is powerful in that many arguments can be passed from the main routine to a subroutine. Take sorting as an example; if a different array address is passed to the sort subroutine, the integers in an array at that particular address can be sorted. Two questions must be answered. First, where should the arguments be stored? The answer is; push an argument on the stack or store it in a register. Second, in what form should the argument be passed? The answer is; pass the bit string value itself or the address of the bit string. The former is also known as call by value, and the latter is also known as call by reference or address. Passing the arguments and issuing the call instruction are known as the calling sequence. Deciding the type or ordering of arguments in a calling sequence is a software design issue.

5.4.2 Call by Value

If the bit string of an argument is pushed on the stack or stored in a register before issuing the call instruction, it is a call by value. Because each constant also occupies a memory location in the calling routine, the subroutine being called can not change the constant in the main routine if it is passed by value. In order to pass an array by value, the entire array needs to be pushed on the stack, which also means overhead. This is especially true if the array is very large. To solve the problem, we can have a call by reference, as introduced next.

5.4.3 Call by Reference

If the address of a data structure is passed to the subroutine before the call instruction, it is a call by reference (address). The data structure can be an integer, an array, or a structure. This design feature is necessary provided that the subroutine is written to interchange two variables or to sort integers in an array. In a typical sort subroutine, the array is passed by address and the array size is passed by value. That is to say, arguments can be passed by value, by address, or a combination of both.

A subroutine does not need to save and restore registers if the main routine does not care whether the registers are modified or not. One example of a calling sequence is shown below:

```
;-----   Data
ans      dw      13bah    ;13ba in hex means 5050 in decimal.
dchar    db      ?, ?, ?, ?, '$'
...
;-----   Calling sequence
         mov     ax, offset dchar+3
         push    ax          ;push destination address
         mov     ax, offset ans
         push    ax          ;push address of integer
         call    convert
;-----   After return, the next instruction is executed.
```

The first line defines a 16-bit integer named ans (answer) in memory that contains 13ba (5050 in decimal). The second line reserves five characters and the last character is the dollar sign delimiter to indicate the end of the ASCII string. Two arguments are pushed on the stack and both are addresses. The stack frame is shown in Figure 5.7a. The first argument is the address of the fourth character in the buffer named Dchar, which stands for destination characters. The second argument is the address of an integer variable named ans. When executing the call, the CPU pushes the return address and passes control to the subroutine return, as shown in Figure 5.7b. The data area and the stack after, are shown in Figure 5.7c provided that the subroutine pops the passing arguments off the stack.

The near subroutine converts the 16-bit positive integer from binary form into 4 decimal characters in ASCII. If a system call is issued to display the string, we will show the answer 5050 on the screen. The coding of the convert subroutine is shown below:

```
          convert proc   near      ;declare near procedure.
          mov            bp, sp    ;bp points to stack frame.
          mov            si, 2[bp] ;si has the source address.
          mov            di, 4[bp] ;di has the dest address.
          mov            bx, 10    ;radix
          mov            cx, 4     ;loop count
          mov            ax, [si]  ;ax = 13ba
loop010:  cwd
          div            bx
          add            dl, 30h   ;add bias
          mov            [di], dl  ;store the remainder char.
          dec            di
          loop           loop010
          ret            4         ;near ret, pop 4 more bytes
convert endp
```

The first line is a pseudo op that declares a procedure whose symbolic address is convert. Note that in the last line, the pseudo op is endp (end procedure), which has the same convert symbol placed in the front as a subroutine delimiter. The mov instruction copies the SP into the BP pointing to the current stack frame. Subsequently, the BP is used as an index register to access the entry in the stack because SS is the default base. The body of the loop uses the divide algorithm to find the remainder digit, add the bias to it, and store the character in the memory location indicated by DI. After exit, we see the ret instruction and the numeric four specified in the operand field, which means four more bytes are popped off the stack as discussed below.

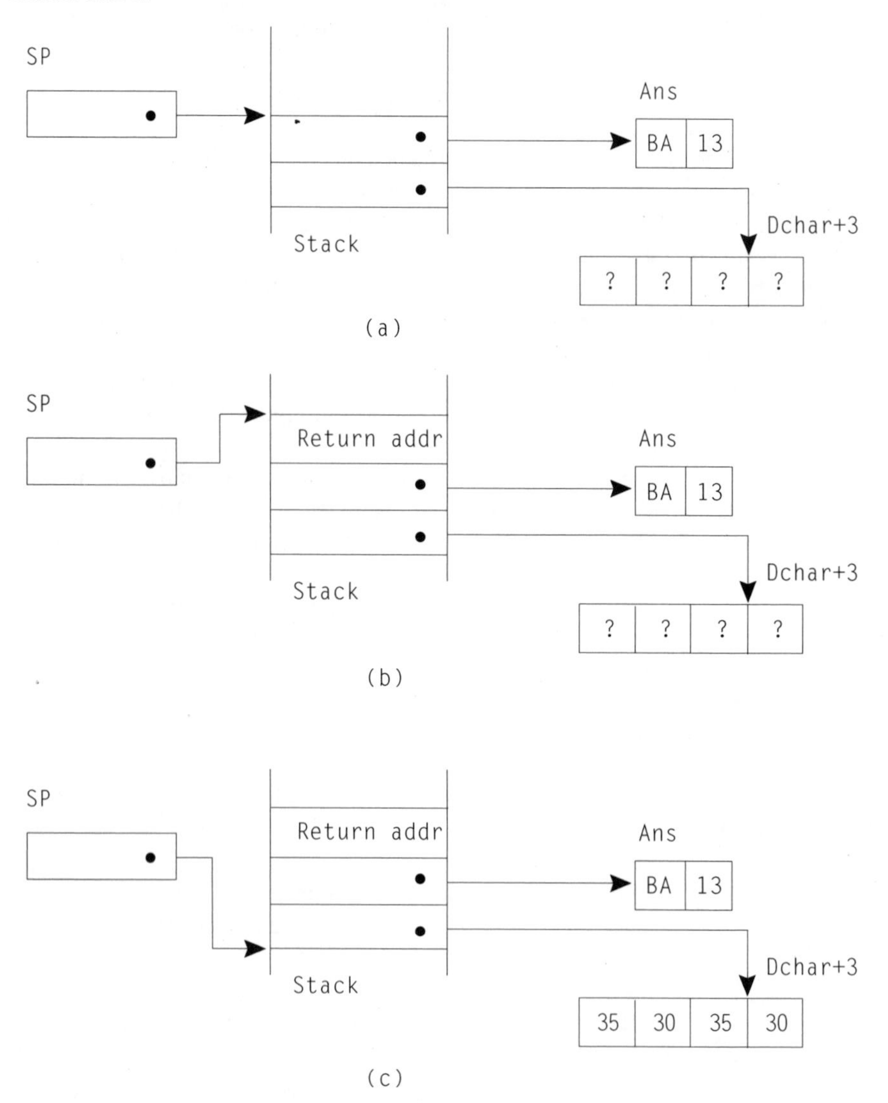

Figure 5.7 Data area and stack frame: (a) before the call, (b) after the call, and (c) after return.

5.4.3.1 Popping Off Arguments

If the passing arguments are pushed on the stack before executing the call instruction, either the calling routine or the subroutine being called can pop the arguments off the stack. However, there is one advantage if the popping job is left to the subroutine being called instead of the calling routine; that is, the subroutine only needs to do it once with one single copy of code. If the calling routine pops the arguments off the stack, the code needs to be placed after the call. Since there may be many calls in a program, the same code is duplicated many times. It is clever that the ret instruction pops the arguments off the stack as shown below:

Code image	Statement		
C2 04 00	ret	4	;near return with extra pop
			;ip <— M[ss:sp]
			;sp <— sp + 2
			;sp <— sp + 0004

Note that the ret instruction has three bytes — a one-byte opcode followed by a two-byte immediate operand.

5.4.4 Recursive Call

A recursive call means that a subroutine can call itself directly or indirectly.

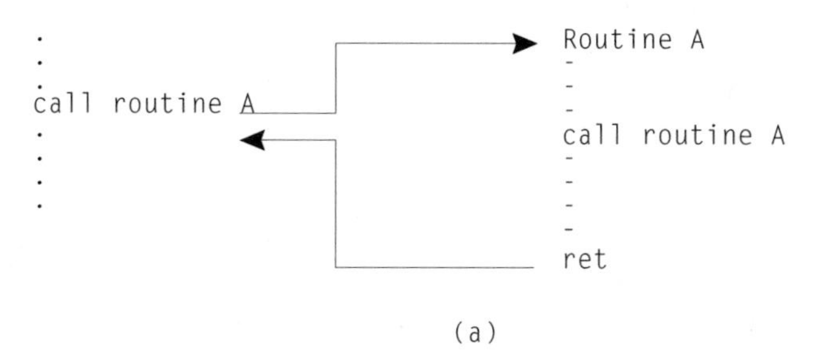

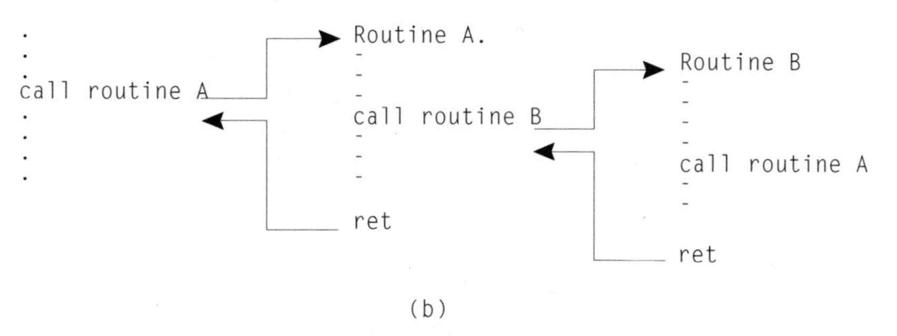

Figure 5.8 Recursive calls: (a) routine A calls routine A directly and (b) routine A calls routine B, and routine B then calls routines A indirectly.

Calling directly means that routine A calls routine A, as shown in Figure 5.8a. Calling indirectly means that routine A calls routine B and then routine B calls routine A as shown in Figure 5.8b.

A requirement for writing a recursive routine is to store the activation record of the subroutine in a stack frame. That is to say, all working registers and local memory variables modified by the subroutine must first be saved on the stack and restored before return. Usually the subroutine pushes the previous environment to save, does the processing, and pops the previous environment to restore. For example, let us study the factorial function as follows:

$$N! = N * (N-1) * (N-2) * \ldots 2 * 1 \quad \text{where } N = 1, 2, 3, \ldots$$
$$= N * (N-1)!$$

We can write a function subroutine under two assumptions. First, the argument N is a positive integer passed in BX. Second, the AX register contains the answer after return. There are two approaches, iterative and recursive, and the following program issues two different calls as shown below:

```
        title       factorial program
DSEG    segment
msg     db          'Job is done.'   ;for display purpose
        db          0dh, 0ah, '$'    ;CR, LF, and terminator
DSEG    ends
SSEG    segment stack
        dw          80 dup (?)
SSEG    ends
CSEG    segment
        assume cs:CSEG, ds:DSEG
main    proc
        mov         ax, DSEG
        mov         ds, ax
        mov         bx, 5            ;call by value
        call        fact1
        nop                          ;use debug to examine result
call                fact2
nop                                  ;ditto
mov     ah, 4ch                      ;system termination
int     21h
; ———   The two subroutines are inserted here.
main    endp
CSEG    ends
        end         main
```

5.4.4.1 Iterative Approach

The iterative routine computes the function in a loop and the assembly code is shown below:

```
;  Fact       = 1
;  DO i       = 1 to N, step 1
;  Fact       = i * Fact
;  ENDDO
; ————-       iterative approach
fact1         proc       near
              mov        ax, 1          ;Fact = 1
              mov        cx, 1          ;CX = i, BX = N

b0010:        mul        cx             ;Fact = i * Fact
              cmp        cx, bx         ;compare i:N
              je         exit020
              inc        cx
              jmp        b0010
exit020:      ret
fact1         endp
```

If N is 5, after return the AX register contains 78 (120 in decimal). Note that if a large passing argument is passed to the factorial subroutine, the result may be too large to be stored in a 16-bit register.

5.4.4.2 Recursive Approach

The second approach is to write a recursive routine as follows:

```
;             Push environment
;             IF N is 1,
;             THEN
;             Fact = 1
;             Goto exit040
;             ELSE
;             Fact = N * Fact( N-1)
;exit040:     Restore environment
;             return
;             ENDIF
; ————-       recursive approach
fact2         proc       near
              push       bx             ;save environment
              push       bp
              mov        bp, sp         ;bp points to stack frame
```

```
          cmp    bx, 1
          jne    skip030              ;If N is 1,
          mov    ax, 1                ;then Fact = 1,
          jmp    exit040              ;pop stack, and return
skip030:  dec    bx                   ;else compute N-1
          call   fact2                ;call Fact( N-1)
          mul    word ptr 2[bp]       ;Fact = N * Fact( N-1)
exit040:  pop    bp                   ;restore environment
          pop    bx
          ret
fact2     endp
```

Note that the same argument five is stored in BX before the call because the fact1 routine never changes BX. The first time, the fact2 routine is called from the main routine and the return address is denoted as return addr 1. The next four times, fact2 is called by itself or from inside. During each inside or recursive call, the return address is denoted as return addr 2 pointing to the mul instruction. Consequently, after pushing the environment and the return address four extra times, the BX register becomes one so the AX register is set to one to return. The stack frame is shown in Figure 5.9a. Note that the mul instruction is executed after each return from an inside call with a different environment for a total of four times. Because the argument in BX is the sec-

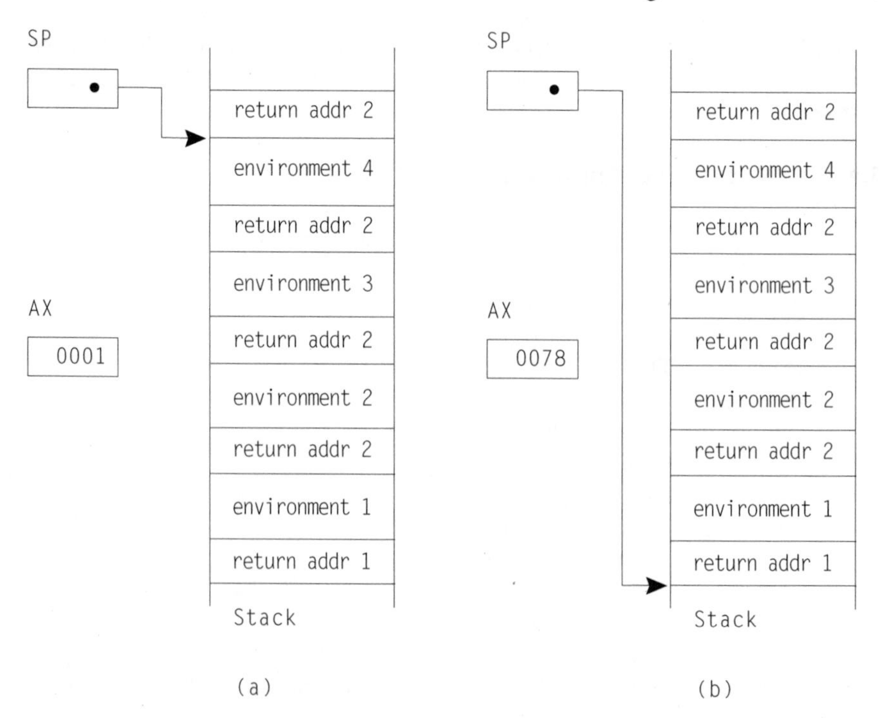

Figure 5.9 Registers and stack frames: (a) after the first return and (b) after the next four returns.

ond word below the top, as indicated by BP, the multiplicand is specified as 2[BP] using SS as the segment base. Following multiply operation, the product is in DX^AX. After the last return, control is passed to the main routine so the AX register contains 0078 in hex, which is 120 in decimal, as shown in Figure 5.9b.

5.5 INTERRUPT MECHANISMS

Interrupt is the fourth major programming concept implemented in third generation computers. As far as the CPU is concerned, an interrupt pending signal may be generated from inside or outside. Either way, the IF (interrupt enable flag) bit in the F register must be set before the CPU can accept the pending interrupt request. When that happens, a hardware mechanism is activated to force the current program to relinquish control. On the other hand, if the IF bit is zero, the CPU just ignores this interrupt signal at the moment. Note that the interrupt pending signal will not disappear until the CPU takes further action. The interrupt hardware mechanism, if activated, causes the CPU to perform context switching. On a microcomputer, the CPU switches the PC (program counter) and SR (status register) referred to as partial context switching. On a mainframe, switching the entire set of working registers is a process known as full context switching.[5] However, the real full running environment still includes the code in memory and the data on disk. A task is defined as the basic computation unit with its resources allocated by the OS. During interrupt, the CPU saves the old environment of the current task and loads a new environment from a low memory address also called an interrupt vector. Interrupting is necessary to protect the system from crashes and to increase throughput. With interrupts, simultaneous I/O operations are possible along with CPU operations.

After an interrupt condition occurs, control is passed to the interrupt handler (IH) in the OS. If the set of working registers is not switched, the IH then saves the working registers on stack, analyzes the cause, and passes control to the ISR (Interrupt Service Routine). While switching the context, further interrupts are temporarily disabled to avoid confusion. After interrupt processing, control is passed to the dispatcher (i.e., task scheduler) which selects the next task to execute.

In the new SR, a special bit is set to indicate that the machine is in supervisor state. A set of privileged instructions can only be executed.[26,28,40,76] That is, the privileged instructions are not accessible to application programmers, and they are designed for system control, e.g., modify SR, interrupt another I/O processor, return from interrupt, etc. If a separate I/O processor is designed to run the task management routines of an OS, then all the instructions on the I/O processor are privileged.

If an interrupt is generated inside the CPU, it is called a trap. A trap triggers the same hardware interrupt mechanism. It is also possible for two or more interrupts to occur at the same time. When that happens, the CPU must honor the interrupt that is assigned with the highest priority. Therefore, each type of interrupt has a priority; the IBM mainframe interrupts are listed in Table 5.1.[26,76]

Table 5.1 Priorities of Interrupts

Type	Priority
Machine check (MC)	4 (Highest)
Program check (PC)	3
Supervisor call (SVC)	3
External	2
I/O	1 (Lowest)

MC (machine check)

Machine check has the highest priority, so when a machine has detected a hardware failure, all computations should be put on hold. The MC ISR displays a message for the operator to run more diagnostic programs and call for help if necessary. All the programs being interrupted should run again when the machine is working properly.

PC (Program Check)

Program check and SVC have equal priority because they are mutually exclusive, i.e., they never occur at the same time. Program Check occurs if a programming error is detected. For example, when a user tries to divide an integer by zero or execute an illegal opcode, the CPU hardware triggers interrupt.

SVC (Supervisor Call)

An SVC means a system call that is a special instruction to trigger the hardware interrupt mechanism. In consequence, control is passed to the OS which in turn, performs the service. Note that the SVC instruction resembles a subroutine call except the SR (status register) is also saved. In a program, if a user needs to read the keyboard input, the program issues a system call to the OS as an I/O request. Hence, system call, supervisor call, trap call, and int (interrupt) are all synonymous. After completing the service, the OS usually returns control to the next instruction after the call. Because a system call and a program check are mutually exclusive, they have equal priority.

External

An external interrupt is a signal generated from an outside device, e.g., real-time clock, another computer, an instrument, etc. If the external signal is important, it should have a higher priority as demonstrated in the microcomputer design. Such a system supports an external interrupt named NMI (nonmaskable interrupt) which has a higher priority than any of the I/O interrupts. This makes a lot of sense if the signal generates from an urgent real-time device.

I/O

The I/O completion interrupt signal is generated by the I/O controller, and it has the lowest priority because the event will be served sooner or later without loss. This design feature enables the CPU and I/O devices to operate in parallel, known as asynchronous processing. For example, when the keyboard input character arrives in a hardware I/O data buffer, it triggers an interrupt to signal I/O completion. After an I/O operation of a task is completed, the CPU processes the event and marks the task ready to execute.

5.5.1 Context Switching

On the 8086, the occurance of an interrupt triggers the hardware interrupt mechanism to perform context switching. For example, a system call to terminate is shown below:

Code image	Statement	
B4 4C	mov ah, 4ch	;function code in ah
CD 21	int 21h	;system call

A function code 4C is passed by value in AH to request program termination. The int instruction has an opcode CD in hex, and the interrupt code 21 in hex is an immediate operand used to compute the address of the interrupt vector in low memory. Each interrupt vector has two words. The top word is the new IP and the bottom word is the new CS. That is to say, the interrupt vector contains the CS:IP pair. As the result of executing the instruction int 21h, the six hardware steps occur as follows:

1.Push the F register.
2.Reset the IF (interrupt flag) and TF (trace flag).
3.Push the CS.
4.CS <— M[21 * 4 + 2].
5.Push the IP.
6.IP <— M[21 * 4].

Before executing the instruction, the stack frame is shown in Figure 5.10a. After executing the int instruction, the F, CS, and IP registers are pushed on the stack as shown in Figure 5.10b by steps 1, 3, and 5. Steps 2, 4 and 6 load a new environment.

In step 4, the new CS is fetched from a low memory address that is computed by hardware. First, shift left the interrupt code by two bits to obtain the 20-bit address is shown in Figure 5.10c. Next, set b1 to one to form the final address as shown in Figure 5.10d. In step 6, shift left the interrupt code by two bits to generate the absolute low address. It is crucial to know that while the CPU executes the first two steps, special hardware is designed to disable any further interrupt. After that, the new F register is the same as the old F register except the two bits, IF and TF, are cleared so no more interrupts are allowed.

5.5.1.1 Interrupt Return

After processing the interrupt, if the OS wants to return to the program being interrupted, it does two things. First it restore the working registers of the program, and then it executes the iret (interrupt return) instruction as shown below:

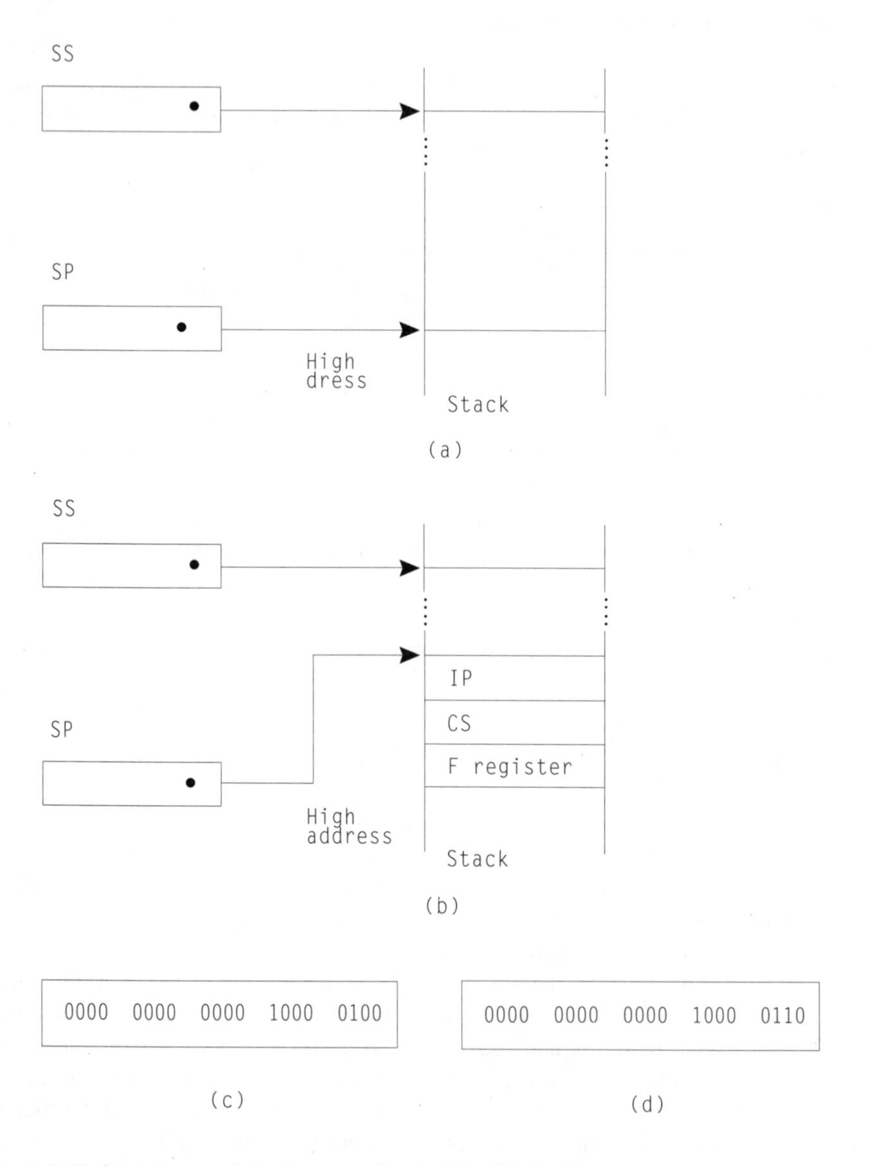

Figure 5.10 Stack frames: (a) before executing int 21h, (b) after interrupt, (c) after shifting left the interrupt code by 2 bits, and (d) the physical address where the new CS is retrieved.

Code image	Statement	
CF	iret	;interrupt return
		;pop ip
		;pop cs
		;pop f

The iret instruction has an opcode CF that is similar to a far ret instruction except it also pops the F register off the stack after popping the return address.

5.5.2 Multitasking

An I/O controller chip is designed to perform an I/O operation. When an I/O operation is completed, if the controller can interrupt the CPU, then the I/O operation can be performed in parallel with the CPU. This design concept is attractive because it allows I/O activities to overlap with CPU activities. After processing an I/O completion event, the OS may decide to place the current program being interrupted in a ready queue and execute the program whose I/O has just been completed. That is to say, the current program is being preempted. As the OS coordinates the execution of many programs from beginning to end, all programs must take turns executing on the CPU. Since the I/O completion event is asynchronous (i.e., unpredictable), running two or more tasks in parallel is known as multitasking or asynchronous processing.

5.5.3 Reentrant Task

With interrupts it is possible to run many tasks concurrently during a time interval. Thus, if one user task or program gets interrupted, another user task may want to execute the same piece of code in memory. If the block of code can be executed the second time by another task, it is reentrant code and the task executing the code is a reentrant task. The requirements of writing reentrant code are similar to writing a recursive subroutine. The main system concept is, do not write into the global addressing space. Reading or executing is fine because it does not change bits in the code. Code sharing means that many user tasks can execute the same code during the same time interval. The requirement is that each task has its own private addressing space, and data can only be written onto the stack or into the registers. This is because the register set is part of the running environment of the current task in execution and it is saved on stack after interrupt and restored before the task resumes its execution. That is to say, if a memory buffer is used for I/O operation, each task must use either its own private memory slot or its own private stack.

If the return address is stored in a memory location, the code is not reentrant. This is because in a multiprogramming environment, the OS supports many tasks during the same time interval. If, right after entering a subroutine, the first user task gets interrupted, the OS suspends the task and executes a second task. If the second task tries to call the same subroutine, the return address of the first task will be damaged. Consequently, when the first task resumes its execution, it can not return properly to its calling program. This is why the return address is usually pushed on a stack or stored in a register — to avoid potential damages from other running tasks.

5.6 I/O STRUCTURE

Each I/O controller can be thought of as a small processor that has its own logic and a set of registers. An I/O controller provides the hardware interface between the CPU and an I/O device. Each register in the controller is called an IOR (I/O register) that can be used to store data, control functions, or status information about the I/O device. In general, the CPU executes an input instruction to transfer bits from an IOR to a CPU register and an output instruction to transfer bits from a CPU register to an IOR. Thus, the I/O instructions enable the CPU to communicate with controllers. In an I/O controller, some registers are read only, some are write only, and others allow both read and write.

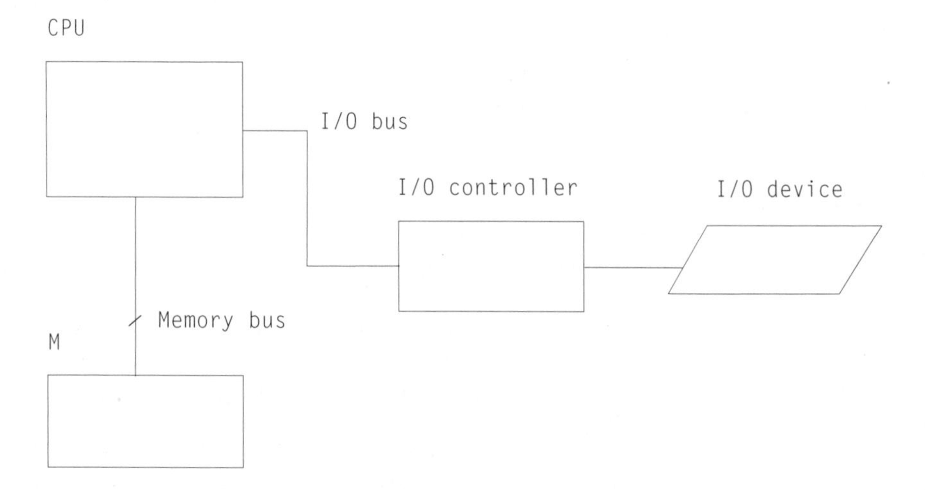

Figure 5.11 A computer system with a separate I/O bus.

5.6.1 Separate I/O Bus

A computer system may use a separate I/O bus to connect an I/O controller to the CPU, as shown in Figure 5.11. Each IOR is assigned a short address (e.g., 12 bits) that uniquely identifies the hardware register in a controller. As a consequence, an I/O instruction must be executed to transfer bits between an IOR and a CPU register. Note that the 8086 chip uses a separate I/O bus. Even though an I/O instruction requires two register addresses to be specified, one for I/O and one for CPU. They may be implied, so the I/O instruction has only a one-byte opcode because the address of an I/O register is implied in DX, and the CPU register is implied in AL or AX as a function of data length. Some coding examples are given below:

Code image	Statement		
EC	in	al, dx	;al ← ior[dx] <7:0>
ED	in	ax, dx	;ax ← ior[dx] <15:0>
EE	out	dx, al	;ior[dx] ← al <7:0>
EF	out	dx, ax	;ior[dx] ← ax <15:0>

As far as addressing mode is concerned, the in (input) instruction has the destination operand specified as register direct, the source operand as register indirect, and the data bits are transferred from an I/O register to a CPU register. For an out (output) instruction, the source operand is register direct, the destination operand is register indirect, and the bits are transferred from a CPU register to an I/O register. Thus, before an I/O instruction is executed, we should place the address of an IOR in DX. These I/O addresses are wired by hardware not subject to change. [27, 77]

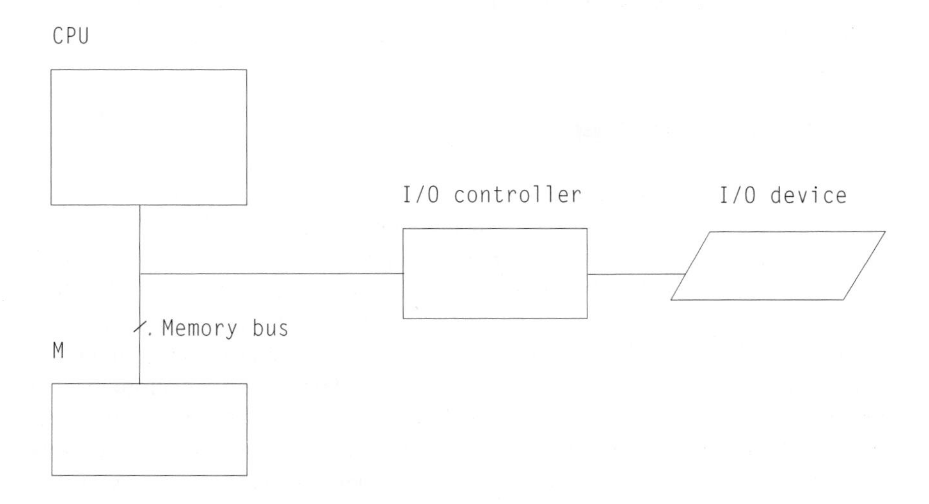

Figure 5.12 A computer system with only one memory bus.

5.6.2 Memory Mapped I/O

A computer system may have only one bus so all the I/O controllers are placed on the same memory bus as shown in Figure 5.12.[40] In other words, such a system has no separate I/O bus. With respect to I/O addressing, the IOR is treated the same as a memory word. Because a regular move instruction can move data between an IOR and a CPU register, there is no need to have special I/O instructions, a technique called memory mapped I/O.

5.6.3 Direct Memory Access Channel

If a controller can perform block I/O functions, then it is an I/O channel or DMA (direct memory access). Thus, a channel is a logic entity that can transfer a block of bytes directly between memory and an I/O device without CPU intervention. A DMA is usually connected to some I/O device, such as disk, tape, etc. A mainframe commonly has 16 or more channels. The micro has a few, one example of which is using the SCSI (small computer system interface) chip for the hard disk. Note that each I/O channel can interrupt the CPU when its I/O operation is completed.

How does the CPU communicate with an I/O channel? The key concept is passing messages back and forth. Typically, a microcomputer provides I/O instructions to write and read I/O registers in the channel. After the CPU sets the command code, memory address, block count in bytes, and other bits, the channel knows what to do. The memory address points to a data block whose size in bytes is the block count. There are other bits that tell the channel what to do after an I/O is completed. After an I/O interrupt, the OS issues an input instruction to read the status register in the channel. The status information includes the completion code set by the I/O channel and device.

5.6.3.1 Cycle Stealing

Since the CPU and I/O operations can be done in parallel, it is possible for the CPU and the I/O channel to request a memory cycle at the same time. When both memory requests are sent to the bus arbitration unit for a decision, the rule is, device first, then CPU. When the CPU gives the memory cycle to the I/O device, it is called cycle stealing because the program on the CPU does not even know of the existence of the event (the I/O device takes its memory cycle). The reason to assign the I/O device a higher priority is twofold. First, the CPU is flexible so it can afford to yield a memory cycle to the I/O device. Second, from a system viewpoint, another task may wait for its I/O to complete. In other words, the task can not proceed unless its I/O operation is successfully completed. If the memory cycle is given to an I/O device, the OS tends to run more programs in a multiprogramming environment.

5.6.3.2 Serial Port vs. Parallel Port

An I/O cable is abbreviated as cable that connects an I/O controller and an I/O device. Each side has a connector on the hardware chassis. If the cable has one data wire (i.e., line), the controller is a serial port because data are transmitted serially on the line. For example, a com (communication) port is serial because only one wire in the cable is used to transmit data from one computer to another. If the I/O cable has many data wires, the controller is a parallel port. For example, a parallel port is used to drive a printer, so eight bits are transmitted on the line at the same time. Note that a serial or parallel port is a physical device, while a programming port is a logical device (i.e., private resource) owned by a user task in a computer network.

5.7 COMMUNICATION PORT

A com (communication) port is a controller, also called a UART (universal asynchronous receiver/transmitter), or ACIA (asynchronous communications interface adaptor).[41,45] The chip acts as a small processor to transmit one byte at a time to the line or receive one byte at a time from the line. As shown in Figure 5.13a, two PCs at two locations are interconnected by a telephone line. Each side has a PC and modem (modulator/demodulator) pair. A computer is a DTE (data terminal equipment), and a modem or data set is a DCE (data communication equipment). Between a DTE and a DCE there is an RS-232 (recommended standard 232) cable that has many wires. One wire is TxD (transmit data), another is RxD (Receive Data), and other lines are for signal ground and control. As far as the PC is concerned, the data bits are transmitted serially on the TxD line and received serially from the RxD line. That is, a com port drives the RS-232 port cable. Because the two PCs can transmit data at the same time, the transmission mode is said to be full-duplex.

Characters are transmitted asynchronously on the line. By using an oscilloscope, we can plot the waveform at the TxD pin of the RS-232 connector. For example, the waveform of a seven-bit ASCII b (62 in hex) is shown in Figure 5.13b. The quiescent line is about -11 volts. After executing an I/O instruction to place a character in the transmit data register, the controller swings the line voltage from -11 v to +11 v to make the leading edge of the start bit. Because inverted logic is used by the telephone industry, the start bit is considered zero. Next, we see seven data bits (b0 to b6), with b0 transmitted first. Next, the controller transmits an even parity bit and one stop bit. The even parity bit one means that after adding all the ones in the character including the parity bit, the sum is an even number (four in this case). If the data rate is 9600 bps, then the line setting is specified as 9600 E71. This is called

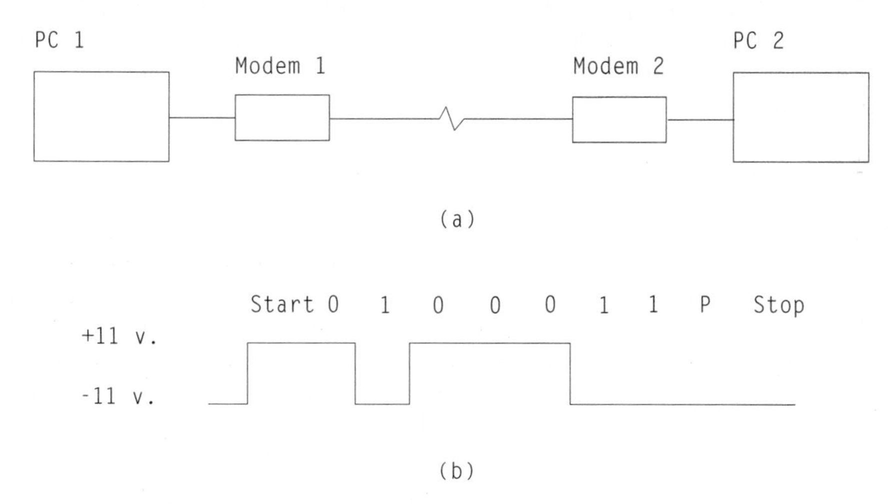

Figure 5.13 Interprocessor communications: (a) two PCs are interconnected by a telephone line and (b) asynchronous character transmission of 7-bit ASCII b.

asynchronous character transmission because the leading edge of the start bit is not predicable. The reason for having one or more stop bits is to restore the line to its quiescent state so as to separate two adjacent characters. In fact, one stop bit is the minimum as required in the spec.

The modem box at the sending side modulates the digital signal to make it suitable for transmission, and the modem box at the receiving side demodulates the voltage back to digital. Note that the controller at the receiving side strips the start bit, parity bit, and stop bit. When the receiving CPU executes an input instruction to read the receive data register, it gets the ASCII b with b7 as zero. Each controller provides various signals to the buffer gates, and each one drives a line in the RS-232 cable. The modem box is connected to the telephone line via a modular jack on the wall.

5.7.1 Com Port Characteristics

The com port has several I/O registers and it has the ability to interrupt the main CPU. The controller can be programmed to set a different data rate, byte size, parity bit, number of stop bits, etc. Because of the complex logic and registers, its size is not much smaller than the CPU chip. Intuitively, after a byte arrives in the receive data register, the receiver generates a receive interrupt signal to the CPU. The transmitter interrupt is generated when the transmit data register is empty. It is clever that the CPU can immediately write the transmit data register after being interrupted.

Each PC has one com port designated as com1 (com port 1), and a second port can be added by inserting a card into the back panel. The port can also be connected to a mouse or any terminal as long as it uses the same serial interface. The port addresses (i.e., I/O addresses) of com1 are listed in Figure 5.14a. The registers are TxD, RxD, Divisor Latch, LCR (line control register), IER (interrupt enable register), LSR (line status register), MCR (modem control register), MSR (modem status register), and interrupt ID register. The interrupt ID register contains an identification code to denote what type of interrupt has occurred. The remaining registers are described below.

5.7.1.1 Divisor Latch

After mapping an I/O address into a three-bit port address due to the limited number of pins, the same address may be used to access a different register depending on the setting of b7 in LCR, which is the DLAB (divisor latch access bit). For example, the address 3F8 (1) with a one in the paren indicates that the address is for the LSB of the divisor when DLAB is one. The address 3F8 (0) with a zero in the parentheses indicates that the address is for the receive data register (RxD) or the transmit data register (TxD) when DLAB is zero. This is because there is a different command line for input for output. Consider the divisor as shown below:

$$\text{line speed} = 1{,}843{,}200 / (16 * \text{divisor}) \text{ bps}$$

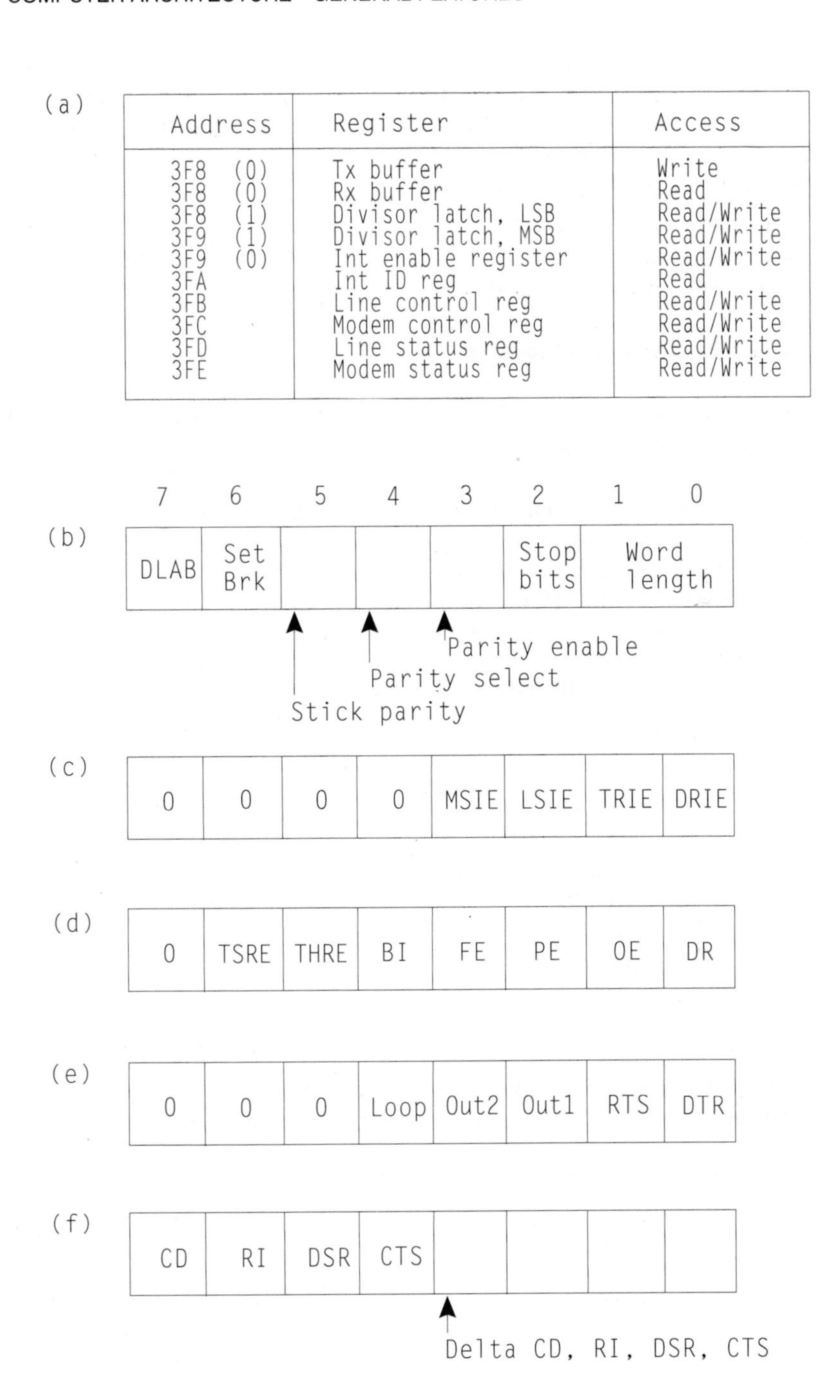
(a)

Address	Register	Access
3F8 (0)	Tx buffer	Write
3F8 (0)	Rx buffer	Read
3F8 (1)	Divisor latch, LSB	Read/Write
3F9 (1)	Divisor latch, MSB	Read/Write
3F9 (0)	Int enable register	Read/Write
3FA	Int ID reg	Read
3FB	Line control reg	Read/Write
3FC	Modem control reg	Read/Write
3FD	Line status reg	Read/Write
3FE	Modem status reg	Read/Write

Figure 5.14 Com port characteristics: (a) the com 1 port addresses, (b) LCR, (c) IER, (d) LSR, (e) MCR, and (f) MSR.

The crystal oscillator provides a 1,843,200 Hz clock to the chip. This basic clock is fed to a counter with a count 16 times a divisor. Thus, the counter output is another clock at a lower frequency that is used to shift the data register for both input and output. By programming the 16-bit divisor register, we can change the frequency of the shift clock, which is the data rate. In other words, we can write the divisor latch LSB (least significant byte) and the divisor latch MSB (most significant byte) to set the shift frequency. If the divisor is 12, the line speed is 9600 bps. If the divisor is changed to six, the speed becomes 192,000 bps.

5.7.1.2 Line Control Register

As demonstrated in Figure 5.14b, there are eight bits in the LCR, denoted as b0 to b7 with the following explanations:

- **b7:** The leading bit is DLAB. The divisor latch LSB and MSB can only be addressed when this bit is set to one.
- **b6:** This is the set break bit. If this bit is set to one, the quiescent voltage of the line is switched from low to high and alerts the remote computer. The line voltage is normal if b6 is zero.
- **b5, b4, b3:** If the stick parity (b5) is set to one, the parity bit never changes. If b5 is zero, then the parity select (b4) and parity enable (b3) bits decide the parity. The three bits jointly determine the setting of the parity bit as follows.

```
CASE b5 b4 b3 of
1 1 1: the parity bit is always 0;
1 0 1: the parity bit is always 1;
0 X 0: no parity exists;          {X is a don't care}
0 1 1: even parity;
0 0 1: odd parity;
ENDCASE;
```

- **b2:** This bit controls the number of the stop bit. If it is set to one, then we have two stop bits, otherwise we have one stop bit.
- **b1, b0:** The last two bits jointly decide the byte size as follows.

```
CASE b1 b0 of
0 0:  5 bits;
0 1:  6 bits;
1 0:  7 bits;
1 1:  8 bits;
ENDCASE;
```

One popular standard on UNIX is to set the LCR to 03 in hex, which indicates 9600 N81. Namely, the data rate is 9600 bps, there is no parity bit, there are eight data bits per byte, and one stop bit. If the receiving side uses the same line setting, it can fetch the data correctly from the line.

5.7.1.3 Interrupt Enable Register

The IER (interrupt enable register) has four leading zeroes and four enable bits as specified in Figure 5.14c. For each port, the interrupts can be divided into four different groups and each enable bit is used to gate the group. As a prerequisite, the enable bit must be on to generate the interrupt signal from the group. Note that after system boot, the IER contains all zeroes meaning all port interrupts are disabled.

- **b0:** This is the DRIE (data ready interrupt enable) bit. If set, an interrupt signal is sent to the CPU after receiving the data in the buffer.
- **b1:** This is the TRIE (transmit ready interrupt enable) bit. If it is turned on, an interrupt signal is sent to the CPU if the transmit buffer is empty.
- **b2:** If the LSIE (line status interrupt enable) bit is set to one, an interrupt signal is sent to the CPU after detecting an abnormal line status condition.
- **b3:** If the MSIE (modem status interrupt enable) bit is set to one, an interrupt signal is sent to the CPU as a result of a modem status condition.

Recall that the CPU has the final say to accept the interrupt if its IF (interrupt enable flag) is set to one. An I/O interrupt causes context switching just like executing a system call. As far as hardware is concerned, the com1 interrupt is equivalent to executing an int 0ch instruction, and the com2 interrupt is like executing an int 0bh.

5.7.1.4 Line Status Register

As shown in Figure 5.14d, the status bits in the LSR reflect the status information of the controller. After a port interrupt, the CPU program needs to examine the LSR to know exactly what has happened. The eight-bit register has a leading bit zero followed by seven status bits as described below:

- **b6:** The TSRE (Tx shift register empty) bit indicates that the entire data byte is shifted out to the line.
- **b5:** The THRE (Tx holding register empty) bit indicates that the data byte is moved down from the holding buffer to the shift register. Therefore, the controller is ready to accept another byte to be transmitted.
- **b4:** The BI (break interrupt) bit indicates that an abnormal line voltage is detected.
- **b3:** The FE (frame error) bit indicates that the received character has no stop bit.
- **b2:** The PE (parity error) bit indicates that a parity error has been detected.
- **b1:** The OE (overrun error) bit indicates that the data register is overlaid. That is, if the receiver has new data arrive before the old data are read, we have a data missing condition known as data overrun.
- **b0:** The DR (data ready) bit indicates that the data is already in the receive buffer.

5.7.1.5 Modem Control Register

The MCR (modem control register) is used to interface with the modem as shown in Figure 5.14e. The upper three bits are all zeroes and the lower five bits are described below:

- **b4:** This is the loop test bit. If b4 is set to one, the transmit shift register is connected to the receive shift register in a loop test.
- **b3:** When the OUT2 signal is set to one, it provides the enable interrupt signal at the second level for the N8250 chip. Recall that the IER provides the first level interrupt enable signals.
- **b2:** The OUT1 signal is not used by the PC.
- **b1:** This is the RTS (request to send) signal to indicate that the computer wants to transmit. If the modem is ready to accept, it activates the CTS (clear to send) signal as set in the MSR (modem status register).
- **b0:** This is the DTR (data terminal ready) signal to indicate that the computer is in a ready state to transmit or receive.

For full-duplex transmission, both b1 and b0 are set to ones all the time. Note that even if b2 is set to 1, no interrupt signal is generated unless the enable bits in IER are set.

5.7.1.6 Modem Status Register

The MSR (modem status register) in Figure 5.14f provides information as described below.

- **b7:** The CD (carrier detection) bit is set when a carrier is detected.
- **b6:** The RI (ring indicator) bit is set if ringing is detected.
- **b5:** The DSR (data set ready) means that the modem is in a ready state to transmit and receive.
- **b4:** The CTS (clear to send) is a response signal that tells the DTE go ahead and transmit.
- **b3:** The delta CD (delta carrier detection) signal indicates that a change in carrier is detected. The Greek letter delta used in a mathematical notation means a change.
- **b2:** The delta RI signal indicates that a change in RI is detected.
- **b1:** The delta DSR signal indicates that a change in DSR is detected.
- **b0:** The delta CTS signal indicates that a change in CTS is detected.

5.7.2 Software Polling

To program a com port, we can either write an interrupt service routine that is active in nature, or write a passive software polling routine that means that the program running on the CPU checks the status of the I/O device in a loop. For example, the CPU program polls the DR (data ready) bit in the line status register in a com port. If it is set to one, then the routine issues an input instruction to read the data. In general, the software polling technique makes the code simple but less efficient because the CPU is tied up in a loop. Nonetheless, we show a polling routine in the following.

```
; ——— define com1 port addresses
RxD1      equu    03F8h       ;receive data buffer
DLLSB1    equ     03F8h       ;divisor latch LSB
DLMSB1    equ     03F9h       ;divisor latch MSB
LCR1      equ     03FBh       ;line control register
LSR1      equ     03FDh       ;line status register
MCR1      equ     03FCH       ;modem control register
; ——— Initialize com1
main:     mov     al, 80h     ;set DLAB in LCR
          mov     dx, LCR1
          out     dx, al      ;DLAB = 1
          mov     al, 0
          mov     dx, DLMSB1
          out     dx, al
          mov     al, 0ch     ;divisor = 000c
          mov     dx, DLLSB1
          out     dx, al      ;9600 bps
          mov     al, 03h     ;DLAB = 0,
          mov     dx, LCR1    ;No parity bit, 8 data bits,
          out     dx, al      ;and 1 stop bit
          mov                 dx, MCR1
          mov     al, 0bh     ;set OUT2, RTS, and DTE
          out     dx, al
; ——— poll receive buffer status of com1
pollAgn:  mov     dx  LSR1    ;poll line status reg 1
          in      al, dx
          and     al, 01h     ;bit 0 => DR (Data Ready)
          jz      pollAgn
          mov     dx, RxD1
          in      al, dx      ;read in character
          . . .
end       main
```

The initialization routine is placed in the front. There is a busy-wait loop to check whether the receiver status of com1 is ready. If so, the routine then reads the character from the receive data register.

5.8 PROGRAMMING INTERFACE WITH C

In embedded system design, it is sometimes desirable to link a piece of assembly code to a C program or insert the assembly code in the C program. For the former approach, the piece of assembly code is a closed subroutine. For the latter approach, the code is an open subroutine. By open, we mean that the code is executed just once. The MSC (Microsoft C) compiler has its own coding convention to pass arguments to the subroutine as described below:

1. Each of the symbol names has an underscore in the front.
2. All the subroutine arguments are pushed on the stack in reverse order.
3. It is the responsibility of the calling routine to pop arguments off the stack after return.

5.8.1 Assembly Code Subroutine

The C program generates 20 pseudo random numbers first in an array named fib. These are the 21st to 40th Fibonacii numbers mod 1024. The main program calls the print subroutine written in C to print out the numbers in unsorted order. It then calls the subroutine written in assembly code to sort the numbers in fib. The sort routine uses the push down method in a double loop to sort the numbers in ascending order. After return, the main routine prints out the sorted numbers in ascending order. It is interesting to note that in the sort(fib, 20) statement, two arguments are passed to the subroutine. The stack frame after the call is shown in Figure 5.15 with the value 20

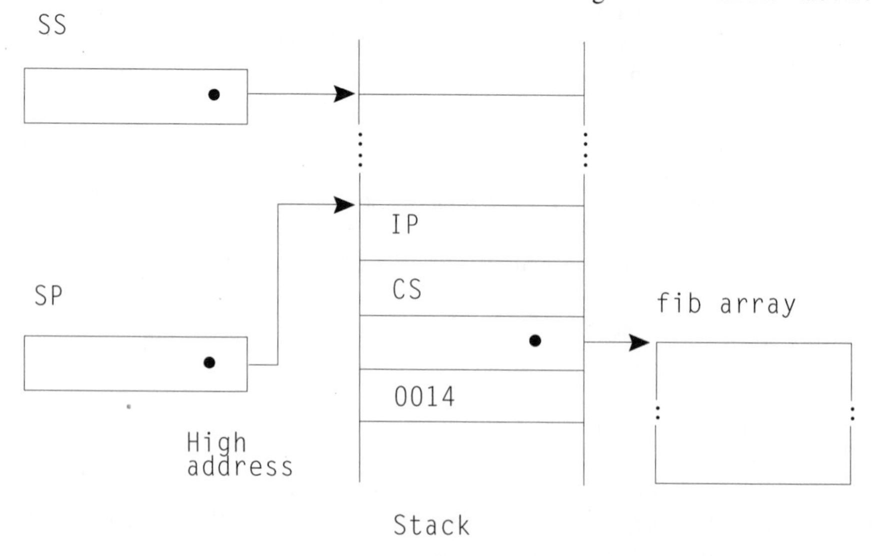

Figure 5.15 The stack frame after executing the call.

pushed first and the fib address pushed next. The far return address is pushed last by executing the call instruction. Note that fib is declared as an array, so the symbol fib alone with no ampersand (&) in front represents an address.

The main routine in C is stored in a disk file named pgm8.c as shown below.

```
/* —- The main program and the print subroutine are in C, but
the sort subroutine is in assembly code.
Name:           John Y. Hsu
Date:           1998-12-14
Version:        1.0
Description:    The C program calls two subroutines, one is written in
                C and the other one in MASM.
Job steps:      1. The pgm8.c file contains the main program in C.
                The pgm8a.asm file contains the assembly code of the
                sort routine whose entry point is _sort.
                To run the C compiler and the assembler, we type:
                msc  pgm8.c
                masm pgm8a.asm
                2. To link, we type:
                link pgm8.obj + pgm8a.obj
                3. To execute, we type
                pgm8.exe
---------------------------------------------------------------------- */
#include <stdio.h>
int   i, j, k, num, fib[20];
void print_array();
void far sort(int*, int);
main()
{
        i = 0;
        j = 1;
        for (k = 2; k < 20; k++)
        { num = i + j;
             if (num >= 1024)
             num = num - 1024;
             i = j;
             j = num; }
        for (k = 0; k < 20; k++)
        { fib[k] = i + j;
             if (fib[k] >= 1024)
             fib[k] = fib[k] - 1024;
             i = j;
             j = fib[k]; }
        printf("This is the unsorted array of random numbers:\n");
        print_array();
```

```
        sort(fib, 20);                /* ——————————————————
                                      Call the assembly subroutine to sort
                                      the array named fib.
                                      ——————————————————*/
        printf("This is the sorted array of random numbers:\n");
        print_array();
        printf("\n\n—- Job completed. —-\n");
        return;
}
void print_array()
                                      /* —— print subroutine ————-*/
{
        for (k = 0; k < 20; k++)
        printf("%d\n", fib[k]);
        return; }
```

The file pgm8a.asm contains the sort subroutine written as a far procedure in assembly code. Therefore, after assembling, a far return instruction is placed at the end as shown in the following:

```
; —— This is the sort subroutine in assembly code. ——-
; Name:              John Y. Hsu
; Date:              1998-12-14
; Version:           1.0
; Description:       This a far subroutine in assembly code to sort
;                    an array. The main program issues:
;                    sort(fib, 20);
;                    According to the C convention, value 20 is
;                    pushed first, and next the address of fib.
;                    Finally, the far return address is pushed on top
;                    during the call. The calling routine is
;                    responsible for popping the two arguments off
;                    the stack after return.
; ————————————————————————————————————————————
CSEG  segment        para "code"
          assume     cs: CSEG, ds: CSEG
          public     _sort           ; define entry point as external symbol
_sort     proc       far             ;declare far procedure and an underscore
                                     ;is needed before any external symbol.
          push       bp
          mov        bp, sp          ;bp points to stack frame
          jmp        f0010
endsi     dw         0               ;upper bound of si
f0010:    push       dx
          push       cx
```

```
                push    bx
                push    ax
                push    si
        mov             cx, 8[bp]       ;fetch value 20
        dec             cx              ;outer loop count
        mov             ax, cx
        shl             ax, 1
        mov             cs:endsi, ax    ;initial value is 38 decimal
        mov             bx, 6[bp]        ;fib near base address
oloop:  mov              si, 0
iloop:  mov             ax, [si+bx]     ;fetch the 1st element
        mov             dx, 2[si+bx]    ;fetch the next element below
        cmp             ax, dx
        jle              skip20
        mov             2[si+bx], ax     ;interchange the two elements
        mov             [si+bx], dx
skip20: add              si, 2
        cmp             si, cs:endsi
        jl              iloop
        sub             cs:endsi, 2     ;no need to compare the sorted
                                        ;portion
        loop            oloop
        pop             si
        pop             ax
        pop             bx
        pop             cx
        pop             dx
        pop             bp
        ret                             ;far return
_sort   endp
CSEG    ends
        end
```

5.8.2 In-Line Assembly

The in-line assembly scheme allows a block of assembly code to be inserted in the middle of a high-level language program. Intuitively, the compiler must be informed that the inserted code is of assembly type, and special keyword symbols are used as delimiters to enclose the assembly code. The C language coding example is shown below to illustrate the concept of in-line assembly.

```
<C statement>
<C statement>
```

```
_asm
 {
 <asm statement>
 <asm statement>
 . . .
 }
```

The keyword is asm preceded by an underscore (i.e., _asm). After the keyword, the assembly code block enclosed in braces just like a block structure. The pair of angular brackets are meta-symbols to mean the class of whatever is enclosed.

5.9 SUMMARY POINTS

1. Indexing is the second major programming concept invented in second generation computers.
2. The effective address of a memory operand is computed as the sum of the address coded in the instruction and the content of a hardware register, as shown below:

   EA = Address coded in the instruction
   + content of an index register if specified

3. The addition operation is done by the adder, and the sum bits are placed on the address bus to fetch the operand in memory.
4. Indexing means that a base address is coded in the instruction and an index is stored in a hardware register. Based addressing means that an index is coded in the instruction and a base address is stored in a hardware register. There is no difference in hardware, but there is a subtle difference in software.
5. Relative addressing means that the program counter is used as an index register to compute the EA (effective address).
6. Register direct means that the operand is in a hardware register.
7. Register indirect means the memory address of an operand is in a register.
8. Double indexing, or based-indexing, means that two index registers are used; one is used to store the base address and the other is used to store an index.
9. Auto-indexing means that the specified index register is automatically modified before or after the operation as the result of instruction execution.
10. If indexing is coupled with memory indirect, we may have pre-indexing or post-indexing. Pre-indexing means that indexing is performed before the first memory read cycle. In contrast, post-indexing means that indexing is performed after the first memory read.
11. A subroutine is a set of instructions to perform a function.

12. A call instruction is similar to an unconditional branch except the CPU saves a return address so as to fetch it into the pc upon executing a return instruction.
13. A call by value means that the value of an argument is passed to the subroutine.
14. A call by reference or address means that the address of an argument is passed to the subroutine.
15. Setting up the passing arguments and issuing the call instruction are known as the calling sequence.
16. A recursive call means that a subroutine can call itself directly or indirectly.
17. The interrupt mechanism, if activated, causes the CPU to perform context switching.
18. An interrupt generated inside of the CPU is also called a trap.
19. A system call triggers the interrupt mechanism.
20. Since the I/O completion event is asynchronous, running two or more tasks in parallel is called multitasking or asynchronous processing.
21. Reentrant code means one copy of code can be executed by two or more tasks during the same time interval.
22. Each I/O controller can be thought of as a small processor in that it has its own logic and a set of registers.
23. If a separate I/O bus exists in a computer system, the CPU must execute an input instruction to transfer bits from an IOR to a CPU register and an output instruction to transfer bits from a CPU register to an IOR.
24. If a computer system has only one bus, its IOR is treated as a memory word, and a move instruction can move data between an IOR and a CPU register, a process called memory mapped I/O.
25. Cycle stealing is when the CPU gives a memory cycle to the I/O device.
26. A com (communication) controller or port is a processor chip called UART (universal asynchronous receiver/transmitter) or ACIA (asynchronous communications interface adaptor) that can transmit or receive one byte at a time.
27. Asynchronous character transmission means that each character contains a start bit, several data bits, a parity bit, and one or more stop bits.
28. A modem is a communication device.
29. Software polling means that the program running on the CPU checks the status of the I/O device in a loop. If it is ready, the program takes further action.
30. An interrupt mechanism makes it possible for CPU executions to overlap with I/O operations.

PROBLEMS

1. What is an addressing mode?
2. In software design, what is the subtle difference between indexing, based addressing, and relative addressing?
3. What is the difference between register direct and register indirect?

4. Explain the difference among double indexing, predecreasing, and post-increasing.
5. If indexing is used in conjunction with memory indirect, what is the difference between pre-indexing and post-indexing?
6. Describe the six hardware steps when the CPU executes the instruction int 21h.
7. What are the requirements of writing recursive code as well as reentrant code?
8. Explain the system concept of multitasking or asynchronous processing.
9. What is the difference between an unconditional jump instruction, a subroutine call instruction, and an interrupt?
10. Describe the characteristics of a com (communication) port.
11. Explain the concept of memory mapped I/O.
12. Using the cycle stealing technique, why does the I/O device have a higher priority than the CPU?
13. The mov statement copies the SI into AX, and it can be translated into a two-byte instruction with a different bit pattern as shown on the LHS. Use the db (define byte) statement to construct either two-byte string in a program. Use debug to verify that they are functionally equivalent.

Code image	Statement	
8B C6	mov ax, si	;ax <— si
89 F0	{same as above}	

14. The mov statement copies AX into a memory word whose address is in the SI. After assembling, verify that the code image on the LHS executes correctly and the destination operand is specified as register indirect.

Code image	Statement	
89 04	mov [si], ax	;M[ds:si] <— ax

15. Run the rdump (register dump) program (below) which displays the content six registers in hex. Before execution starts, the OS sets DS pointing to PSP, so the user program must set DS to the data segment. Using address constants, we write the AL program in a single loop. After each execution of the logical and shift instructions, we obtain four bits in AX. Call the convert subroutine to convert it to ASCII in hex. At the end, a system call is issued to display all six registers on one line. If the divide conversion algorithm is used, the divisor should be 16 and the program becomes much shorter. Note that a dw statement is used to declare a pointer variable array that has six near addresses to conform with the following C statement:

```
char *pointr[6] = {str1, str2, str3, str4, str5, str6};
```

The keyword char * specifies that the object indicated is of char (character) type. The array named pointr has a 6 in the square brackets, defined as size or dimension. Since str1, str2, etc. are declared as arrays of char, the symbol itself with no & in the front represents an address. Therefore, the pointr array is statically assigned with six initial values as enclosed in braces.

```
;-----------------------------------------------------------
; Name:          John Y. Hsu
; Date:          1998-09-02
; Version:       1.0

; Description: This register dump program displays the initial
;   setting of six registers on one line, in particular, the
;   original DS when the program is first entered.
;------------------------------------------------------------------
        title  Register dump program
DSEG    segment
msg     db     `cs='
str1    db     4 dup(?),', `
        db     `ip='
str2    db     4 dup(?),', `
        db     `ss='
str3    db     4 dup(?),', `
        db     `sp='
str4    db     4 dup(?),', `
        db     `original ds='
str5    db     4 dup(?),', `
        db     `program ds='
str6    db     4 dup(?)
        db     `$'                ;end of string mark
pointr  dw     str1, str2, str3, str4, str5, str6
acon    dd     CSEG:main          ;far address constant
reg     dw     6 dup(0)
temp    dw     0
DSEG    ends
SSEG    segment stack
        dw     80 dup(?)          ;160-B stack
SSEG    ends

CSEG    segment
        assume cs:CSEG, ds:DSEG, ss:SSE
main    proc   far
        mov    cx,ds              ;save original ds
        mov    ax,DSEG
        mov    ds,ax              ;set new ds
        mov    reg,cs             ;save cs
```

```
            mov   ax,acon
            mov   reg+2,ax          ;save ip
            mov   reg+4,ss          ;save ss
            mov   reg+6,sp          ;save sp
            mov   reg+8,cx          ;save original ds
            mov   reg+10,ds         ;save program ds
            mov   si,0              ;initialize index
again:      mov   ax,reg[si]
            mov   temp,ax           ;move word operand to temp
            mov   al,byte ptr temp+1
            mov   cl,4
            shr   al,cl             ;decode upper 4 bits in high order byte
            mov   di,pointr[si]     ;move destination address to di
            call  convert
            mov   al,byte ptr temp+1
            and   al,0fh            ;decode lower 4 bits in high order byte
            inc   di                ;increase destination address by 1
            call  convert
            mov   al,byte ptr temp
            mov   cl,4
            shr   al,cl             ;decode upper 4 bits in low order byte
            inc   di
            call  convert
            mov   al,byte ptr temp
            and   al,0fh            ;decode lower 4 bits in low order byte
            inc   di
            call  convert
            add   si,2              ;increase index pointing to next operand
            cmp   si,10
            jle   again             ;jump if less or equal — 6 times total
            mov   dx,offset msg
            mov   ah,09h
            in    21h               ;display message
            mov   ah,4ch
            int   21h               ;system termination
; —— The convert subroutine is as follows.
convert     proc  near
            add   al,30h            ;convert to ASCII
            cmp   al,39h
            jbe   skip
            add   al,7              ;make A to F by adding 7 more
skip:       mov   [di],al           ;store at destination address in di
            ret                     ;near return
convert     endp
main        endp
CSEG        ends
            end   main              ;entry point
```

CHAPTER 6

Microprogrammed CPU Design

6.1 HISTORY OF MICROPROGRAMMING

This chapter discusses the basic principles of microcode and the microprogrammed CPU that executes a subset of 8086 instructions. That is to say, the execution of microinstructions generates the various timing signals required to execute a target instruction. This innovative idea was proposed by Wilkes as an alternative to hardwired logic.[68]

A microinstruction is a string of bits in ROM (read-only memory) and when executed, it generates many enable signals to control the timing of executing a target instruction.

Each bit in an microinstruction provides a single control function. Due to the fact that less hardwired logic is required to execute microinstructions, it is easier to debug the logic. A microinstruction means microcode, and it is occasionally just abbreviated as instruction.

The microprogrammed CPU has the obvious advantage of being flexible. The trade-off is slower speed because more overhead is required to fetch, interpret, and execute microinstructions. Thus, the microprogrammed CPU is slower than its counterpart using hardwired logic. This is also the reason why in a family of computers, low-end models are microprogrammed but high-end models are hardwired. A microprogrammed CPU has its own merit for many reasons. When a new computer is developed, it is preferably backward-compatible so that all the instructions of an old processor can execute on the new processor. By using microprogrammed logic, we can always install an emulator for the old processor. An emulator is a microcode interpreter in ROM along with the one for the new machine. This is referred to as emulation because the interpreter in ROM can not be changed without special hardware tools. To add or delete a target instruction, we simply modify the microcode in ROM without any wiring change. This chapter shows that a complex target instruction, such as multiply, divide, or floating point, can be interpreted and executed by microcode.

6.1.1 Computer in a Computer

By storing microcode in ROM, we obtain a computer in a computer, as shown in Figure 6.1. The inner computer is the microcode engine that has a microinstruction register (μIR), microinstruction address register (μIAR), and hardwired logic to execute microinstructions. That is to say, a microcode interpreter in ROM drives the engine. The engine does not have an ALU, but it has its own μIR and μIAR that are different from IR and IAR. Because ROM is known as the control store, the μIR is the control instruction register containing the current microinstruction, and the μIAR is the control instruction address register pointing to the next microinstruction in ROM. The μIR and μIAR are designed for executing the microcode in ROM, while the IR and IAR are designed for executing the target machine code in central memory.

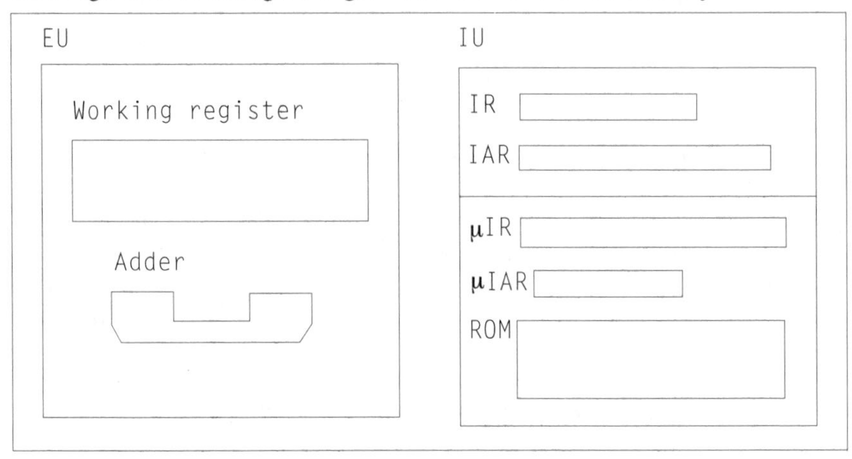

Figure 6.1 Microcode engine — computer in a computer.

The time spent debugging a hardware unit is exponentially proportional to its complexity. If the complexity of the unit is doubled, its debug time is increased three to four times. Since the engine uses much less hardwired logic, its debug time is shortened. We trade speed for flexibility because the target instruction is interpreted by a microprogram.

6.1.1.1 Microcode Execution

The microcode interpreter is an infinite loop implemented in ROM. The microcode length is fixed in words, so the μIAR or μPC (micro program counter) is incremented by one after obtaining a microinstruction. The bits in the microinstruction may be routed into the μPC as a result of execution. To illustrate this concept, we can use a sequence of five microinstructions to interpret a target instruction, as shown below:

```
add  sopd                ;acc ◄— acc + M[sopd]
```

1. The first microinstruction is fetched into the μIR and the μPC is increased by one. Enable bits in the first microinstruction route the address of the target instruction (add) from PC to the address bus and initiate a memory read request.
2. The second microinstruction reads the add instruction from the data bus into the IR. At this point, the target instruction fetch phase is completed. An enable bit of the second microinstruction in the μIR copies the opcode of the target add instruction from IR into the μPC in order to branch. That is, three more microinstructions jointly interpret this add operation.
3. The third microinstruction has bits to route the operand address from the IR to the address bus, initiate a memory read request, and increase the μPC by one.
4. The fourth microinstruction has bits to clock the operand from the data bus into a temp register and increase the μPC by one.
5. The fifth microinstruction has bits to route an accumulator to adder input A, a temp register to adder input B, and adder output to acc. Meanwhile, a fixed ROM address is routed from the μIR to the μPC so control is passed to the beginning of the loop.

6.2 TWO BASIC MICROCODE TYPES

A microinstruction has the following characteristics:

1. All microinstructions have a uniform instruction length between 22 to 150 bits.
2. Each bit is an enable signal to open a group of gates or to activate a control function.
3. An address field can be used to pass control to another microinstruction in ROM. That is, there are bits set in the microinstruction to route an address into the μIAR based on a condition. Thus, it is possible to write a sequence of microinstructions to interpret a complex target instruction.
4. The size of ROM is determined by the address field, usually between 256W (word) and 4KW (kilo-word).

There are two basic types of microcode: horizontal and vertical. It is possible to have a hybrid design by mixing horizontal microcode with vertical microcode in a microinstruction. In a family of computers sharing the same architecture, the microcode can be different from model to model. Yet, the main objective, to interpret target instructions, remains the same. Some engines may use two sets of microinstructions, and in each set the instructions have a uniform length. For example, the interpreter of the IBM 3033 mainframe is stored in 4KW ROM divided into two halves. The first half, from address 000 to 7FF, contains 108-bit instructions, and the second half, from address 800 to FFF, contains 126-bit instructions. Even with two sets, the engine still increases its μPC by one to fetch the next instruction in ROM.

6.2.1 Horizontal Microcode

A horizontal microcode is longer, and each bit is a direct enable signal for control. Usually, the bit serves as an enable signal for a group of gates and no decoding circuit is required. A 16-bit horizontal microcode is shown in Figure 6.2a. The bits are denoted as E0, E1, E2, ..., and E15, and only one bit can be set at a time. Usually, if a bit is set, one out of 16 registers is selected as an adder input. That is, the engine uses the bit to open a group of gates directly without decoding circuit delay. In practice, horizontal microcode is popular because much less hardwired logic is used.

6.2.2 Vertical Microcode

In Figure 6.2b, a four-bit vertical microcode field is shown to provide up to 16 enable signals after a four-bit decoding circuit. Obviously the microcode is short, but it requires extra logic to decode the four-bit field. That is, the decoder maps the four-bit input signal into 16 output signals with only one output on. Consider the following examples. If the input is 0000 in binary, the E0 signal is on. If the input is 1111 in binary, the E15 signal is on. That is, with four bits we have 16 possible combina-

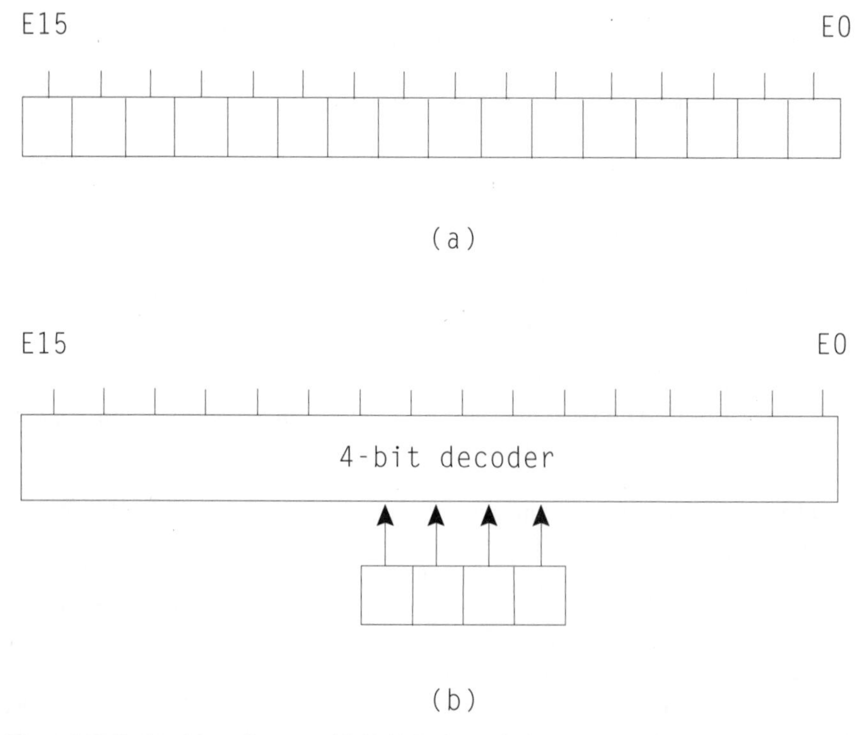

Figure 6.2 Basic microcode types: (a) 16-bit horizontal microcode and (b) 4-bit vertical microcode.

tions. Because of encoding, the ROM size is reduced. The encoding method is commonly used in opcode design for a processor to interpret and execute.

Vertical microcode is attractive provided that fully debugged logic is available from a previous generation of processor design. For example, the PDP-11 was a popular 16-bit minicomputer in the 1970's and its instruction uses a three-bit register address to select one of eight GPRs. Its successor, the LSI-11, uses a vertical microcode interpreter in ROM which is 2KWin size The 22-bit microinstruction has a uniform length as shown in Figure 6.3a.[56]

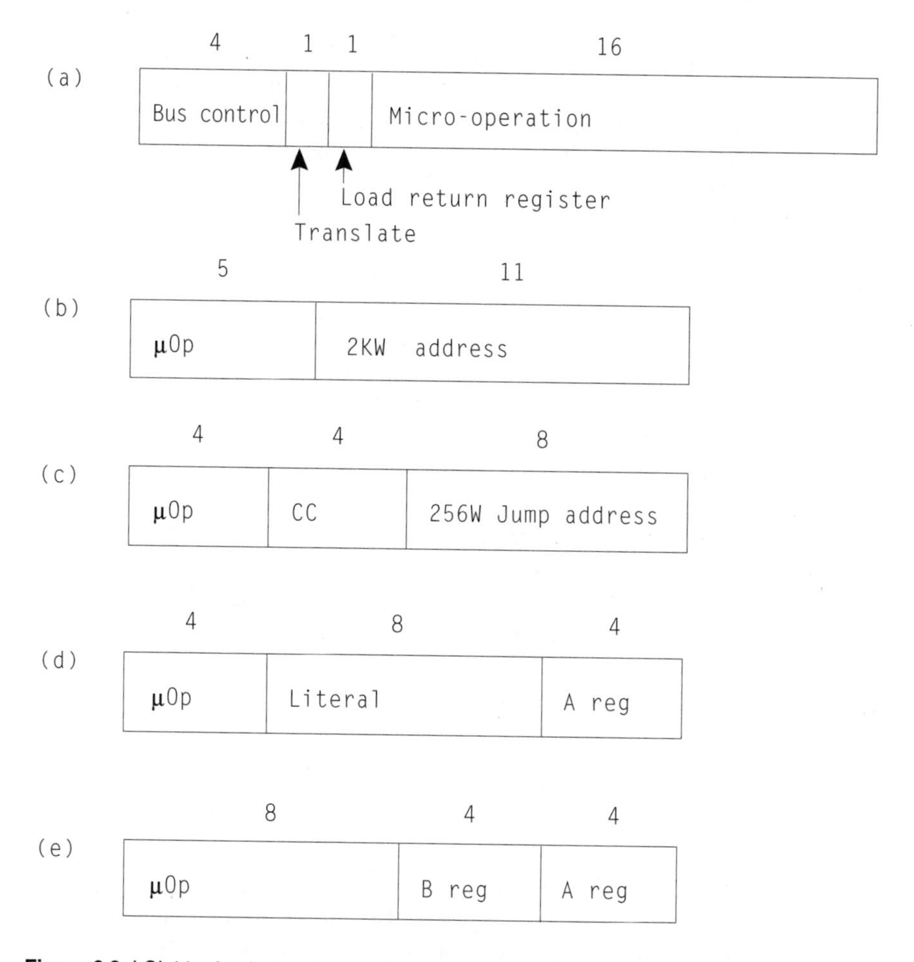

Figure 6.3 LSI-11 microinstructions: (a) general format, (b) unconditional jump, (c) conditional jump, (d) memory immediate to register, and (e) register to register.

The upper four bits are for bus control, the next two are bits for special control, and the lower 16 bits constitute a micro-operation. A bus operation means that a memory cycle is initiated by executing a few microinstructions in a row. First, the address is placed on the memory bus with the read or write bit set to initiate the cycle. Taking the read cycle as an example, after a delay another microinstruction reads the data bus into a temporary register. Next to the bus control field we see two special bits. The T (translate) bit checks whether any pending interrupt request exists. If so, control is passed to a fixed address in the control store to activate the interrupt mechanism. In the case of an interrupt pending condition, this bit allows the engine to abort a tedious microinstruction sequence immediately. To the right of the T bit, the LRR (load return register) bit allows the μPC to be loaded from the return register. When the LRR bit is set in the last microinstruction of a subroutine, control is returned to the next microinstruction after the call. The lower 16 bits represent a micro-operation that has a micro-opcode (μop) in the front, divided into four groups.

The unconditional jump operation has a five-bit μop followed by an 11-bit ROM address as shown in Figure 6.3b. A conditional jump operation as shown in Figure 6.3c has a four-bit μop followed by a four-bit CC (condition code) plus an eight-bit microcode address. The literal micro-operation as shown in Figure 6.3d has a four-bit μop followed by a one-byte literal plus a four-bit register address. A literal is an immediate operand of eight bits, same as the adder width.

The last group consists of register to register operations as shown in Figure 6.3e. Each operation has an eight-bit μop followed by two four-bit register addresses, reg A and reg B. Each address selects a byte of an internal register or GPR. This means that it takes two passes to perform a 16-bit word operation via the eight-bit adder. If reg A or reg B contains 0000 or 0001 in binary, it means indirect mode. If the bit pattern is 0000, the three-bit GPR address in the target instruction is further decoded to select the low-order byte of a GPR. Similarly, if the bit pattern is 0001, the three-bit GPR address in the target instruction selects the high-order byte of a GPR. Besides GPRs, other bit patterns mean direct mode and the pattern itself is decoded directly to select the low-order byte or the high-order byte of an internal register, such as bus address (BA), instruction register (IR), source operand (SRC), destination operand (DST), program status register (PSW), PC, etc. With proper bits set in the bus control field, the micro-operation may initiate a memory cycle, read or write.

6.3 MICROCODE ENGINE

The microcode engine has hardwired logic to support the execution of an interpreter in ROM. That is to say, the microcode in ROM drives the engine. The interpreter is written in microassembly language. The notion is that a microassembler must execute in order to translate each statement into a bit string. Two approaches can be used in designing a microassembler. The low-level approach simply sets all bits as they are explicitly specified in a statement. The high-level approach makes the statement look like a register transfer language. Therefore, the microassembler has intelligence to translate the English-like syntax into a bit string.

Our goal is to describe the internal operations inside a CPU. Therefore, the

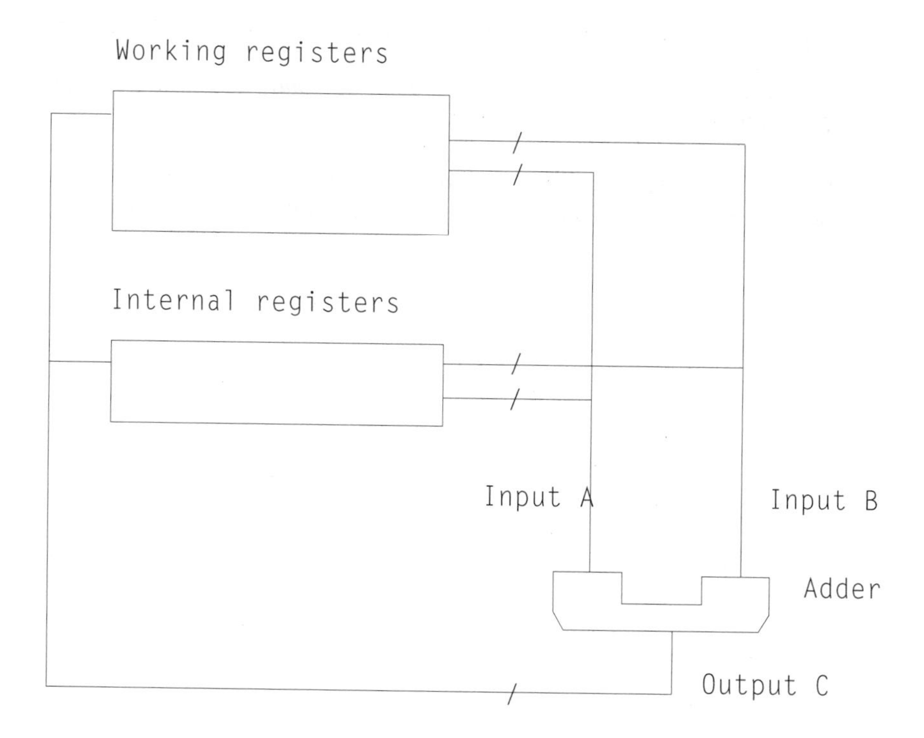

Figure 6.4 Adder internal bus structure.

microcode interpreter is written under two conditions. First, there is an instruction pipe, so instruction stream fetch is automatic. Second, there is no operand pipe, so target instruction executions are sequential. That is to say, the engine does not support parallel operand executions. Those performance topics are left to the next chapter. The adder has three internal buses, input A, input B, and output C, as shown in Figure 6.4. The multiplexer used to select a particular register is omitted in the block diagram. The engine is designed for two-address target instructions. The adder input A is for sopd1 (source operand 1), input B is for sopd2 (source operand 2), and the adder output C is routed to the destination, i.e., sopd1.

The format of the 103-bit microinstruction is shown in Figure 6.5a. From the rightmost side, the bit is numbered from 0 to 102. Each microinstruction has six fields: address, sequence control, adder control, input control A, input control B, bus, and general control.

6.3.1 Address and Sequence Control

The Addr (address) and SC (sequence control) fields are shown in Figure 6.5b and described below:

Addr <11:0>: The rightmost 12-bit field contains an address of the microcode in ROM that is 4KW in size.
SC <18:12>: The seven-bit field is for sequence control as to where to fetch the next microinstruction.

The b12 is opc (opcode control). If opc is set, the upper four bits in Addr concatenated with the opcode in the IR are routed into the μPC. The upper four bits des-

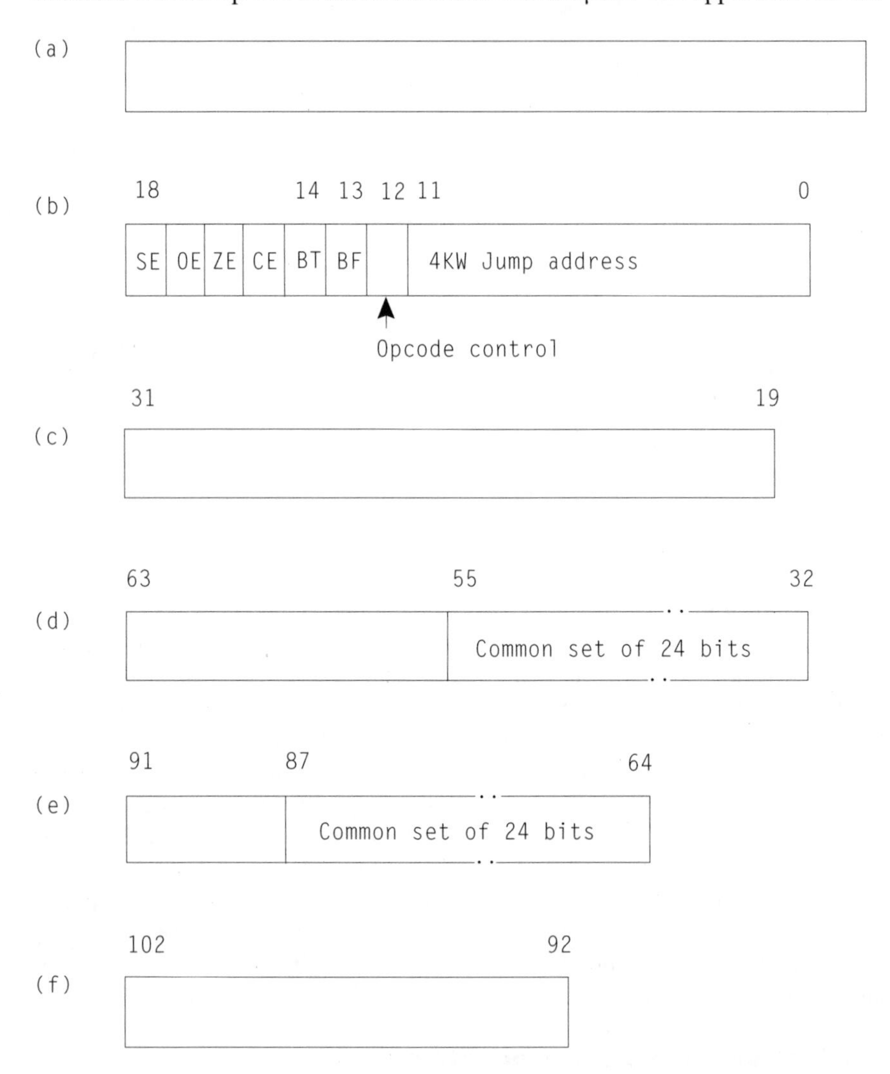

Figure 6.5 Horizontal microcode: (a) 103-bit general format, (b) address and sequence control, (c) adder operation control, (d) adder input control A, (e) adder input control B, and (f) bus and general control.

ignated as SC <18:15> are condition enable bits, the b13 is BF (branch on false); and b14 is BT (branch on true). If b12 is 0, either b13 or b14 must be set for sequence control based on a condition. That is, depending on the condition, the 12-bit Addr is either copied into the μPC to branch, or the μPC is increased by one. Routinely, the μPC is incremented by one after retrieving the current microinstruction, so if there is no change, the next microinstruction in ROM is fetched.

6.3.2 Conditional Branches

Before exploring in further detail, the four ways for sequence control are summarized as follows:

1. If b12 is set, the upper four-bit field in Addr and the eight-bit opcode byte in the IR are concatenated and copied into the μPC to branch by opcode.
2. If the specified condition is true, copy the 12-bit Addr from the μIR into the μPC to branch.
3. The specified condition is always true as an unconditional branch.
4. The specified condition is never true as an unconditional no-branch.

In the case that opc is zero, the six-bit <18:13> condition specifier determines whether or not to branch. The four condition enable bits <18:15> are SE (sign enable), OE (overflow enable), ZE (zero enable), and CE (carry enable). The four enable bits are followed by BT and BF. Each enable bit selects the condition flag in the SR (status register) as input, and the logic schematic is shown in Figure 6.6.

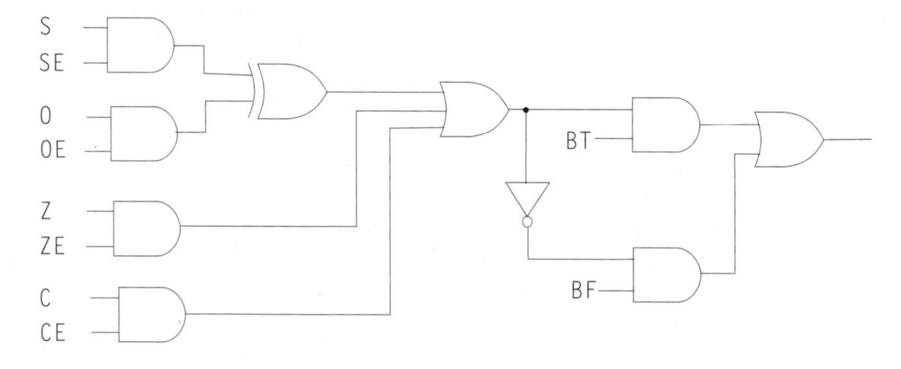

Figure 6.6 Condition logic schematic.

The selected S and O are two inputs to an EOR gate whose output is further fed to a three-input OR gate. The other two inputs to the OR gate are the selected Z bit and C bit. Amazingly, the equation below covers all the compares, signed and unsigned.

$$
\begin{aligned}
\text{condition} &= BT\,(((S\;SE) \oplus (O\;OE)) + Z\;ZE + C\;CE) + \\
&\quad BF\,\backslash(((S\;SE) \oplus (O\;OE)) + Z\;ZE + C\;CE)
\end{aligned}
$$

where $\oplus$ means EOR (exclusive or) and the branch condition is the sum of two products. The first product is the test condition gated with BT, while the second is the complement of the test condition gated with BF. Consequently, we have two cases of branch as listed below:

1.If BT is set and the test condition is true, then there is branch.
2.If BF is set and the test condition is false, then there is branch.

6.3.2.1 Unconditional Branch vs. No Branch

To make an unconditional branch, clear SE, OE, ZE, CE, and BT to zeroes, but set the BF bit to one. Obviously, the first product is false, but the second product is true (i.e., not false). Suffice it to say, branch occurs unconditionally. It is also possible to have a no branch condition. That is, if we reset SE, OE, ZE, CE, and BF to zeroes, but set BT to one, the selected condition is never true, so after being increased by one, the μPC points to the next microinstruction.

6.3.2.2 Branch Addresses in Microcode and Target Machine Code

The 12-bit address in a microinstruction is different from the address in a target machine branch instruction. However, a conditional branch microinstruction is executed to interpret a conditional branch target instruction. They are related, so the engine always maps the opcode of the target machine conditional branch instruction into a ROM address where a conditional branch microinstruction is fetched. Interestingly, the microinstruction has the same test condition specified. At the target machine level, a central memory branch address is placed in the IAR if the condition is true. Henceforth, the conditional branch microinstruction passes control to another microinstruction to copy the target instruction branch address from the IR to the IAR. If the condition is false, the next microinstruction is executed to increase the IAR by the length of the conditional branch target instruction. That is to say, if the microinstruction branches, the target instruction also branches. Let us study an example on 8086, as shown below:

Code image	**Loc**	**Statement**	
90	0013	b0020:	nop
. . .			
3B C3	001F	cmp	ax, bx
7D F0	0021	jge	b0020

On each line, the code image is followed by the location of the instruction and the assembly statement. Assume that the nop statement has a symbol b0020 defined as 0013 in hex with respect to the segment base. The compare instruction sets CC in the F register based on the result of (AX - BX). If the GE (greater or equal) condition is true, branch to b0020 (0013). Otherwise, the IP is incremented to 0023 pointing to the next instruction. This is because the location of jge is 0021, and the instruction length is two. Two questions must be answered. First, how is the logical test performed? Second, how is the new branch address computed? The specified condition is (\S \O + S O) which means that the sign and overflow bits are identical (both zeroes or ones). Comparing AX with BX (AX:BX), we have four possibilities to branch, as listed below:

1. AX is positive and BX is negative, e.g. +5 and -4.
2. AX is positive and BX is also positive but less than AX, e.g., +5 and +4.
3. AX is negative and BX is also negative, but the absolute value of the former is less than that of the latter, e.g., -4 and -5.
4. AX is equal to BX, e.g., +5 and +5.

As shown in the fourth case, we question whether the variable Z should be included in the sum. Boolean algebra proves that Z can be absorbed into the first two products as follows:

condition = \S \O + S O + Z
= \S \O + S O + Z (S O + \S O + S \O + \S \O)
= \S \O (Z + \Z) + S O (Z + \Z) + S O Z + \S \O Z
+ \S O Z + S \O Z {Don't cares}
= \S \O + S O
= \(S $\oplus$ O)

Interestingly, because the O and Z conditions can not be true at the same time (likewise for S and Z), the last two minterms are "do not matters". To interpret such a condition, the enable bits SE, OE, and BF are set in the microinstruction. However, in a microcode assembly language statement, we can use an IF-THEN-ELSE construct to determine branch or not. That is, if \(S $\oplus$ O) is false, we execute a microinstruction to not branch. Otherwise, a microinstruction to branch is executed. That is, the target machine branch address is placed into the IP. In the case of a no branch, the IP is incremented by two, the length of the jge instruction. The instruction has one-byte opcode 7D followed by an offset F0 which is sign extended. As the target instruction is interpreted, the IP is increased at the same time. By the time the interpretation is complete, the IP is already 0023, so the offset is relative to the next target instruction. The branch address is computed below:

$$
\begin{array}{lr}
 & 0023 \\
+) & FFF0 \\
\hline
 & 0013
\end{array}
$$

The one-byte offset is a signed integer ranging from 127 to -128. In this particular example, the relative offset is -16 in twos complement notation (FFF0). In fact, the next instruction after the conditional branch is the reference point, so the CPU can jump ahead up to 127 bytes, or jump behind up to 128 bytes wrapped-around. That is to say, whenever a relative offset is added to the IP, the address grows until it reaches 64 K - 1. Beyond that, the next IP address is zero so in order wrap around 64 K. It is essential to understand the system concept that the address is an unsigned integer. After sign extension, the 16-bit unsigned integer is added to the IP, and the final branch address is within 64 K, wrapped around. In fact, the adder adds the two unsigned integers mod 2^{16} (i.e., 64 K). This notion is important because in a jump instruction, a 16-bit unsigned relative offset is used to compute the branch address that covers the entire 64 KB segment. All the branch instructions and their conditions are listed in Table 6.1.

Obviously, the branch condition for JL (jump on less than) is the complement of JGE, so the bits SE, OE, and BT are set in the microcode. The condition for GT (greater than) is derived from the GE (greater or equal) condition by excluding the equal condition. Since the Z (zero) flag indicates an equal condition, we obtain the branch condition for JG (jump greater) as shown below:

$$
\begin{aligned}
\text{condition} &= \backslash(S \oplus O)\ \backslash Z \\
&= \backslash((S \oplus O) + Z) \quad \{\text{De Morgan theorem}\}
\end{aligned}
$$

The unsigned compare group is easy. Recall that the C flag in the SR (status register) is really a borrow after internal subtract, so JB (jump below) means jump on borrow, which is same as JC.

In practice, a microcode loop can be used to interpret a complex target instruction such as float mul, float div, sine, string move, string search, etc. At the end of the loop, a conditional branch microinstruction is executed to perform the test/branch step. If the condition is true, a new address is placed in the μPC so control is passed to the beginning of the loop. Otherwise, execution is completed in order to exit.

6.3.3 Adder Operation Control

Figure 6.5c shows the 13-bit adder operation control field <31:19> to specify ALU operations. The lower three bits <21:19> are for basic adder control: Clk (clock), Cin, and Addop (adder operation).

Table 6.1 Conditional Branch Instructions

Opcode	Description	Condition	Specifier Bits
{Unconditional}			
JMP	jump	\ False	BF
{Single bit compare group}			
JE	equal	Z	ZE, BT
JNE	not equal	\ Z	ZE, BF
JZ	zero {JE}	Z	ZE, BT
JNZ	not zero {JNE}	\ Z	ZE, BF
JC	carry	C	CE, BT
JN	no carry	\ C	CE, BF
JO	overflow	O	OE, BT
JNO	no overflow	\ O	OE, BF
JS	sign	S	SE, BT
JNS	no sign	\ S	SE, BF
{Signed compare group}			
JL	less	(S ⊕ O)	SE, OE, BT
JLE	less or equ	(S ⊕ O) + Z	SE, OE, ZE, BT
JG	greater	\((S ⊕ O) + Z)	SE, OE, ZE, BF
JGE	greater or equal	\(S ⊕ O)	SE, OE, BF
{Unsigned compare group}			
JB	below {JC}	C	CE, BT
JBE	below or equal	(C + Z)	CE, ZE, BT
JA	above	\(C + Z)	CE, ZE, BF
JAE	above or equal {JNC}	\ C	CE, BF

- **Clk (b21):** This bit is set to clock the adder output to the destination. If this bit is zero, the destination operand remains unchanged.
- **Cin (b20):** This is the carry input signal to the adder at bit position zero.
- **Addop (b19):** This bit indicates a basic adder operation, add or subtract, and it is also an NS (no shift) signal to the one-bit shifter.

For a binary operation, two adder inputs A and B are specified. For an unary operation, only adder input B is specified while the adder input A is zero. Take the neg (negate) instruction as an example. A string of zeroes is fed to adder input A, the ones complement of the unary operand is fed to adder input B, and at the same time the Cin bit (b20) is turned on. Consequently, the twos complement of the unary operand is clocked into the destination by the Clk bit. Recall that the CPU does an internal subtract when executing a cmp (compare) instruction. That means the Clk bit is reset to zero so the engine sets CC in the F register and the destination is not changed.

The upper 10 bits in the adder operation control field are for logical and shift operations. Precisely, the three bits <24:22> are for logical operations: AND, OR, and EOR. Note there is no bit for the NOT operation because it is similar to negate except the Cin bit is turned off. The next seven bits <31:25> are for shift control: SHL, SHR, ARI, ROL, ROR, RCL, and RCR. Recall that in Chapter 3, the ARI (arithmetic) bit was set in conjunction with SHR (b30) in order to achieve the SAR (shift arithmetic right) operation. The shifter design is a simple one because it can only shift one bit at a time in one clock cycle.

6.3.4 Adder Input Control

The engine uses a 16-bit adder, and each of the two input control fields has a common set of 24 bits. Input A has eight more bits in the front and input B has four more bits. The adder input control A has 32 enable bits as shown in Figure 6.5d.

The common bits <55:32> include 12 accumulators. This is because the low-order byte, the high-order byte, or the full word can be selected from any accumulator, so three enable bits are necessary for each set of AX, BX, CX, and DX. Each register signal has E (enable) appended to its name, e.g., ALE, AHE, and AXE. The other 12 bits include four index registers: SIE, DIE, SPE, and BPE; three temp registers, T0, T1, and T2, IP, BDL (bus data low-order byte), BDH (bus data high-order byte), IODL (I/O data low-order byte), and IODH (I/O data high-order byte). The BDH and BDL jointly constitute a 16-bit register BDW (bus data word) in the hardware bus interface unit. If a target instruction has a 16-bit memory operand specified, the BDW (i.e., set BDH and BDL) contains the operand before a memory write cycle or after a memory read cycle. Similarly, because the processor uses a separate I/O bus, the IODW (I/O Data Word) contains two bytes, IODH and IODL.

The eight extra enable bits <63:56> include four segment registers, CSE, DSE, ESE, and SSE; plus F, BA (bus address), IOA (I/O address), and Zero (zero operand). Note that the BA register has 20 bits residing in the hardware bus interface unit that can place the address bits on the memory address bus to initiate a memory cycle. By the same token, the IOA register is used to store the I/O address during an I/O request. The zero bit is designed for unary operations as well as register to register transfers. If the zero bit is one, the engine acts like a three-address machine. That is to say, no register is selected for adder input A. As a result, an operand zero or low voltage is fed to adder input A. A register operand, however, is enabled to adder input B. The destination register is specified in the adder input control A field. For example, we can transfer AX to DS as shown below:

$$DS \longleftarrow 0 + AX;$$

In the microinstruction, both DSE and zero are set in the adder input control A field, and AXE is set in the adder input control B field.

The adder input control B has a total of 28 bits as shown in Figure 6.5e. The lower 24-bit <87:64> field has the common bits, but the leading four bits <91:88>

namely Comp (complement), IRB (IR byte), IRW (IR word), and Imm (immediate), are different.

- **Comp (b91):** This bit is set for complement control. If it is one, the ones complement of the operand is selected as adder input B; otherwise just the operand is selected.
- **IRB (b90):** Select the leading byte of the IR as adder input B.
- **IRW (b89):** Select the leading two bytes of the IR as adder input B. The leading byte in the IR is the low-order byte in a word, and the next byte is the high-order byte.
- **Imm (b88):** This bit indicates that a 16-bit immediate operand is right justified in the input control B field <79:64> which becomes the adder input B. That is to say, we can implement a constant in the microcode so to speak. Usually, this constant represents +1, -1, 16, 0, etc. In a hardwired CPU, such a constant is hardwired without using flip-flops. In a microprogrammed CPU, a register-immediate microinstruction allows an operation between the two. That is to say, any 16-bit constant can be copied, masked, or padded into a register so as to eliminate hardwired constants altogether.

6.3.5 Bus and General Control

The leftmost 11 bits <102:92> are for bus and general control as shown in Figure 6.5f. Before discussing each control bit, we need to understand the three hardware design features described below:

1. The IR is really a part of an instruction queue or pipe. Between the IR and the central memory bus, there is an on-chip hardware BIU (bus interface unit), abbreviated as BU (bus unit). The BU resembles an IU (instruction unit) that can initiate a memory request to fetch a target instruction stream into the instruction pipe. Conceptually, the instruction bits flow from right to left into the pipe on a continuous basis. The BU saves the 20-bit linear memory address in a counter, so after retrieving a two-byte instruction the BU increments the counter by two. The BU issues another read request as long as there is empty space to the left of the instruction stream. With the design of an instruction pipe, instruction fetch is overlapped with operand executions. That is, the BU and EU may operate at the same time, as shown in Figure 6.7.

As the BU starts to fetch an instruction stream, the EU waits for the arrival of the first instruction. When the instruction reaches the head of the queue, the EU is activated to execute the instruction. Meanwhile, the BU fetches the next instruction in sequence. If the memory data bus is only 16 bits, it takes two physical memory cycles to retrieve of a four-byte instruction. During execution, if the instruction has a memory operand specified, the BU must take a break to fetch the operand first. The EU must be stalled until the operand arrives in the bus data register so the EU may

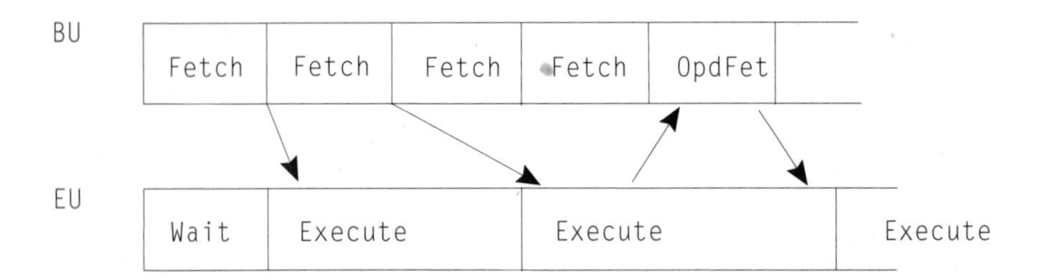

Figure 6.7 Overlapped operations between the BU and EU.

continue. That is to say, the BU has intelligence to coordinate instruction fetch, operand retrieval, and operand execution.

2. When the BA bit is set in a microinstruction, the engine acts like a three-address machine for the purpose of computing a memory address since a memory cycle is in order. For a branch, the CS (code segment) is fed to adder input A, the upper 12-bit field in the IP is fed to input B, and the adder output is clocked into the upper 16-bit field of BA. Thus, we compute the 20-bit linear address of the target instruction as follows:

$$BA \leftarrow CS + IP$$

In the transfer language, the CS really means (CS * 16) as the lower four bits of IP flow directly into the lower four bits of BA. Remember two things. First, the InsF bit in the microinstruction must be zero to indicate an instruction fetch so the BU knows to place them in the instruction pipe. Second, the instruction pipe must be flushed because all the instructions below the branch are no longer valid.

If an instruction contains a 16-bit offset of a memory operand, during execution the CPU must perform a memory cycle to fetch the operand whose 20-bit linear address is computed below:

BA <— DS + offset;

Again DS really means (DS * 16) because the lower four bits of the offset flow directly into the lower four bits of BA. The upper 12-bit field in the offset is fed to input B, the 16-bit DS is fed to input A, and the output is clocked into the upper 16-bit field of BA.

3. The IP (instruction pointer) is a counter. After interpreting one byte or two bytes in the IR, the engine does two things. First, it shifts the IR left by one byte or two bytes. Second, the IP increases by one or two accordingly. That is to say, if one byte is shifted, the IP is increased by one due to the setting of SHLIR1 (SHL IR 1 byte) in the microinstruction. If two bytes are shifted, the IP is incremented by two because SHLIR2 (SHL IR 2 bytes) is set. Interestingly, the IP always points to the next instruction byte to be interpreted.

In the 11-bit bus and general control field, the upper four bits are for memory bus and I/O bus control, namely Wr (write), Rd (read), In, and Out. The remaining seven bits are for general control: Int (interrupt), SHLIR1, SHLIR2, Link, Ret (return), InsF (instruction fetch), and Ill (illegal), as introduced below:

- **Wr (b102):** Initiate a memory write cycle.
- **Rd (b101):** Initiate a memory read cycle.
- **In (b100):** Initiate an input operation.
- **Out (b99):** Initiate an output operation.
- **Int (b98):** This bit checks any interrupt pending condition after executing the current microinstruction. If there is one, the engine activates the hardware interrupt mechanism to abort the execution of the current microinstruction and the target instruction as well.
- **SHLIR1 (b97):** Shift left the IR by one byte and the IP is incremented by one. Therefore, the next instruction byte is left justified in the instruction pipe.
- **SHLIR2 (b96):** Shift left the IR by 2 bytes and IP is increased by two. Therefore, the third byte in the instruction is left justified in the instruction pipe.
- **Link (b95):** This bit is set in a branch instruction to pass control to a micro-subroutine. That is, the return address in the μPC is copied into T2, and then the branch address is copied into the μPC.
- **Ret (b94):** The Return bit in the last instruction in the micro-subroutine is used to copy the return address from T2 into μPC. As a consequence, the next instruction in ROM after the call is fetched next.
- **InsF (b93):** This InsF bit is of particular importance because it changes the flow of the instruction stream. When InsF is set, the hardware BU (bus unit) flushes the instruction queue and starts retrieving the instruction stream from memory based on the new address in BA. This signals a continuous operation until another bus request microinstruction is executed with the InsF bit on. If the InsF bit is zero in a bus request microinstruction, then the BU writes or reads the operand as a one shot deal.
- **Ill (b92):** This bit is set to indicate that the target instruction has an illegal opcode. The interrupt hardware mechanism is triggered to abort the target routine.

6.4 MICROCODE ASSEMBLY LANGUAGE

A microassembly statement has many fields separated by a semicolon. Each microinstruction may be extended to a second line with a period as the end delimiter. Each statement is translated into a 103-bit microinstruction by the micro assembler. After that, all the microcode bits must be placed in ROM. The statement is case insensitive and each line may have a label defined in column one.

An equate file is shown in the front so mnemonics can be used to designate bit positions. For example, Addr (address) represents the 12-bit address field starting at b0. As each microinstruction is cleared to all zeroes before assembling, the default value in the Addr field is zero. Either Cin or b20 can be used to denote the 21st bit in the microinstruction counting from the rightmost end. The basic rules are:

1. A comma is used to separate bits, a semicolon is used to separate phrases, and a period is the end delimiter. It is possible for one microinstruction to spread over two lines.
2. The equal sign is an assign operator and its LHS specifies the enable bits to be set by the micro assembler. The mnemonic helps to remember the meaning of the bit.
3. The transfer language can be used to mean a run-time action so the microassembler knows how to interpret the language and set the enable bits accordingly. Some microcode phrases are explained below:

AX ← AX - BX; {Set Addop, Clk, Cin, AXE in input A; Comp, BXE in input B.}

AX ← AX .AND. BX; {Set And, Clk, AXE in input A, BXE in input B.}

AX ← 0 + BX; {Set Addop, Clk, Zero, AXE in input A, BXE in input B.}

AX ← BX; {Same as above.}

μPC ← μPC + 1; {Set BT with no condition code. This also means that μPC is incremented by 1.}.

μPC ← Addr^Op; {Set b12 for branch control by the 8-bit opcode.}

In, Zero, Clk = 1; IOA ← 0 + DX; {Set In = 1; Zero = 1; Clk = 1; Addop = 1; IOA = 1; DXE = 1 in input B for an input operation.}

Addr = 100; {Sets the Addr field to 100 in hex.}

Addr = Interp; μPC ← Addr.

{Set BF with no condition code to branch. Set Addr field to the address whose symbolic name is Interp so control is passed to Interp.}

Goto Interp. {A short cut same as above.}

The IF-THEN-ELSE statement can be used to describe three microinstructions. For example, we describe the logic below to interpret the adc (add with carry) instruction.

IF C .EQ. 1, (Test Carry flag in SR.}
THEN Int, BF, Clk, Cin = 1; AX <— AX + BX + 1;
Goto Interp. {Same as Addr = Interp; μPC <— Addr.}

ELSE Int, BF, Clk = 1; AX <— AX + BX;
Goto Interp.

The micro assembler can translate this IF statement into three microinstructions as follows:

CE, BF = 1; Addr = skip010; μPC <— Addr.
Int, BF, Clk, Cin = 1; AX <— AX + BX + 1; Goto Interp.
skip010:
Int, BF, Clk = 1; AX <— AX + BX; Goto Interp.

The first microinstruction tests the carry flag. If it is false, branch to the third microinstruction (the ELSE clause). Otherwise, execute the next instruction in sequence (the THEN clause). After executing either clause, control is passed to Interp, which is the beginning of the interpretation loop.

A microassembler translates a statement into a 103-bit string. How can we ensure that the bits are set correctly in the microinstruction? One solution is to write a micro-disassembler. That is, the micro-disassembler can print out all the bits with their mnemonics in the microcode for further verifications.

6.4.1 Subset of Target Instructions

A subset of 18 instructions is listed by the lexicographical order of its opcode in Table 6.2.

Table 6.2 A Subset of 8086 Instructions

Code image	Statement	Description
03 C3	add ax, bx	;ax ← ax + bx
13 C3	adc ax, bx	;ax ← ax + bx + C
1B C3	sbb ax, bx	;ax ← ax - bx - C
2B C3	sub ax, bx	;ax ← ax - bx
3B C3	cmp ax, bx	;Compare ax:bx and set CC.
7C XX	jl xx	;IF (S ⊕ O), ;THEN ip ← ip + 2 + xx sign ext. ;ELSE ip ← ip + 2 ENDIF
7D XX	jge xx	;IF \(S ⊕ O), ;THEN ip ← ip + 2 + xx sign ext. ;ELSE ip ← ip + 2 ENDIF
8B D0	mov dx, ax	;dx ← ax
8B D8	mov bx, ax	;bx ← ax
8E D8	mov ds, ax	;ds ← ax
90	nop	;No operation
99 XXXX	jmp xxxx	;ip ← ip + 3 + xxxx
A1 XXXX	mov ax, ds:xxxx	;ax ← M[ds*16 + xxxx]

Code image	Statement	Description
B8 XXXX	mov ax, xxxx	;ax ← xxxx
EC	in al, dx	;al ← IOR[dx]
EE	out dx, al	;IOR[dx] ← al
F7 D0	not ax	;ax ← .NOT. ax
F7 D8	neg ax	;ax ← -ax

6.4.2 Microcode Interpreter

The interpreter is stored in on-chip ROM and the target instructions are stored in off-chip central memory. The execution of the very first microinstruction initiates target instruction stream fetch. The engine is stalled until the very first target instruction is retrieved into the IR. The execution of the second microinstruction maps the first byte, i.e., the opcode of the target instruction, into a ROM address and meanwhile shifts the IR to the left by one byte. If more actions need be taken, the subsequent microinstruction again maps the next byte of the target instruction into another ROM address to proceed.

The central memory has BIOS and the system boot routine is embedded in ROM at a high-address. The microcode engine ROM contains an interpreter as shown in Figure 6.8. After boot, the μPC is initialized to FFE to fetch the very first microinstruction in the μIR. The very first microinstruction in μIR provides enable signals to add the IP to the CS to form a 20-bit linear address in BA. As specified in the system reference manual, the initial value of the CS is FFFF and the IP is 0000. Therefore, the start address of the system boot routine is at memory address FFFF0. Because the Rd (read) bit and the InsF (instruction fetch) bit are set in the microcode, the BU starts retrieving the target instruction stream. That is to say, the instruction fetch and operand execute are overlapped.

The target instruction execution is still sequential because the execution of the next target can not start until the current instruction is completed. The interpreter listing has an equate file in the front that defines mnemonics for bit positions. The org (origin) statement tells the microassembler where the microinstruction is to be loaded in ROM. For example, the address is denoted as x'ffe'. The letter x denotes the hex data type and the pair of single quotes are delimiters of the bit pattern ffe. Because the microinstruction at this address has both the BA and InsF bits set, the bus unit starts obtaining the target instruction stream. Execution stalls until an instruction arrives in the IR. Then, the next microinstruction at address x'fff' executes. Logically, this is the beginning of the interpretation loop. Physically, it is at the bottom of ROM. After this microinstruction is loaded in the μIR, it has enable bits to route the opcode from the IR into the μPC in order to form the address of the next microinstruction in low ROM. Because the opcode is eight bits, the formed ROM address is less than 256. That is to say, the first 256 words in ROM are reserved for the microcode sequence to start. Each of those microinstructions is indicated by an org statement in the front so its ROM address is fixed. Due to different target instructions, the engine may take further actions.

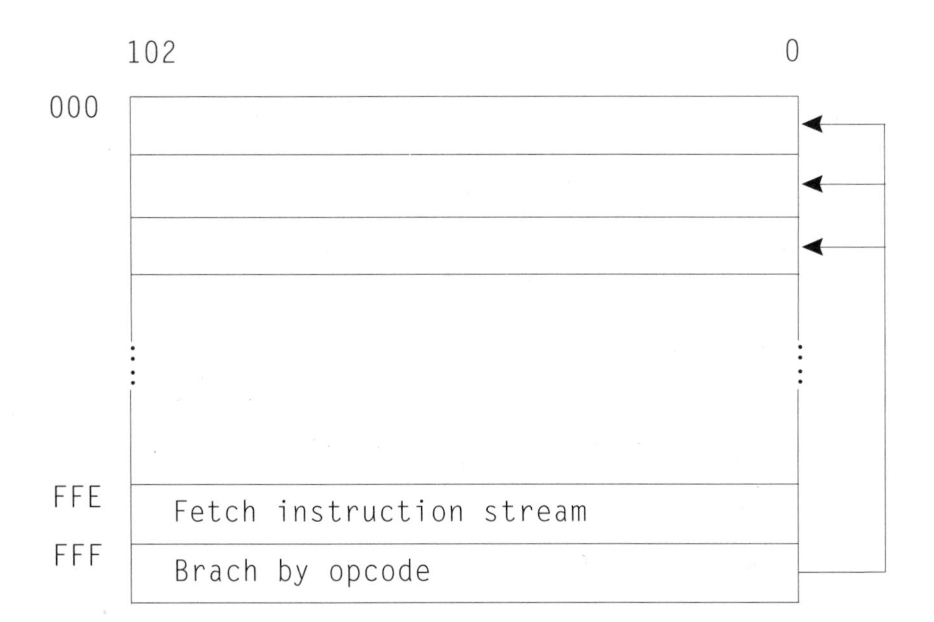

Figure 6.8 Microcode in ROM.

This branch control by opcode concept is crucial because a microinstruction or group of microinstructions is needed to interpret the target operation. In the listing, any microcode block with a defined label resides in a ROM address higher than 256. The microcode sequence or block is equivalent to a micro-order, and we enclose it in a comment block for easy comparisons.

```
; ——————————————- Microcode interpreter listing
  start
Wr equ  102            {Equate file: Wr is bit 102.}
Rd equ  101
. . .
org x'ffe' {Address FFE is where the instruction retrieval microinstruction
            resides.}
InsFetch:
  BT, Rd, InsF, Clk = 1; BA <— CS + IP; μPC <— μPC + 1.
        {This first microinstruction at ROM address x'ffe' initiates instruc
        tion stream retrieval.  After each execution of a branch target
        instruction, this microinstruction receives control to start retrieve
        ing a new instruction stream.}
Interp:
  SHLIR1 = 1; Addr = 000; μPC <— Addr^Op.
```

```
        {This microinstruction resides at address x'fff',i.e., the bottom of
        ROM. Because instruction retrieval is automatic, the microcode
        Interp (interpreter) actually starts here.}
org x'03'       ;This is for the target machine code:
                ;03 C3   add ax, bx
                ;The address x'03' is as same as the opcode.
                ;It receives control after opcode interpretation
                ;and it passes to sa03C3 (symbolic address for
                ;machine code 03C3)for further action.
  SHLIR1, BF = 1; Goto sa03C3.
{ —— This microinstruction resides at a high ROM address labeled as
sa03C3. After executing the microcode below, control is passed to the inter-
pretation loop.
sa03C3:         ;ax ← ax + bx
  Int, BF, Clk = 1; AX ← AX + BX; Goto Interp. ——— end}
org x'13'       ;13 C3   adc ax, bx
  SHLIR1, BF, Clk = 1; Goto sa13C3.
{ ——— microinstruction block:
sa13C3:         ;ax ← ax + bx + C
  IF C .EQ. 1,
  THEN Int, BF, Clk, Cin = 1; AX ← AX + BX + 1; Goto Interp.
  ELSE
  Int, BF, Clk = 1; AX ← AX + BX; Goto Interp. ——— end}
org x'1B'       ; 1B C3   sbb ax, bx
  SHLIR1, BF, Clk = 1; Goto sa1BC3.
{ ——— microinstruction block:
sa1BC3:         ;ax ← ax - bx - C
  IF C .EQ. 1,
  THEN Int, BF, Comp, Clk = 1; AX ← AX - BX - 1; Goto Interp.
  ELSE Int, BF, Comp, Clk, Cin = 1; AX ← AX - BX; Goto Interp.}
org x '2B'      ;2B C3   sub ax, bx
  SHLIR1, BF = 1; Goto sa2BC3.
{ ——— microinstruction:
sa2BC3:         ;ax ← ax - bx
  Int, BF, Comp, Clk, Cin = 1; AX ← AX - BX; Goto Interp. }
org x'3B'       ;3B C3   cmp ax, bx
  SHLIR1, BF = 1; Goto sa3BC3.
{ ——— microinstruction:
sa3BC3:         ;compare ax:bx
  Int, BF, Comp, Cin = 1; AX ← AX - BX; Clk = 0;
  Goto μPC ← Addr. —- }
org x'7C'       {7C XX   jl xx}
  SHLIR1, BF, Zero = 1; T1 ← 0 + IRB sign extended;
  Goto sa7C.
```

```
{ ——— microinstruction block:
sa7C:
   IF (S ⊕ O),
   THEN Int, BF, Clk = 1; IP ◄— IP + T1;
   Goto InsFetch.        ;Flush instruction queue
   ELSE Int, BF = 1; Addr = Interp; μPC ◄— Addr.  ;No branch}
org x'7D'                ;7D XX    jge xx
   SHLIR1, BF, Zero = 1; T1 ◄— 0 + IRB sign extended;
   Goto sa7D.
{ ——— microinstruction block:
sa7D:
   IF \(S ⊕ O),
   THEN Int, BF, Clk = 1; IP ◄— IP + T1;
   Goto InsFetch.        ;Flush instruction queue
   ELSE Int, BF = 1; Addr = Interp; μPC ◄— Addr.  ;No branch}
org x'8B'                ;8B for Move register
   SHLIR1 = 1; Addr = 100; μPC ◄— Addr^Op.
org x'1D0'               ;8B D0 for mov dx, ax
   Int, Zero, BF, Clk = 1, DX ◄— 0 + AX; Goto Interp.
org x'1D8'               ;8B D8    mov bx, ax
   Int, Zero, BF, Clk = 1; BX ◄— 0 + AX; Goto Interp.
org x'8E'                ;Move into segment register
   SHLIR1 = 1; Addr = 200; μPC ◄— Addr^Op.
org x'2D8'               ;8E D8    mov ds, ax
   Int, Zero, BF, Clk = 1; DS ◄— 0 + AX; Goto Interp.
org x'90'                ;90 for No operation
   Int, BF = 1; Goto Interp.
org x'99'                ;99 XXXX  jump direct
   SHLIR2, Zero, BF, Clk = 1; T1 ◄— 0 + IRW; Goto sa99.
{ ——— microinstruction block:
sa99:                    ;ip <— ip + 3 + xxxx
   Int, BF, Clk = 1; IP ◄— IP + T1;
   Goto InsFetch.  ;Flush instruction queue ——}
org x'A1'                ;A1 XXXX  mov ax, ds:xxxx
   SHLIR2, Zero, BF, Clk = 1; T1 ◄— 0 + IRW;
   Goto saa1.
{ ——— microinstruction block:
saa1:          ;ax ◄—     M[ds*16 + xxxx]
   Rd, BA, BT, Clk = 1; BA ◄— DS + T1; μPC ◄— μPC + 1.
                         ;Wait for memory operand to arrive
   Int, Zero, BF, Clk = 1; AX ◄— 0 + BDW; Goto Interp. ——}
org x'B8'                ;B8 XXXX  mov ax, xxxx
   SHLIR2, Zero, BF, Clk = 1; T1 ◄— 0 + IRW;
Goto sab8.
```

```
{ ——— microinstruction block:
sab8:                    ;ax ◄— xxxx
   Int, Zero, BF, Clk = 1; AX ◄— 0 + T1; Goto Interp. —— }
org x'F7'                ;F7 XX for unary operation
   SHLIR1 = 1; Addr = 100; μPC ◄— Addr^Op.
org x'1D0'         ;F7 D0    not ax
   Int, Zero, Comp, BF, Clk = 1; AX ◄— .NOT. AX; Goto Interp.
org x'1D8'         ;F7 D8    neg ax
   Int, Zero, Comp, BF, Clk, Cin = 1; AX ◄— 0 - AX;
   Goto Interp.
org x'EC'          {EC      in al, dx     ;al ◄— IOR[dx]}
   BF = 1; Goto saec.
{ ——— microinstruction block:
saec:                      ;Input byte and wait
   In, Zero, Clk = 1; IOA ◄— 0 + DX; μPC ◄— μPC + 1.
   Int = 1; AL ◄— 0 + IODL; Goto Interp. ——— }
org x'EE'          {EE      out dx, al    ;IOR[dx] ◄— al}
   BF = 1; Goto saee.
{ ——— microinstruction block:
saee:                      ;Output byte
   Clk, Zero = 1; IODL ◄— 0 + AL; μPC ◄— μPC + 1.
   Out, Int, Zero, Clk = 1; IOA ◄— 0 + DX; Goto Interp. —- }
End.
```

6.5 MULTIPLY OPERATIONS VIA ONE ADDER

Our society uses a decimal number system, and a negative number is represented by a signed-magnitude notation. Given two positive integers M and N, their product is shown as (M * N), where * is the multiply operator as used in a high-level programming language. The symbol M is the multiplicand and the symbol N is the multiplier. If positional notations are used to represent M and N, the multiply operation (M * N) result is the sum of many partial products. Assuming that M = 15 and N = 15, we show the decimal multiply operation as follows:

$$
\begin{aligned}
P &= 15 * 15 \\
&= 15 * (10 + 5) \\
&= (15 * 10) + (15 * 5) \\
&= 150 + 75 \\
&= 225
\end{aligned}
$$

where the final product 225 is the sum of two partial products, 75 and 150. The multiply operation can be decomposed into a sequence of add operations provided that the next partial product is shifted one digit to the left as shown below:

```
  15
* 15
-----
  75
 15
-----
 225
```

The first partial product is 75 which is (15 * 5). The second partial product is 15, which is (15 * 1). Because the multiplier digit one carries a weight of 10, we need to shift the partial product one digit to the left to get 150. The trailing zero is not shown in the partial product for simplicity. Because the multiplier has two digits, by adding the two partial products, 75 and 150, we obtain the final product as 225.

If the multiplier N has more than two digits, we repeat the process by adding more partial products. If the digit of the multiplier is a zero, its associated partial product is zero. Therefore, the adding step can be skipped, so perform a shift operation only. The same concept works perfectly for negative numbers because our society uses the signed-magnitude notation. That is, if M or N is negative, we place a minus sign (-) in front of the number to mean negative. The product is negative if the signs of M and N differ, otherwise the product is positive.

The binary multiply operations are similar with some differences. First, the number is in binary. Second, instead of shifting the partial product to the left, we shift the sum of partial products to the right. Third, we have two types of integers, signed and unsigned.

6.5.1 Unsigned Multiply

The symbolic representation of an n-bit integer is shown below:

$$C_{n-1}\ C_{n-2}\ C_{n-3}\ .\ .\ .\ .\ C_2\ C_1\ C_0$$

where C_{n-1} is the most significant bit of an unsigned integer. The same bit string can represent a signed integer and C_{n-1} is the sign bit. As intricate as it may be, we must determine the meaning of the sign bit.

6.5.1.1 Sign Bit

If an n-bit integer is unsigned, it has a value as computed by the following positional notation:

$$C_{n-1}\ 2^{(n-1)} + C_{n-2}\ 2^{(n-2)} + ... + C_2\ 2^2 + C_1\ 2^1 + C_0\ 2^0$$

where Ci (coefficient i) is bi (bit i) for i = 0, 1, ..., (n-1). Because each bit carries a positive weight as determined by its position in the notation, the value of an unsigned integer is always positive. As a simple example, if n = 4, the unsigned 1111 has a value computed below:

$$
\begin{aligned}
\text{value} &= 2^3 + 2^2 + 2^1 + 2^0 \\
&= 8 + 4 + 2 + 1 \\
&= 15
\end{aligned}
$$

where the sign bit carries a positive weight of 2^3. For the sake of simplicity, assume that the hardware register is only four bits long. Thus, a four-bit unsigned integer has a value ranging from 0 to 15. Hence, multiplying an unsigned 15 by an unsigned 15, we obtain an unsigned 225, as shown below:

```
        1111          {M: unsigned 15}
  *     1111          {N: unsigned 15}
  ----------
        1111
       1111
      1111
     1111
  ----------
    11100001          {P: unsigned 225}
```

where the multiplicand M is an unsigned 15, the multiplier N is an unsigned 15, and the product P is an unsigned 225. Just like multiplying two decimal numbers, an unsigned binary multiply is also comprised of a sequence of add and shift operations. As we can see, it is not intuitive to add four binary partial products at the same time. However, the adding job becomes easier if we perform four iterations and we merely add two numbers per iteration. The partial product for the next bit in the multiplier is shifted one bit to the left before adding as shown below:

```
      1111
  +  1111
  --------
    101101
  + 1111
  --------
   1101001
  +1111
  --------
  11100001        {P: unsigned 225}
```

Assume that M is any four-bit unsigned integer and N is 1111; the algebraic value of the product is computed below:

$$P = M * N$$

$$= M\,(2^3 + 2^2 + 2^1 + 2^0)$$

$$= M\,2^3 + M\,2^2 + M\,2^1 + M\,2^0$$

From the above equation we notice that each bit in the multiplier carries a positive weight. Assume that two four-bit registers by the name of DX and AX are used. Before the unsigned multiply operation, the multiplier is placed in AX and the multiplicand M can be placed in a register or in memory, as shown in Figure 6.9a. After the operation, the DX register concatenated with the AX register (i.e., DX^AX) constitutes an eight-bit double register to contain the unsigned product.

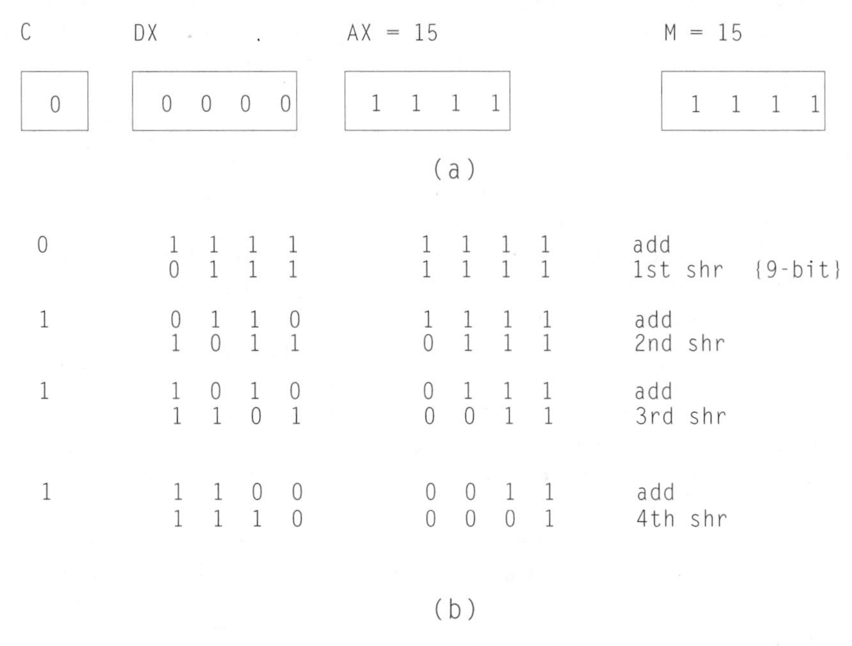

Figure 6.9 Four-bit unsigned multiply operation, 15 * 15 = 225: (a) initial register condition and (b) register contents after each iteration.

Two clever variations are incorporated in the design. First, instead of shifting the partial product to the left, we shift the sum to the right. That is, during each iteration, test the LSB in AX first. If AX <0> is one, then add M to DX, which is the high-order half of the product register; otherwise take no action. After the test with either outcome, perform a double logical right shift on DX and AX by one bit. This test and shift operation sequence is repeated a total of four times. If M is any four-bit unsigned integer and N is 1111, the algebraic value of the unsigned product is computed below:

$$
\begin{aligned}
P &= (M\,2^4 + (M\,2^4 + (M\,2^4 + (M\,2^4)/2\,)/2\,)/2\,)/2 \\
&= M\,2^3 + M\,2^2 + M\,2^1 + M\,2^0 \\
&= M\,(\,2^3 + 2^2 + 2^1 + 2^0) \\
&= M * N
\end{aligned}
$$

Note that adding M to the high-order half of the product register is equivalent to adding M 2^4 (M with four trailing zeroes). If the unsigned M has a large magnitude, a carry may be generated by the adder. When that happens, a total of nine bits is needed to store the sum correctly. That is to say, after adding the unsigned multiplicand to DX, if there is a carry generated, a nine-bit shift logical right (SHR) operation is in order. Note that if the C bit is zero, a nine-bit SHR is no different from an eight-bit SHR, provided that the lower eight bits are of interest. Adding one extra bit to the adder design would mean more hardware. A clever solution is to take advantage of the RCR (rotate with carry) operation. To achieve a uniform design, the second variation is done in two steps. First, before the iteration begins, clear the C bit. After the add or no add operation, always perform a nine-bit SHR. This is accomplished by performing an RCR on DX followed by an RCR on AX. Shifting logical right both DX and AX by one bit means that the sum of all the partial products is divided by two. Not surprisingly, this shift logical right operation always produces a perfect quotient because the sum is an even number. The old LSB in the multiplier is shifted out, and its left neighbor bit takes over its position as the new LSB. After the divide operation, the multiplier bits are completely shifted out of AX.

Figure 6.9b shows the contents of DX^AX after each iteration. Whether there is an add operation depends on the LSB in AX, but there is always a shift operation. If the LSB in AX is one, the first line shows the register contents after the add operation, and the second line shows the register contents after the shift operation. If the LSB in AX is zero, the add zero operation is skipped and there is only one line to show the result after the shift. After four iterations, the DX^AX pair contains the eight-bit unsigned product.

The basic concept works for 16-bit, 32-bit, or 64-bit unsigned multiply operations. Take the 8086 as an example; before the 16-bit unsigned multiply instruction executes, the 16-bit multiplier N is placed in AX and the 16-bit multiplicand M is in a register or in memory.

6.5.1.2 16-bit Unsigned Multiply Algorithm

```
          Clear DX;
          Retrieve M into a temporary register;
          Set loopCount to 16;
loop010:  Clear the C flag;
          IF the LSB of AX is 1,
          THEN add M to DX; ENDIF;
          RCR DX, 1 bit;
          RCR AX, 1 bit;
```

```
                Decrement loopCount by 1;
                IF loopCount is not equal to 0,
                THEN goto loop010; ENDIF;
exit:           . . .
```

The 16-bit unsigned multiply instruction is shown below:

```
mul   bx          ;unsigned mul
```

After executing the instruction, the unsigned product (BX * AX) is in DX^AX. This instruction can be implemented by microcode or simulated by a software routine. The symbol mask in the code represents an address of a memory word.

```
          mov    dx, 0      ;clear dx
          mov    cx, 16     ;loop count
loop010:  clc               ;Clear the C flag
          mov    mask, 1    ;test the LSB of AX
          and    mask, ax
          jz     skip015    ;skip add if 0
          add    dx, bx     ;add multiplicand
skip015:  rcr    dx, 1      ;rotate with carry right dx 1 bit
          rcr    ax, 1      ;rotate with carry right ax 1 bit
          loop   loop010    ;decrease cx, loop if cx .NE. 0
exit:     .. .
```

6.5.2 Signed Multiply

A signed multiply instruction can multiply two signed integers. Using four-bit arithmetic, a signed integer has a value ranging from -8 to 7 in twos complement notation. Multiplying -1 by -1, we obtain a product that is +1. To illustrate the concept, assume that the registers DX and AX are four bits long. After a signed multiply operation, the DX^AX pair contains a signed product that is eight bits long. If M is 1111 in binary and N is also 1111, the product is +1 (0000 0001 in binary).

The signed multiply instruction can be implemented by microcode via one single adder or by a multiplication tree using combinational logic. First, consider the mathematical theory behind an n-bit signed integer in twos complement notation. Recall that the integer is negative if its sign bit is one. In fact, the sign bit carries a negative weight as determined by its bit position in the positional notation. Thus, the value of an n-bit negative integer is computed as follows:

$$\begin{aligned}\text{Value} &= -(2^n - (C_{n-1}\, 2^{(n-1)} + C_{n-2}\, 2^{(n-2)} + \\ &\qquad \ldots + C_2\, 2^2 + C_1\, 2^1 + C_0\, 2^0)) \\ &= -C_{n-1}\, 2^{(n-1)} + C_{n-2}\, 2^{(n-2)} + \\ &\qquad \ldots + C_2\, 2^2 + C_1\, 2^1 + C_0\, 2^0\end{aligned}$$

The C_{n-1} or sign bit carries a negative weight of $2^{(n-1)}$ and each bit to its right carries a positive weight. That is to say, the n-bit negative integer can be viewed as a binary code whose sign bit carries a negative weight. For example shown below, the arithmetic value of four-bit 1111 is computed as -1.

$$\begin{aligned} \text{Value} &= -2^3 + 2^2 + 2^1 + 2^0 \\ &= -8 + 4 + 2 + 1 \\ &= -8 + 7 \\ &= -1 \end{aligned}$$

Because the adder generates the signed sum of two signed numbers, the CPU can interpret each bit in the multiplier from right to left and add the multiplicand accordingly. Before the signed multiply operation begins, the four-bit multiplier N is placed in AX, and the four-bit multiplicand M can be in a register or in memory. The multiplier is 1111 (-1) and the multiplicand is also 1111 as shown in Figure 6.10a. The top line shows the initial register condition followed by the register contents after each iteration. During each iteration, test the LSB in AX first. If it is zero, perform an eight-bit shift arithmetic right (SAR) operation on DX^AX by one bit. If the LSB of AX is one, add M to DX. After the operation, if the Overflow (O) bit is set, we need to perform a nine-bit shift arithmetic right (SAR), otherwise perform an eight-bit SAR. Again, the nine-bit SAR is accomplished by performing an RCR on DX one bit followed by another RCR on the AX one-bit. If there is an add operation, the first line shows the register contents after the operation and the second line shows the register contents after the shift. Otherwise, only one line exists to show the register contents after the shift.

After three iterations, the sign bit of the multiplier that needs special action is reached. If the sign bit is zero, perform an eight-bit SHR to exit. If the sign bit is one, subtract M from DX because the bit carries a negative weight. The subtract operation is actually done by adding the twos complement of M to DX. After that, perform an eight-bit SAR to exit. After the signed multiply, the double word signed product is in DX^AX. If the multiplicand M is any signed integer and the multiplier N is -1, the algebraical value of the signed product is shown below:

$$\begin{aligned} P &= (-M\,2^4 + (M\,2^4 + (M\,2^4 + (M\,2^4)/2\,)/2\,)/2\,)/2 \\ &= -M\,2^3 + M\,2^2 + M\,2^1 + M\,2^0 \\ &= M\,(\,-2^3 + 2^2 + 2^1 + 2^0) \\ &= M * (-1) \end{aligned}$$

Figure 6.10b shows another example — when -7 is multiplied by +3, the signed product is -21. After the second add operation, because there is an overflow condition, a nine-bit SAR is performed to take care of the sign. It should be mentioned that this design always works, except in one extreme case when N is -2^3 and M is -2^3 as shown in Figure 6.10c.

```
     C      DX          AX = -1      M = -1
(a) [?]   [0 0 0 0]   [1 1 1 1]    [1 1 1 1]

     0     1 1 1 1     1 1 1 1     add
           1 1 1 1     1 1 1 1     1st  sar  {8-bit}

     1     1 1 1 0     1 1 1 1     add
           1 1 1 1     0 1 1 1     2nd  sar  {8-bit}

     1     1 1 1 0     0 1 1 1     add
           1 1 1 1     0 0 1 1     3rd  sar  {8-bit}

     1     0 0 0 0     0 0 1 1     subtract
           0 0 0 0     0 0 0 1     4th  sar  {8-bit}
```

```
     C      DX          AX = 3       M = -7
(b) [?]   [0 0 0 0]   [0 0 1 1]    [1 0 0 1]

     0     1 0 0 1     0 0 1 1     add
           1 1 0 0     1 0 0 1     1st  sar  {8-bit}

     1     0 1 0 1     1 0 0 1     add     {overflow}
           1 0 1 0     1 1 0 0     2nd  sar  {9-bit}

           1 1 0 1     0 1 1 0     3rd  sar  {8-bit}
           1 1 1 0     1 0 1 1     4th  sar  {8-bit}
```

```
     C      DX          AX = -8      M = -8
(c) [?]   [0 0 0 0]   [1 0 0 0]    [1 0 0 0]

     0     0 0 0 0     0 1 0 0     1st  sar  {8-bit}

           0 0 0 0     0 0 1 0     2nd  sar  {8-bit}

           0 0 0 0     0 0 0 1     3rd  sar  {8-bit}

     0     1 0 0 0     0 0 0 0     subtract

           0 1 0 0     0 0 0 0     4th  shr  {8-bit}
```

Figure 6.10 Four-bit signed multiply operations: (a) -1 * -1 = 1: (b) -7 * 3 = -21, and (c) -8 * -8 = 64.

The problem arises after subtracting the multiplicand M from DX during the final iteration. That is because when we take the twos complement of -2^3, the result is still the same negative number. Therefore, after performing the final 8-bit SAR, we obtain a product -2^6 instead of $+2^6$. To solve this anomaly, we need to test DX first after the final subtraction. If DX contains the bit pattern 1000, then perform an eight-bit SHR; otherwise perform an eight-bit SAR.

The 16-bit signed multiply instruction can be simulated by the following algorithm provided that the 16-bit multiplier N is in AX and the 16-bit multiplicand M is in a register or in memory.

6.5.2.1 16-bit Signed Multiply Algorithm

```
                Clear DX;
                Retrieve M into a temporary register;
                Set loopCount to 15;
      loop020:  IF the lsb of AX is 1,
                THEN add M to DX;
                   IF Overflow, THEN
                   RCR DX, 1 bit;
                   RCR AX, 1 bit;
                   Goto f0024; ENDIF; ENDIF;
                SAR DX, 1 bit;
                RCR AX, 1 bit;
      f0024:    Decrease loopCount by 1;
                IF loopCount is not 0,
                THEN goto loop020; ENDIF;
                IF the lsb of AX is 1,
                THEN subtract M from DX;
                   IF DX is not 8000 in hex,
                   THEN goto skip026;
                   ELSE
                   SHR DX, 1 bit;
                   RCR AX, 1 bit;
                   goto exit028; ENDIF; ENFIF;
      skip026:  SAR DX, 1 bit;
                RCR AX, 1 bit;
      exit028:  . . .
```

After executing the instruction, the 32-bit signed product is in DX^AX. For example, the following instruction,

```
        imul    bx    ;integer, i.e., signed multiply
```

can be simulated by the assembly code below:

```
                mov     dx, 0        ;clear dx
                mov     cx, 15       ;loop count
    loop020:    mov     mask, 1      ;test the lsb of AX
                and     mask, ax
                jz      skip0022     ;skip add if 0
                add     dx, bx
                jno     skip022      ;jump on no overflow
                rcr     dx, 1        ;33-bit sar
                rcr     ax, 1
                jmp     f0024
    skip022:    sar     dx, 1        ;32-bit sar
```

```
            rcr     ax, 1
f0024:      loop    loop020
            mov     mask, 1      ;test the divisor sign
            and     mask, ax
            jz      skip026
            sub     dx, bx
            cmp     dx, 8000h
            jne     skip026
            shr     dx, 1        ;32-bit shr
            rcr     ax, 1
            jmp     f0028
skip026:    sar     dx, 1        ;32-bit sar
            rcr     ax, 1
f0028:      . . .
```

After iterating 15 times, the original sign bit of the multiplier becomes the LSB in AX. If the bit is one, the multiplier is negative, so we subtract M from DX. After the final subtract, if the DX contains the bit pattern 8000 in hex, perform a 32-bit SHR, otherwise perform a 32-bit SAR before exiting.

In early computers, DX was considered to be the accumulator and AX was the MQ (multiplier/quotient) register. Before a multiply operation, the MQ register contained the multiplier. After the operation, the accumulator and MQ register pair contained a double word product. The next chapter demonstrates that a multiplication tree using combinational logic can add all the shifted multiplicands at once. As a result, it takes only one clock to do a four-bit multiply: unsigned or signed.

6.6 DIVIDE OPERATIONS VIA ONE ADDER

Both divide operations, unsigned and signed, are intricate. We perform a decimal divide operation on paper, as shown below:

```
        14
      _____
  15 | 224
       15
      _____
        74
        60
      _____
        14
```

The dividend 224 is placed on the right and the divisor 15 is placed on the left. Starting from the leftmost digit of the dividend, we compare a group of leading digits of the dividend with the divisor. The group of digits in the dividend is treated as a partial dividend. If the value of the partial dividend is less than the divisor, the quotient digit is zero, otherwise we divide the divisor into the partial dividend to obtain the quotient digit. In our example, after examining the first two digits, the partial dividend 22 is greater than the divisor 15, so a quotient digit one is placed on top of the second most significant digit of the dividend. Multiply the divisor by the quotient digit to obtain the partial product. Subtracting the partial product from the partial dividend, we obtain a difference of seven as the leading digit of the remaining dividend. Pull down another digit from the dividend and compare the group of digits with the divisor again. The same operation is repeated until the remaining dividend (remainder) is less than the divisor. In this example, after the decimal divide operation, the quotient is 14 and the remainder is also 14 as a coincidence.

An unsigned binary divide operation is similar with the understanding that the quotient in each iteration is a bit instead of a decimal digit. Thus, if the quotient bit is one, the partial product is simply the divisor. The operation of dividing 15 into 224 in binary arithmetic is shown below:

```
                 1 1 1 0   <-- quotient
          _____________________
  1 1 1 1 |  1 1 1 0  0 0 0 0
             1 1 1  1
          _____________________
               1 1 0  1 0
                 1 1  1 1
          _____________________
                  1 0 1 1 0
                    1 1 1 1
          _____________________
                    1 1 1 0   <-- remainder
```

For simplicity reasons, the dividend is an eight-bit unsigned integer 224 and the divisor is a four-bit unsigned integer 15. After the unsigned divide operation, the quotient has a decimal value of 14 and the remainder also has a decimal value of 14. Starting from the leftmost bit of the dividend, compare the leading bits of the dividend with the divisor. If the value of the leading bits of the dividend is less than the divisor, we pull another bit of the dividend down. Because the value of the leading five bits is greater than the divisor, a quotient bit one is placed on top of the fifth significant bit of the dividend.

In a case where the value of the leading bits is not greater than the divisor, a quotient zero is placed on top. However, all the leading zeroes in the quotient are not shown. Just like in decimal divide, we multiply the quotient by the divisor to compute the partial product. Intuitively, if the quotient bit is one, the product is the divisor, otherwise the product is zero. In the above example, the subtract operation is shown as an explicit binary subtract. In a computer, the twos complement of the divisor is added to the leading bits of the dividend. Pull down the next trailing bit from the dividend and compare again. The process is repeated a total of four times. At the end, a four-bit partial dividend is left as the remainder.

6.6.1 Unsigned Divide

During an unsigned divide operation, an interrupt on divide overflow means that the quotient is too big for a CPU register to store it correctly. As a result, the divide operation is aborted. Testing an overflow condition is quite easy because the dividend and the divisor are both unsigned integers. Assume that the DX and AX registers are four bits long. Before the unsigned divide operation begins, the DX^AX contains the eight-bit dividend, and the divisor D is in a four-bit register or in memory, as shown in Figure 6.11a. First, compare the divisor with DX, which contains the four leading bits of the dividend. If there is no borrow, it is an unsigned divide overflow because the quotient requires five bits. For example, a divisor zero will generate an overflow. If there is no overflow, iterate four times to generate a remainder in DX and the quotient in AX.

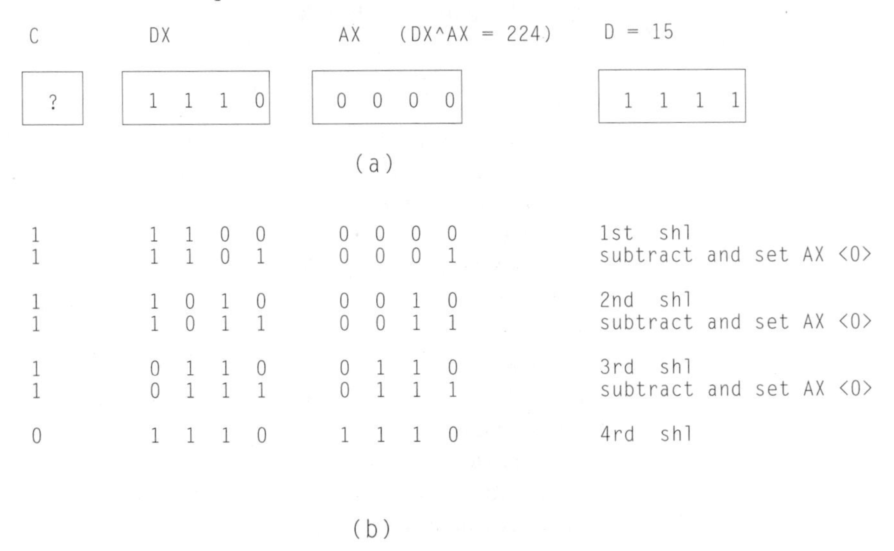

Figure 6.11 Four-bit unsigned divide operation, 224/15 = (14) + (14)/15: (a) initial register condition and (b) register contents after each iteration.

Recall that in each iteration of an unsigned multiply, we perform a sequence of add and shift logical right operations. For unsigned divide, we perform a sequence of shift logical left and subtract operations. After each shift, the engine compares DX with the divisor. It then performs a subtract operation only if the partial dividend in DX is greater than or equal to the divisor. The register contents after each iteration, are shown in Figure 6.11b. At the end, we obtain a quotient 14 in AX and a remainder in DX that is also 14. In the equation, each unsigned integer is enclosed in a pair of parentheses. During each iteration, a nine-bit shift logical left is performed. The carry flag is the sign bit of the partial dividend. If the C flag is one, then we subtract the divisor from DX because the dividend is definitely greater than the divisor. If the C flag is zero, the engine needs to compare DX with the divisor. Recall that the internal subtract operation is implemented by adding the twos complement of the divisor to DX, so the C flag is a borrow. Only if the C flag is 0 (no borrow), does the engine perform the subtract operation.

In practice, both DX and AX and are 16-bit registers. Before the 16-bit unsigned divide instruction executes, we place the 32-bit dividend in DX^AX, the divisor in a 16-bit register or in memory.

6.6.1.1 16-bit Unsigned Divide Algorithm

```
            Fetch divisor D into a temp register;
            Compare DX with D;
            IF C flag is 0, {no borrow means above or equal}
            THEN interrupt on overflow and abort; ENDIF;
            Set loopCount to 16;     {no overflow}
loop030:    SHL (Shift Logical Left) DX^AX by 1 bit;
               {shl ax, 1 followed by rcl dx, 1}
            CASE C flag of,
            1: Subtract D from DX;   {DX above divisor}
               Set the lsb of AX to 1;   {AX <0> <— 1}
            0: Compare DX with D;
               IF C flag is 0,       {no borrow}
               THEN Subtract D from DX;
                  Set the lsb of AX to 1;
               ENDIF;
            ENDCASE;
            Decrease loopCount by 1;
            IF loopCount is not 0,
            THEN Goto loop030; ENDIF;
               {After 16 iterations, DX contains remainder
                and AX contains quotient.}
            Exit;
```

6.6.2 Signed Divide

Signed divide is more intricate. After studying for several hundred hours, the author failed to develop a direct signed divide algorithm without too many patches. The trouble arises when the divisor is one or most negative and the dividend is negative. However, by converting a negative dividend or divisor to positive, we turn the problem into unsigned divide with minor modifications because the overflow condition is a little different. After the iterations, we negate the quotient and/or the remainder as determined by its sign combinations. The dividend sign and the divisor sign are decoded into a two-bit sign flag as shown below:

00: Both DX and D are positive.
01: DX is positive but D is negative.
10: DX is negative but D is positive.
11: Both DX and D are negative.

With sign flag 00 or 11, the quotient is positive because the two signs are equal. With sign flag 01 or 10, the quotient is negative because the two signs are different. The remainder obeys the sign of the dividend. Thus, sign flag 00 or 01 means a positive remainder while sign flag 10 or 11 means a negative remainder.

We use four-bit arithmetic to illustrate the concept so DX and AX are four-bit registers. The eight-bit signed dividend is in DX^AX, and the signed divisor D can be in a four-bit register or in memory. The key ideas are described below:

1. Decode the sign of DX and D into a two-bit sign flag. For DX^AX or D, if its sign is one, we negate the operand.
2. Using four-bit arithmetic, an interrupt on overflow occurs if the quotient is greater than 7 or less than -8. The quotient can not be correctly stored in a four-bit register, so the operation is aborted. For signed divide, the overflow condition is tested in two phases.
3. Before the first SHL (shift logical left), if DX is greater than or equal to D, it is an overflow.
4. After the first SHL, if DX is greater than or equal to D and the two signs are the same, it is an overflow. However, it is valid if DX is equal to D and the two signs are different. Such a design allows the most negative integer as a quotient.

Four cases of signed divide are shown in Figure 6.12. The dividend has a magnitude of 23 and the divisor has a magnitude of three, but their sign flags are different. Because any negative number is converted to positive, all the shift and subtract operations look the same as shown in (a). The final quotient and remainder are negated according to the 2-bit sign flag. As shown in (a), the quotient is 7 and the remainder is 2. In (d), the quotient is 7 and the remainder is -2.

This algorithm is attractive for three reasons. First, it handles overflow correctly. Second, it allows the most negative number as a divisor because taking the twos com plement of -2^3 means no change. Third, it allows the most negative integer as quotient so long as the two signs are different. As shown in Figure 6.13, when -56 is divided by +7, the quotient is -8 and the remainder is zero. After the loop, the unsigned eight is in AX, but after negation we obtain the same quotient treated as -8. Using a 32-bit adder, the engine can handle both signed divide operations, and the operand can be 16 bits or 32 bits. If DX and AX are 16-bit registers, we have the following algorithm.

```
 C      DX              AX   (DX^AX = 23)       D = 3

[?]   [0 0 0 1]       [0 1 1 1]              [0 0 1 1]

SignFlag = 00                                Both DX and D are positive.
                                             Compare DX:D and pass.

0       0 0 1 0         1 1 1 0              1st shl

0       0 1 0 1         1 1 0 0              2nd shl
0       0 0 1 0         1 1 0 1              subtract, set AX <0>

0       0 1 0 1         1 0 1 0              3rd shl
0       0 0 1 0         1 0 1 1              subtract, set AX <0>

0       1 0 0 0         0 1 1 0              4rd  shl
0       0 0 0 0         0 1 1 1              subtract, set AX <0>

                          (a)

SignFlag = 01                                DX is positive and D is negative.

. . .

0       0 0 1 0         0 1 1 1              subtract, set AX <0>
        0 0 1 0         1 0 0 1              negate AX

                          (b)

SignFlag = 10                                DX is negative and D is positive.

. . .

0       0 0 1 0         0 1 1 1              subtract, set AX <0>
        1 1 1 0         1 0 0 1              negate DX and negate AX

                          (c)

SignFlag = 11                                Both DX and D are negative.

. . .

0       0 0 1 0         0 1 1 1              subtract, set AX <0>
        1 1 1 0         0 1 1 1              negate DX

                          (d)
```

Figure 6.12 Four-bit signed divide operations: (a) 23/3= (+7) + (+2)/3, (b) 23/-3 = (-7) + (+2)/-3, (c) -23/3 = (-7) + (-2)/3, and (d) -23/-3 = (+7) + (-2)/-3.

```
 C       DX                AX    (DX^AX = 56)      D = 8

[ ? ]   [ 0 0 1 1 ]       [ 1 0 0 0 ]             [ 1 0 0 0 ]

 SignFlag = 11                                  Both DX and D are negative.
                                                Compare DX:D and pass.

 0        0 1 1 1          0 0 0 0              1st shl

 0        1 1 1 0          0 0 0 0              2nd shl
 0        0 1 1 0          0 0 0 1              subtract, set AX <0>

 0        1 1 0 0          0 0 1 0              3rd shl
 0        0 1 0 0          0 0 1 1              subtract, set AX <0>

 0        1 0 0 0          0 1 1 0              4rd  shl
 0        0 0 0 0          0 1 1 1              subtract, set AX <0>
          0 0 0 0          0 1 1 1              negate DX
```

Figure 6.13 Four-bit signed divide, -56/-8 = (+7) + (0)/-8.

6.6.2.1 16-bit Signed Divide Algorithm

```
          Retrieve divisor D into a temp register;
          Decode the signs of DX and D into 2-bit sign Flag;
          CASE signFlag of,
          01: negate divisor D;
          10: negate dividend in DX^AX;
          11: negate both dividend and divisor; ENDCASE;
          Compare DX with D;
          IF C flag is 0,            {no borrow}
          THEN interrupt on overflow and abort; ENDIF;
          SHL DX^AX by 1 bit;
          Compare DX with D;
          IF Z (Zero) flag is 1 and the two signs differ,
          THEN subtract D from DX;
               set the lsb of AX to 1;
               {This patch allows the most negative integer as quotient.}
          ELSE
          IF C flag is 0,            {no borrow}
          THEN interrupt on overflow and abort; ENDIF; ENDIF;
          Set loopCount to 15;       {Divide loop}
loop040:  SHL DX^AX by 1 bit;
          CASE C flag of
          1: Subtract D from DX;  {DX above divisor}
               Set the lsb of AX to 1;
          0: Compare DX with D;
```

```
        IF C flag is 0,
        THEN Subtract D from DX;
        Set the lsb of AX to 1;
        ENDIF;
ENDCASE;
Decrease loopCount by 1;
IF loopCount is not 0,
THEN Goto loop040; ENDIF;
CASE signFlag of,
01: negate quotient in AX;
10: negate quotient in AX;
        negate remainder in DX;
11: negate remainder in DX;
ENDCASE;
Exit;
```

6.7 OTHER TYPES OF MICROINSTRUCTIONS

Some other microinstructions are worth mentioning, and examples include microcode subroutine linking, interrupt, floating point instructions, etc.

6.7.1 Microcode Subroutine Linking

The discussion of micro-subroutine linking is important as a concept. The reason to have a micro-subroutine is that the small set of code can be used repeatedly by other microinstructions. Micro-subroutine implementation is simpler than the subroutine at the target machine level.

6.7.1.1 Microcode Subroutine Call

Recall that during a target subroutine call, the return address in the IP is pushed on the stack before the IP is replaced by the branch address. At the microcode level, because a one-level call is adequate, the return address can be placed in a particular temp register, e.g., T2. Where to store the return register is a hardware decision because the return instruction must know where to find the address. If a branch microinstruction has the link bit set, it is a call instruction. Thus, the microcode engine implements the call in two steps. The first step is to store the μPC in T2 because it has already been increased by one when the branch microinstruction was obtained from the control store. The second step is to store the branch address in the μPC.

6.7.1.2 Microcode Return

If any microinstruction has the return bit set, it becomes the micro-return instruction, and it is usually the last instruction, in a micro-subroutine. After a micro-return instruction is executed, the return address in T2 is copied into the μPC which points to the next microinstruction in the control store after the call.

6.7.2 Microcode Interrupt Routine

When the Int bit is set in a microinstruction, the engine checks if there is any interrupt pending condition. If so, the engine passes control to a particular ROM address that contains the microcode interrupt routine. As a result of executing the microcode interrupt routine, the CS and IP are saved on the stack, the IF (interrupt flag) and TF (trace flag) in the F register are reset to zero, and the CS and IP are loaded from the interrupt vector in low central memory.

6.7.3 Microcode Floating Point Routines

The floating point instruction can be implemented by microcode. In a modern computer, a scientific function can be written as a routine of target machine instructions or implemented by microcode. If one single target instruction is executed to obtain the result of a scientific function, the microcode routine is usually a loop that performs floating point arithmetic. That is to say, a microcode sequence can be used to compute a complex scientific function. Therefore, at the target instruction level, we can implement functions, such as square root, sine, cosine, tangent, arctangent, natural log, decimal log, etc.

6.7.3.1 Floating Point Add/Subtract

Assume that the first source operand has its biased exponent denoted as bexp1 and the second source operand has bexp2. After the operation, the destination operand has its biased exponent denoted as bexp3. The floating add and subtract operations are similar except for a subtract operation; the sign bit of the subtrahend is flipped as a pre-operation step. The four sign bit combinations are 00, 11, 10, and 01, and a sign bit zero means positive. The floating point add or subtract operation is done in three steps:

1. Locate the float number with a smaller exponent and shift its significand to the right. After each shift of the significand by one bit, increase its biased exponent by one. This shift right operation continues until the two exponents are equal.
2. Compute the sum or difference of the two significands as determined by the two signs. If the two signs are equal, positive or negative, compute the sum of the two significands; otherwise compute the difference. This is done by feeding the positive significand to one adder input and the twos complement of the negative significand to the other input.
3. Negate the adder output if it is negative, normalize the significand, adjust its biased exponent, and determine its final sign.

6.7.3.2 Floating Point Multiply

The floating point multiply operation is done in four steps, as shown below:

1. Add the biased exponent of the multiplier (bexp2) to the biased exponent of the multiplicand (bexp1) so the result has one extra bias.
2. Subtract the bias from the result to obtain a biased exponent for the destination operand.
3. Multiply the two significands to generate a product.
4. Normalize the product, adjust its biased exponent (bexp3), and determine the overall sign.

6.7.3.3 Floating Point Divide

The floating point divide operation is done in four steps, as shown below:

1. Subtract the biased exponent of the divisor (bexp2) from the biased exponent of the dividend (bexp1) so the result has one less bias.
2. Add the bias to the result.
3. Divide the significand of the divisor into the significand of the dividend to generate a quotient.
4. Normalize the quotient, adjust its biased exponent (bexp3), and determine the overall sign.

6.8 SUMMARY POINTS

1. The microprogrammed CPU approach has the major advantage of flexibility.
2. If the microcode interpreter is stored in ROM in the CPU, we have a computer in a computer.
3. A horizontal microcode is longer because each bit is a direct enable signal for control. In contrast, a vertical microcode is shorter because the field is encoded.
4. In general, each microinstruction can be divided into six fields: address, sequence control, adder control, input control A, input control B, and bus control.
5. The sequence control field decides where and how to fetch the next microinstruction.
6. After a micro-assembler is developed, we can write a micro-disassembler to print out the bits in the microcode for verification.
7. The instruction queue or pipe is designed to allow overlapped operations between the BU and EU.

8. When the BA bit is set in the microinstruction, the engine acts like a three-address machine.
9. After interpreting the opcode in the IR, the engine passes control to different places in ROM to take further action.
10. A binary multiply may be implemented as a sequence of add and shift operations.
11. A binary divide may be implemented a sequence of shift and subtract operations.
12. A target instruction may compute a scientific function, such as square root, sine, cosine, tangent, arctangent, natural log, log, etc.

PROBLEMS

1. Explain the concept of a computer in a computer?
2. What is the difference between horizontal microcode and vertical microcode?
3. Describe the functions of the μIR and μIAR in a microcode engine, and justify that the μIAR is really a program counter.
4. Explain why the branch condition for the jge instruction is (\S \O + S O) instead of (\S \O + S O + Z).
5. What is the logical equation of the branch condition for the jl instruction?
6. What is the condition specifier for jae, i.e., jump above or equal for unsigned compare?
7. What are the differences among cmp (compare), sub (subtract), and sbb (subtract with borrow) instructions?
8. What is the difference between executing a neg (negate), and a not instruction?
9. Some instructions, such as nop, increase AX, decrease AX, etc., have a single byte opcode. Design the microcode sequence to execute the three following instructions.

Code image	Statement	
40	inc ax	;ax <— ax + 1
48	dec ax	;ax <— ax - 1
90	nop	;ip <- ip + 1

10. What is the advantage of designing an instruction queue or pipe?
11. Why is the IP (instruction pointer) designed as a counter?

The next five problems deal with extreme cases under the assumption that both DX and AX are four-bit registers.

12. Show the sequence of the unsigned multiply operations of

    a. (15 * 14) b. (15 * 15)

13. Show the sequence of the signed multiply operations of

    a. (7 * +7) b. (7 * -7) c. (-7 * +7) d. (-7 * -7)

14. Show the micro sequence of unsigned divide operations as follows.

    a. (225 / 15) b. (225 / 14)

15. Show the quotient and remainder of the following signed divide operations.

    a. (49 / +7) b. (49 / -7) c. (-49 / +7) d. (-49 / -7)

16. Show that the following four signed divide operations are all valid using four-bit arithmetic.

    a. (-56 / 7) = (-8) + (0) / 7 b. (56 / -8) = (-7) + (0) / -8
    c. (-57 / 7) = (-8) + (-1) / 7 d. (55 / -8) = (6) + (-7) / -8

17. How can we implement a micro-subroutine call?
18. What is the advantage of designing a target instruction that can compute a scientific function?
19. On most microcomputers, after a subtract operation, the C (carry) flag is same as the complement of the carry output from the adder. Therefore, the C flag bit is truly a borrow. Let us change our design such that if C is zero, it means a borrow. By doing so, we can copy the carry output of the adder into the C flag bit regardless of the operation, add or subtract. Based on this new definition, answer the following questions:
    a. What should be the microinstruction routine to implement the adc (add with carry) and sbb (subtract with borrow) instructions?
    b. If we design a hardwired logic CPU, can we feed the Carry flag as Cin (carry in) to the adder for adc or sbb?
20. Exponent overflow or underflow means that the result is incorrect after a floating point operation: add, subtract, multiply, or divide. We make the following assumptions:
    a. Given two floating point numbers, the magnitude of each number is one.
    b. The exponent field has four bits, so the excess is seven (0111 in binary). Note that the first floating point number has bexp1 (biased exponent 1), the second operand has bexp2 (biased exponent 2), and the result has bexp3 (biased exponent 3).

If bexp1 is +7 (unsigned 14) and bexp2 is +2 (unsigned 9), then bexp3 is too big a positive number after a floating point multiply, and it generates an exponent overflow interrupt, as shown below.

```
      bexp1  +7        1110
+     bexp2  +2        1001
    ----------------------------
                       0111   (intermediate result)
+            -7        1001
    ----------------------------
    bexp3 (overflow)   0000
```

The signs of both bexp1 and bexp2 are ones, but after adding the twos complement of seven to the intermediate result, the sign of bexp3 changes to zero to signal an exponent overflow.

An exponent underflow condition may occur after a divide operation. If the dividend has bexp1 as -2 and the divisor has bexp2 as +7, subtracting bexp2 means adding its twos complement. The sign bits of both adder inputs are zeroes, but after adding seven to the intermediate result, the sign of bexp3 changes to one to signal underflow.

Verify that an exponent underflow occurs after a multiply operation if bexp1 is -2 and bexp2 is -7. When bexp1 and bexp2 are fed to the adder, both sign bits are zeroes, but after subtracting seven from the intermediate result, bexp3 has its sign bit changed to one.

21. On a Pentium, verify the 16-bit imul instruction can handle the product of the two most negative integers. Write a program named imul.asm with the following instructions:

```
mov  ax, 8000h
mov  bx, 8000h
imul bx
```

After assembling and linking, execute the debugger by entering:

```
debug  imul.exe
```

Enter the t (trace) command to verify that the product in dx^ax is 4000^0000.

22. Write the program named idiv.asm that has the following instructions:

```
mov  dx, 4000h
mov  ax, 0000h
mov  bx, 8000h
idiv  bx
```

After executing the debugger, use the r command to change ip, dx, ax, and bx before executing the idiv instruction. Verify the following extreme cases. On the left, we show the dividend, divisor, and product. On the right, using four-bit arithmetic we enclose the equivalent decimal numbers in braces.

a. 4000 0000 / 8000 = (8000) + (0000) / 8000 {64/-8}
b. 4000 7fff / 8000 = (8000) + (7fff) / 8000 {71/-8}
c. c000 8001 / 8000 = (7ffe) + (8001) / 8000 {-55/-8}
d. c000 7fff / 8000 = (7fff) + (ffff) / 8000 {-57/-8}

CHAPTER 7

Superscalar Machine Principles

If we view the instruction flow and data flow in a computer system, all the machines can be grouped into three classes: SISD (Single Instruction Single Data), SIMD (Single Instruction Multiple Data), and MIMD (Multiple Instruction Multiple Data).[13] A SISD machine contains a scalar processor — each instruction can operate on one piece of data. A SIMD machine contains a vector processor as each instruction can operate on a vector, i.e. array of data. A MIMD machine means that many processors are interconnected. This chapter discusses the design principles of a superscalar machine. The discussions of SIMD and MIMD machines are found in the next chapter.

7.1 PARALLEL OPERATIONS

A superscalar processor is a processor that allows the executions of many instructions at the same time, but each instruction still operates on only one piece of data.

In terms of instruction functions, a superscalar processor is not different from a scalar processor except in speed and complexity. That is, a superscalar processor is able to run much faster because all the operations are performed in parallel. This parallelism includes memory too, so the CPU speed is balanced with its operand access time. A powerful CPU is usually supported by a memory system that has many components operating at different speeds. The goal is to achieve parallel operations at all levels. That is to say, the CPU and all the memory components must be kept busy. In practice, there are two approaches to designing a fast CPU, a pipe or a decoupled pipe. Let us explore the memory structures before discussing CPU design in a balanced system.

7.1.1 Storage Hierarchy

Memory and storage are synonymous. As shown in Figure 7.1, a hierarchical storage system commonly has four levels: the registers, cache, central memory, and disk. The first two levels are on chip and the next two levels are off-chip. As the level is closer to the CPU, the storage is smaller in size but faster in speed. The goal is to keep the storage system busy at all levels.

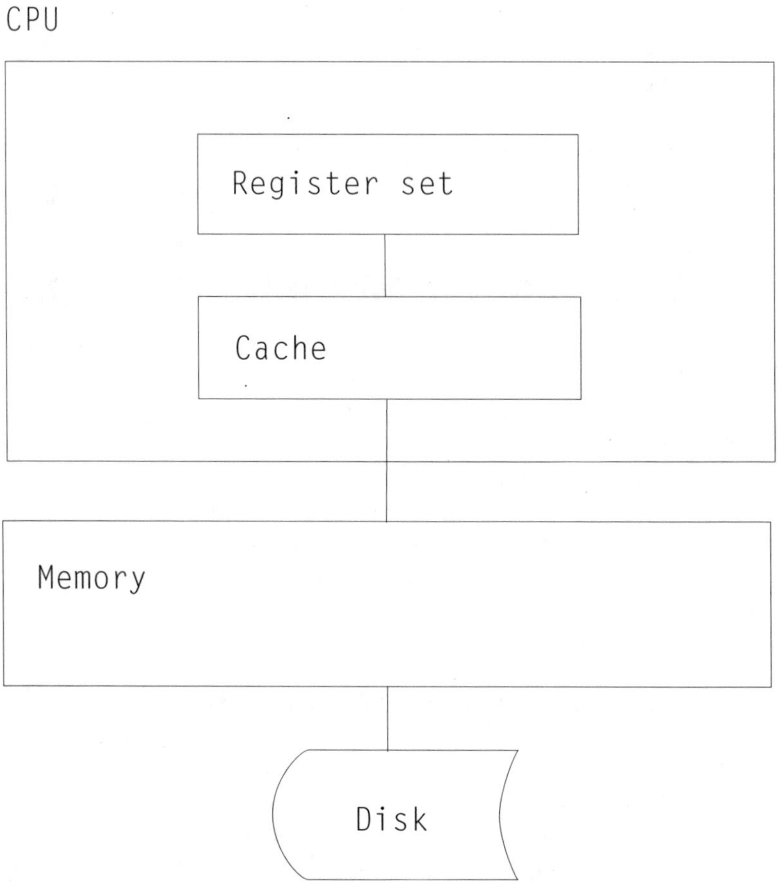

Figure 7.1 The four-level storage structure includes register set, cache, memory, and disk.

The first level includes registers and its access time is one clock, e.g., 2 ns at 500 Mhz. The second level is the cache memory, or cache for short. A cache on-chip is comprised of many high-speed temporary registers. Based on its design, the access time of a cache usually varies from one to two clocks. In concept, the cache is an extra layer containing the duplicate copy of a portion of memory that is transparent to a user program. The cache was first employed by the IBM 360/85 to shorten the memory access time. Generally, the bigger the cache, the slower the speed. In practice, the instruction cache is separated from the data cache in order to improve performance. That is to say, instructions are fetched from the instruction cache while data operandi are fetched from the data cache. Cache memory design will be discussed later in this chapter.

The third level storage is central memory where results are stored during computation. In general, a mainframe computer has 4 GB or more of memory, but a PC has about 128 MB to meet the requirements of graphic applications. The memory data bus can be 64 bits, but its bandwidth can be increased as described below.

7.1.1.1 Memory Address Interleaving

Figure 2 shows a central memory that has two banks. In the memory, the left bank contains blocks with even addresses and the right bank contains blocks with odd addresses. We say that the memory is two-way interleaved in regard to addresses. The purpose is to allow both banks to be busy at the same time. As a result, if the memory cycle time is shortened, its bandwidth is increased. Assume that each bank can provide 64-bit data in one memory cycle. That is, upon a memory request, a block address is placed on the address bus to access the eight-byte block. In order to fully utilize the two-way address interleaved feature, the memory system has two ports as each has its own memory bus. One port is connected to the CPU while the other is connected to an I/O processor.

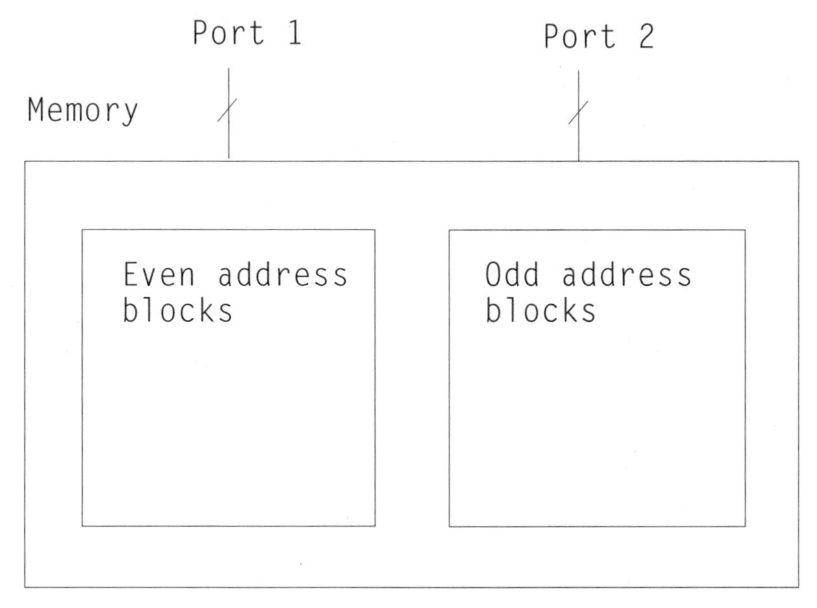

Figure 7.2 Two-way memory address interleaving. The left bank contains even address blocks and the right bank contains odd address blocks.

The memory is designed to operate in two modes as determined by the control line named opdsize (operand size): zero for 64 bits and one for 128 bits. That is, the CPU can retrieve a 64-bit block in one atomic cycle. Because of the two-way address interleaving feature, the CPU can also retrieve a 128-bit block on the same data bus in one atomic cycle. In doing so, the memory provides two data available signals: one for the first 64-bit block and one for the second 64-bit block. The CPU can issue a 128-bit data request by setting the opdsize bit to one while placing a block address on the bus. If the block address is even, the memory reads the even block first and the odd block after it. If the block address is odd, the memory reads the odd block first and the even block after it. When the first 64-bit block of data is ready, the memory places data on the bus and asserts its first data available signal. When the second block of data becomes ready, the memory places it on the same data bus and asserts its second data available signal.

Because the two banks can operate at the same time, the memory keeps a busy flag for each bank. Thus, the memory can allow two 64-bit data requests at the same time, and the arbitration logic in memory can grant a cycle as long as the bank is not busy. After receiving a request, the memory bank becomes busy and any subsequent request must be deferred until the previous cycle is completed. In sum, a two-way interleaved memory can deliver 128 bits in one cycle to double its bandwidth.

The fourth level storage is hard disk. A hard disk spins at a much faster speed than a floppy and has a much larger capacity. A hard disk with six stacked platters is depicted in Figure 7.3a. Each platter is a plastic plate coated with magnetic material. Except the top side and the bottom side, each platter in the middle uses both sides to store digital information. The six platters spin as one unit, so on the side there are 10 disk heads mounted on a mechanical arm to select a cylinder, i.e., a group of tracks with the same radius. This disk has many cylinders, each consisting of 10 tracks, as shown in Figure 7.3b. In order to select a particular cylinder, the arm and 10 heads move as one unit with precision. There is no arm movement to access the tracks in the same cylinder.

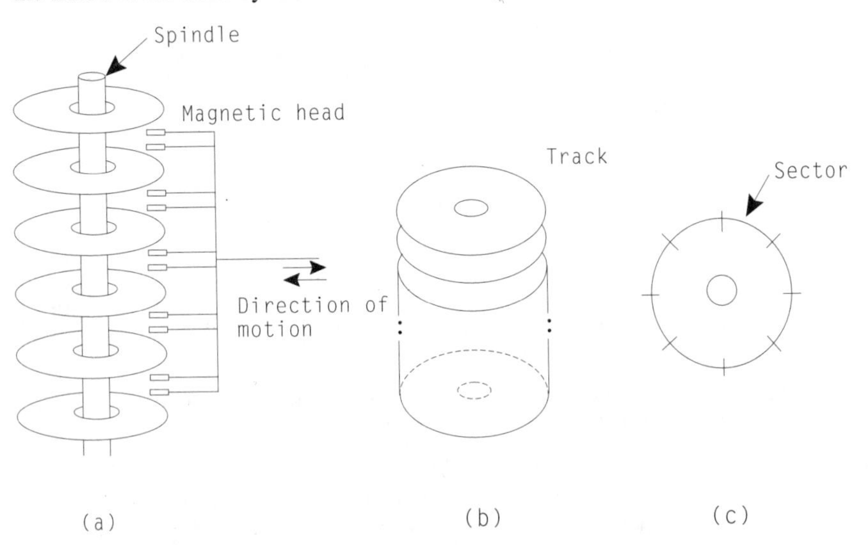

Figure 7.3 Physical layout of a hard disk: (a) six platters with one mechanical arm, (b) a cylinder with 10 tracks, and (c) a track with eight sectors.

A track or concentric ring on the disk surface is further divided into sectors, as shown in Figure 7.3c. A sector is a physical record of fixed size. In general, the physical record size ranges from 512B to 4KB, which is also the allocated I/O buffer size. Each time the I/O routine can read a physical record from disk or write a physical record onto disk. Usually, a high-performance disk has a bit density of 4 gigabit or more per square inch.

The latency time is the time required to move the mechanical arm from the current cylinder to the addressed cylinder, which obviously depends on the relative distance between the current track and the destination track. If the destination track is the same as the current track, the latency time is zero because it is in the same cylinder. From the innermost track to the outermost track, or vice versa, the latency time may be about .25. There are also high-speed disks with fixed heads, i.e., each track has a head at a fixed position, so the latency time is always zero.

The time to access the sector on a track varies as a linear function of the relative distance between the head position and the accessed record. The access time on average is the time to spin half a revolution. If a disk spins at 10,000 Rpm (revolutions per minute), the access time is 3 milliseconds:

$$\begin{aligned}\text{Access time} &= (1/2) \,/\, 10{,}000 \text{ Rpm} \\ &= (1/2) * 60 * 1000 \,/\, 10{,}000 \text{ ms} \\ &= 3 \text{ ms}\end{aligned}$$

7.1.2 Data Dependencies

A high-speed CPU may be simply designed as a three-stage pipe: instruction retrieval, decode, and execute. The third stage may contain multiple functional units, each of which has a different execution time. Thus, many instructions can enter the third stage as long as there are no data dependency problems known as hazards. The three data dependency conditions are listed below:

1. RAW (read after write).
2. WAR (write after read).
3. WAW (write after write).

The RAW data dependency problem occurs when the source operand of the current instruction is the destination of the previous instruction, as shown below:

```
R[2] <— R[0] * R[1]    ;I0 (instruction zero)
R[3] <— R[2] - R[1]    ;I1
```

Instruction I1 can enter the execution stage if the data in R[2] is ready, and the required resource, a functional unit, must be ready. Because it takes longer to execute I0, which is a multiply instruction, I1 can not enter the execution stage until I0 is completed, i.e., the data in R[2] is ready.

The WAR data dependency problem occurs when the destination operand of the current instruction is the source of the previous instruction, as shown below:

```
R[0] <— R[2] * R[3]    ;I2
R[3] <— R[2] - R[1]    ;I3
```

The execution of I3 changes R[3] which is the sopd1 of I2. Therefore, the machine state after executing I3 can not be made permanent until I2 is completed. To solve the WAR problem, use a temporary register to store the result of I3 in the execution stage. Next, retire instructions in program sequence so I3 is after I2.

The WAW data dependency problem occurs when the destination operand of the current instruction is also the dopd of the previous instruction, as shown below:

```
R[3] <— R[0] * R[2]    ;I4
R[3] <— R[2] + R[1]    ;I5
```

As far as programming is concerned, this is not correct because a register, after being written, should be referenced at least once before being written again. Nonetheless, the race problem is eliminated by using a temporary register to store the result of I5 in the execution stage. If I4 is retired after I5, the result in R[3] is the sum instead of the product.

7.1.3 Control Dependencies

The control dependency problem occurs when the instruction execution sequence changes. Therefore, any operations after this instruction in the pipe must be abandoned. Since an instruction changes the program counter, all the instructions retrieved after the instruction must be discarded. There are two types of branches: unconditional and conditional. A conditional branch, if not detected early, causes a bigger headache than an unconditional branch.

7.1.3.1 Unconditional Branches

Normally, the CPU preretrieves instructions in a queue so the instruction is ready when needed. This idea works fine without a branch condition. Because a branch condition changes the flow of instructions, any instructions in the queue after the branch must be discarded. In the CPU there are many different ways to trigger an unconditional branch. Fortunately, the event is deterministic, so actions can be taken immediately. That is, as soon as an unconditional branch condition is detected, the CPU flushes all the instructions in the instruction queue. Usually, the decode unit can signal this condition quickly. There are four ways to trigger a branch condition, as described below:

1. Unconditional branch instruction
2. Subroutine call and system call
3. Subroutine return and interrupt return
4. Interrupt

A system call is an instruction that triggers interrupt. There are many other types of interrupts. Whenever an interrupt occurs, the CPU saves environment, flushes instruction queue, and abandons all the operations after the current instruction. As an example, if a program check occurs, all the operations after the instruction that triggers the interrupt must be discarded because they are no longer valid.

7.1.3.2 Conditional Branches

When a conditional branch instruction executes, the CPU performs a logic test to determine whether or not to branch. The CPU can preretrieve instructions right after the conditional branch instruction. Because there is a greater chance of branching occurring than not occurring, the CPU should preretrieve the instructions at the predicted branch address. If the test condition is true, all the operations after the predicted branch address are valid. If the test condition is false, the program counter is increased by the length of the branch instruction, and any mispredicted operations must be abandoned. This is why in a programming loop, we should minimize the use of branch instructions, conditional or unconditional.

7.1.4 Multiple Functional Units

A functional unit (i.e., subunit) performs an operation on one or two registers. As shown in Figure 7.4a, many of the design concepts were pioneered by CDC6600.[65] First, execute instructions to load data from memory into registers. Next, execute instructions to operate on one or two registers. After that, execute store instructions to store results from registers to memory. As each instruction is retrieved in the IR, the decoder prepares information in decoded forms in an operation register. The decoded information includes opcode, addresses of source registers, and addresses of destination registers that may be temporary. If there is no data dependency problem and the functional unit is available, the operation register is passed to the functional unit. Consequently, the unit executes and places the result in a temp register. If there is no control dependency problem, the result is clocked into the destination register to make the machine state permanent.

Each functional unit may have a different execution time. If the unit is frequently used, it may be duplicated in quantities. In addition, many functional units can be grouped into one execution unit, so a fixed unit can handle integer add, subtract, shift, and logical operations. A multiply unit can handle multiplications, either fixed point or floating point. The function units are described below:

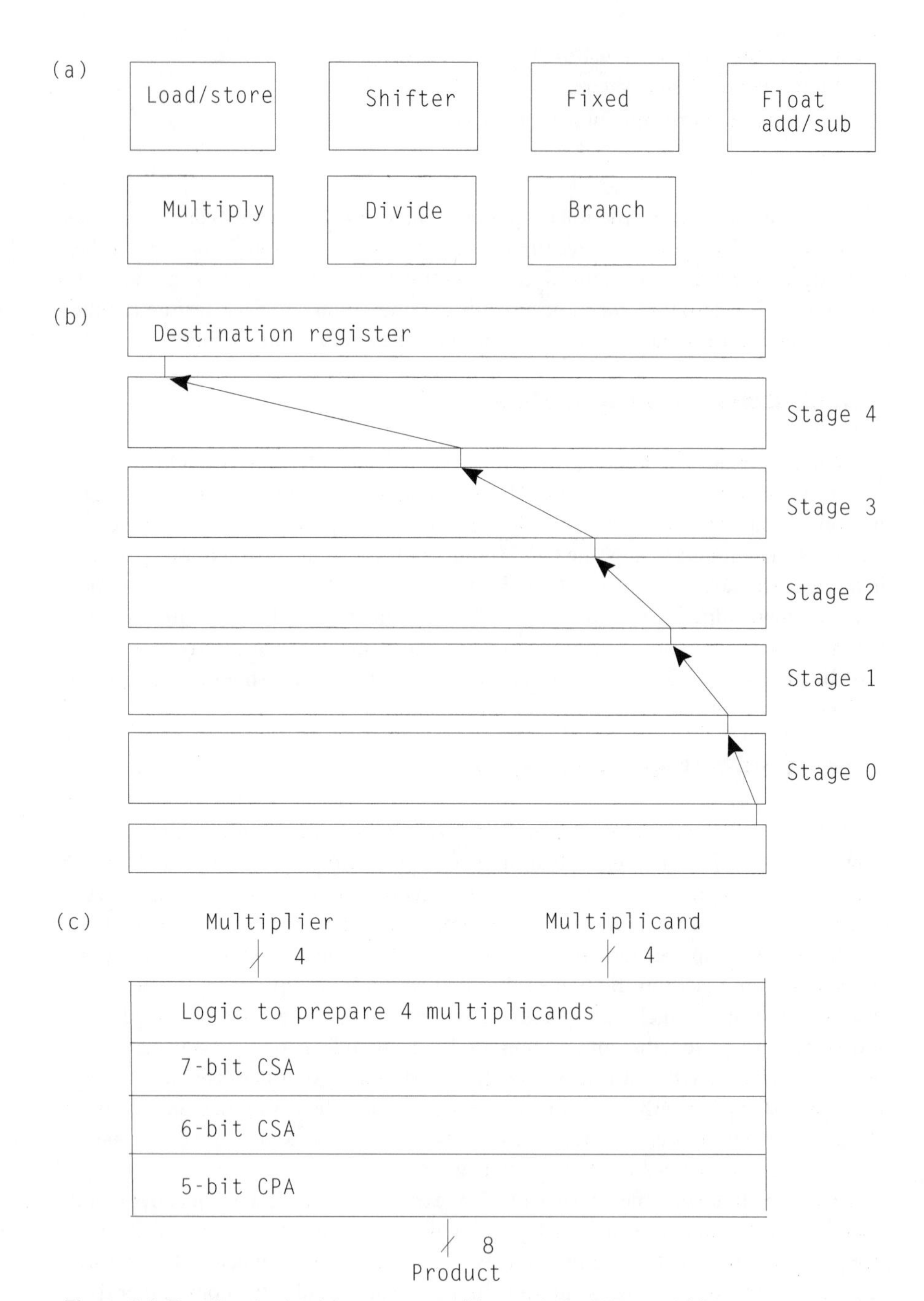

Figure 7.4 Functional units: (a) seven types, (b) one-clock shifter, and (c) four-bit multiplication tree.

1. Load/store

The load/store unit computes the EA (effective address) of the operand, initiates a read or write memory request, and handles the data transfer between a register and memory.

2. Shifter

If a five-bit shift count is given as (C4 C3 C2 C1 C0), where Ci is bit i, its count value ranges from 0 to 31 in decimal. The shifter logic can perform the shift operation, arithmetic or logical, in one clock. The shifter needs five stages of logic gates with no flip-flops. Based on the shift type, source operand, shift count, and destination operand, the shifter propagates the source operand through five stages of combinational logic and clocks the shifted result into the destination register. For illustration purpose, we use a DO loop to explain the logic below:

```
DO i = 0 to 4, step 1
   IF the Ci bit in the shift count is 0,
   THEN transfer stage i to stage i+1 in parallel with no shift;
   ELSE shift stage i into stage i+1 by a count of 2^i; ENDIF;
ENDDO;
```

If a count is (1 1 1 1 1) for a left shift, stage zero shifts left one (20) bit, stage one shifts two (21) bits, stage two shifts 4 (22) bits, stage three shifts eight (23) bits, and stage four shifts 16 (24) bits. Altogether, the source is shifted 31 bits to the left as indicated by the five skewed arrows shown in Figure 7.4b.

3. Fixed

The fixed unit handles fixed point add and subtract as well as logical operations in one clock. Since twos complement notation is used to represent a negative integer, the unit mainly consists of a carry look-ahead adder with logic to handle AND, OR, and EOR operations. The unit accepts two inputs and clocks the result into a temp register that is retired into the destination register at a later time.

4. Float add/sub

The float add/sub unit handles floating point add or subtract operations. The unit consists of a carry look-ahead adder plus logic to perform many operations in parallel. Before feeding the two operandi to the adder, the biased exponents are compared. Shift right the significand with a smaller exponent and increase its exponent. Repeat this process until the two exponents are equal. Then perform the add operation and normalize the result.

5. Multiply

The multiply unit has a multiplication tree that can add all the unsigned or signed multiplicands in five clocks. The idea was derived from a Wallace tree that uses many cascaded CSAs (carry save adder) followed by a CPA (carry propagated adder)67. Each CSA comprises many independent one-bit full adders with no interconnections. At each bit position the CSA accepts three input bits and generates two output bits, i.e., sum and carry. In other words, each CSA can add three binary numbers and produce the bit-wise logical sum vector and the bit-wise logical carry vector. The term carry save really means that the carry vector is saved for a subsequent operation. The key concept is that in the final stage, a CPA with carry look-ahead logic must exist in order to add the two binary numbers into an arithmetic sum.

As shown in Figure 7.4c, the four-bit multiplication tree has a logic box, a seven-bit CSA, a six-bit CSA, and a five-bit CPA. Except for the first one, each multiplicand is properly shifted to the left as determined by the bit in the multiplier. If the multiplier bit is zero, the multiplicand is not enabled, so it is zero. Because the LSB of the first multiplicand is the rightmost product bit, the next adder width is reduced by one bit. In fact, each CSA accepts three inputs and generates two outputs, so it reduces one row. In order to obtain an arithmetic sum, a CPA in the final stage adds the final bit-wise carry after being skewed one bit to the left to the final bit-wise sum. In this case, the CPA in the final stage needs only five bits. For example multiply the unsigned 15 by an unsigned 15, yielding a product of 225.

	0	0	0	0	1	1	1	1	1st multiplicand	(15)
	0	0	0	1	1	1	1		2nd multiplicand	(30)
	0	0	1	1	1	1			3rd multiplicand	(60)
+	0	1	1	1	1				4th multiplicand	(120)
	1	1	1	0	0	0	0	1	eight-bit unsigned product	(225)
	P7	P6	P5	P4	P3	P2	P1	P0		

The binaries are on the LHS, while their decimal equivalents are on the RHS. We examine the CSA output stage by stage. Because P0 (product bit zero) is the LSB of the first multiplicand, the first seven-bit CSA generates the two vector outputs as shown below:

0	0	1	0	1	1	0	SV1 (sum vector 1)
0	0	1	1	1	1		CV1 (carry vector 1)
0	1	1	1	1			4th multiplicand

After the first CSA reduces one row, the second six-bit CSA can handle the fourth multiplicand. Because P1 (Product bit 1) is the rightmost sum bit from the seven-bit CSA, the second CSA needs only six bits. Similarly, because P2 (product bit two) is the rightmost sum bit from the six-bit CSA, the final CPA needs only five bits. The two vector outputs from the six-bit CSA are added by the five-bit CPA to generate the five-bit arithmetic sum as shown below:

	0	1	1	0	1	0	SV2 (Sum Vector 2)
+	0	1	1	1	1		CV2 (Carry Vector 2)
	1	1	1	0	0		Arithmetic sum from CPA
	P7	P6	P5	P4	P3		Upper five-bit product

As we can see for unsigned mul, the logic box on top prepares four multiplicands under two conditions. First, each multiplicand is zero extended. Second, each enabled multiplicand is properly shifted to the left as determined by the bit position in the multiplier. Interestingly, the same tree can handle four-bit signed multiply in one clock. Recall that all the shifted multiplicands should be sign extended. In addition, the sign bit in the multiplier indicates that, in concept, the twos complement of the multiplicand should be input to the six-bit CSA. A four-bit signed multiply example is left as an exercise. We say that a 32-bit multiplication tree has 30 cascaded CSAs and the final stage has a 33-bit CPA.

For floating point multiply, the same multiplication tree can be used in addition to an adder and temporary registers. The tree has a data path of up to 64 bits to support single, double, and extended precisians. Again, many operations are done in parallel. Add the two biased exponents and subtract one bias from the result. At the same time, multiply the two unsigned significands. Finally, normalize the product and adjust the biased exponent accordingly.

6. Divide

It is interesting that the fixed divide instruction is not supported by Alpha and CDC6600. On such a machine, the programmer must convert the integers to float first, perform float divide, and convert the quotient back to integer. However, we can design a fixed divide unit that is hardwired using gates and flip-flops. The unit employs a loop, and the subtract and shift operations are performed in one clock each time they complete the loop.

For a float divide operation, the fixed divide unit is used along with logic to handle exponent arithmetic. In consequence, the divide time is fixed, and for example, the 32-bit float divide operation requires 40 clocks on a Pentium.28

7. Branch

If the test and branch executions are grouped as one instruction, the branch functional unit compares the test condition first. If the tested condition is true, the unit sends the target branch address to the IU. Consequently, the IU fetches the instruction stream at this new address. If the test condition is false, the conditional branch instruction is treated as a nop, so the next instruction below the branch is retrieved.

Since the functional units in the EU (execution unit) can operate in parallel, we can design a decoupled pipe so the CPU can perform out-of-order executions regardless of whether it is hardwired or microprogrammed. Before discussing this a clever idea, we introduce the pipelined CPU, which handles in-order executions.

7.2 PIPELINED CPU

For the sake of simplicity, a hardwired five-stage pipe is shown in Figure 7.5a with the following assumptions:

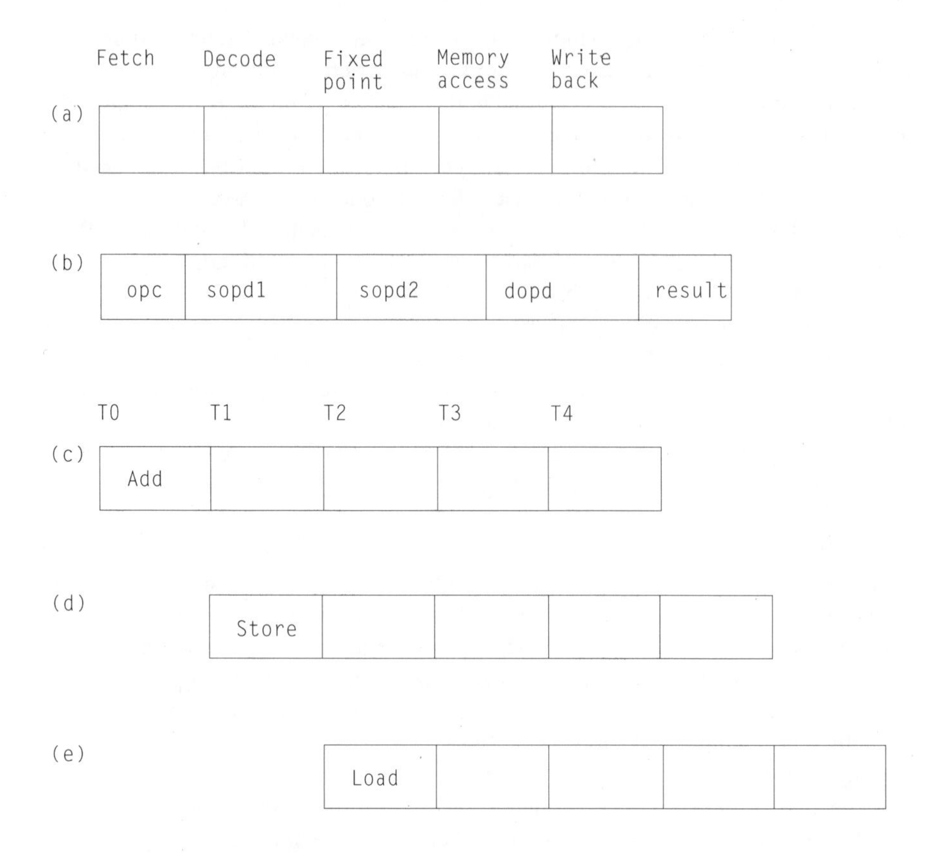

Figure 7.5 Pipelined CPU: (a) a hardwired five-stage pipe, (b) the mop register, (c) add instruction in pipe at T0, (d) store instruction in pipe, and (e) load instruction in pipe.

1. The CPU is hardwired so that only register operations are allowed. That is, a load instruction must be issued to load a memory operand into a register. After an operation, a store instruction may be issued to store the result from register to memory.
2. Each instruction is executed in five phases or less.
3. Each phase executes in one clock.

The five stages are retrieval, decode, fixed point, memory access, and write back.[19] Only in-order operations are allowed, so each stage indicates a progress point regarding the execution of an instruction. Load and store instructions actually go through the memory access phase but other instructions just bypass it. Five instructions can coexist in the pipe with in-order logic flow.

The retrieval unit brings an instruction into the IR. The decode unit prepares a mop (micro operation) register that has many fields, as shown in Figure 7.5b. The opc (operation control) is in decoded form, but the addresses can be either in decoded or encoded forms as dictated by semiconductor technology. For RR (register-register) operations, sopd1 (source operand 1) and sopd2 (source operand 2) are used to select two registers as inputs to the fixed point unit (i.e., ALU). For a load or store operation, sopd1 and sopd2 are also inputs to ALU, but sopd1 is the index register and sopd2 represents a register to store the memory address. That is to say, the decode unit copies the address field from the IR into a temp register that is named sopd2. The result field in mop specifies a temp register to store the result from ALU. The dopd (destination operand) is a different register that is used to store the result from ALU, the MD (memory data) before write, or the MD after read. The three mop registers are lined up in a queue, and each one is passed to the next stage that may use different fields coded in the mop. By executing three instructions, add, store, and load, we can describe what happens in the pipe.

1. At T0 (time 0), i.e., the first clock cycle, the add instruction is retrieved into the IR as shown in Figure 7.5c. At T1, the decode unit prepares opc, sopd1, sopd2, result, and dopd in a mop register. At T2, the mop register is passed to the fixed point unit that performs integer add. At T3, the result remains unchanged (i.e., nop). At T4, the write back unit copies result to dopd to make the machine state permanent.
2. At T1, the store instruction is retrieved in the IR, as shown in Figure 7.5d. At T2, the decode unit prepares opc, sopd1 (index), sopd2 (address), result, and dopd. At T3, after adding the index register to the address, the pipe places the EA (effective address) in the result. At T4, the memory access stage issues a memory write by placing the result on the address bus and dopd on the data bus. At T5, the write back unit does nothing.
3. At T2, the load instruction is retrieved in the IR, as shown in Figure 7.5e. At T3, the decode unit prepares opc, sopd1 (index), sopd2 (address), result, and dopd. At T4, the fixed point stage adds the index to the address to form the EA in the result. At T5, the memory access stage issues a memory read by placing result on the address bus and by obtaining the MD from the data bus after the cycle. At T6, the write back unit copies the MD to dopd to make the machine state permanent.

Conceptually, a pipelined CPU is like an automobile assembly line. The line receives a car body frame from one end, assembles different components at different stages, and outputs a car at the other end. Because the cars are of the same type, all the operations can be performed on a continuous basis. The assembly line is full all the time except for occasional interruptions. In contrast, a pipelined CPU can not be full all the time due to data dependency or branch dependency problems. In other words, holes are developed in the pipe from time to time. Therefore occasionally, certain stages are left idle for a few clock cycles. In an ideal case, the pipe has five instructions all the time and each instruction is progressing in a stage.

7.2.1 CPU Speed

If instructions and operandi are retrieved in advance into the CPU, as discussed later in this chapter, it is reasonable to assume that each stage takes one clock cycle to perform its function in the pipe. Thus, a total of five clocks are needed to execute one instruction on a unpiped CPU, but it takes one clock on average to execute the same instruction on a piped CPU. This is because five instructions are executed in parallel to achieve the ideal execution rate, as computed below:

$$\text{Ideal CPU rate} = \frac{\text{Number of clocks issued to execute an instruction on a unpiped CPU}}{\text{Number of stages in the pipe}}$$
$$= 5 / 5$$
$$= 1 \text{ clock/instruction}$$

7.2.2 Clocks per Instruction

We define CPI (clocks per instruction) as the average number of clocks required to execute an instruction on piped CPU. Assume that when the pipe is 100% full, it takes one clock cycle to execute one instruction, defined as the ideal CPU rate. If the pipe is 80% full on average, its utilization factor is 80%. As a result, it takes 1.25 clocks on average to execute one instruction, as computed below:

$$\text{CPI} = \text{ideal CPU rate} / \text{utilization factor}$$
$$= 1 / .8$$
$$= 1.25$$

7.2.3 Million Instructions per Second

The raw speed of a CPU is often measured in terms of MIPs (million instructions per second) as defined below:

$$\text{MIPS} = \text{Clock rate in Mhz} / \text{CPI}$$

If the CPU supports a clock rate of 500 Mhz with a CPI equal to 1.25, it is a 400 MIPS machine, as computed below:

$$\text{MIPS} = 500 / 1.25$$
$$= 400$$

7.2.4 Million Floating Operations per Second

If computations are numerically oriented, it is necessary to measure the speed in terms of MFLOPS (million floating operations per second) as defined below:

MFLOPS = clock rate in Mhz/clocks per FLOP

That refers to how fast its floating point unit can operate. If it takes an average of two clocks to execute a FLOP (floating point operation), add or subtract, it is a 250 MFLOPS machine, as computed below:

MFLOPS = 500 / 2
= 250

We can use a stop watch to time a benchmark that is composed of fixed point, floating point, logic, load, store, and branch instructions in a loop. Benchmark results[70] and tools[71] are free as they are available from the Internet. It should be stressed that CPU speed alone can not decide the performance or sales of a computer system. Other factors include reliability, system software, application software, cost, marketing, support, momentum, etc.

7.3 CACHE MEMORY

In a conventional memory, the address is not stored as part of the word, so the full address is decoded to select the word or block. In contrast, the cache memory or cache is an associative memory, i.e., content addressable. That means each cache entry, either a word or a line, contains both address and data as shown in Figure 7.6a. Depending on its design, the address field may contain a full memory address or part of an address. Its associated data field contains the copy or mirror image of a block in memory. Conceptually, the cache serves as an extra layer between registers and central memory not seen by the programmer. As shown in (b), the fully associative cache memory has four entries: E0, E1, E2, and E3. The entry E0 contains A0 (Address 0) and D0 (Data 0). For entry E0, the key is said to be A0 and its associated data block is D0.

7.3.1 Cache Hit vs. Cache Miss

During a read cycle, the cache compares the given address with the key in a cache entry. If a match is found, it is defined as a cache hit so its data field is retrieved. If a match is not found, it is defined as a cache miss, meaning the data block of interest in not ready in cache. As a result, the cache memory reads the central memory. Any old data to be replaced must be written into central memory. The data read from central memory and its address are inserted into the cache entry to complete the cache read cycle.

(a)

Address	Data

(b)

	Address	Data
E0	Address 0	Data 0
E1	Address 1	Data 1
E2	Address 2	Data 2
E3	Address 3	Data 3

(c)

	Address	Data
E2	Address 2	Data 2
E0	Address 0	Data 0
E1	Address 1	Data 1
E3	Address 3	Data 3

(d)

	Address	Data
E4	Address 4	Data 4
E2	Address 2	Data 2
E0	Address 0	Data 0
E1	Address 1	Data 1

Figure 7.6 Cache memory entries: (a) the format of an entry, i.e., slot, word, or line, (b) a fully associative cache has four entries: E0, E1, E2, and E3, (c) pull out the matched entry E2 and push it on top, and (d) push the new entry E4 on top.

7.3.2 Fully Associative

Fully associative means that the address fields of all the cache entries are searched at the same time so the key field contains a full blown address. If a central memory has 64 GB, its physical address is 36 bits. To read a block of eight bytes, its block address is 33 bits (36 - 3). During a cache read cycle, the 33-bit address is used to match the key field in each cache entry. Depending on the applications, a cache entry may contain some extra bits for special control functions.

7.3.3. Least Recently Used Algorithm

Because a program has the property of locality, a cache should not contain the LRU (least recently used) entries for they are not likely to be accessed again. If the cache has four entries, the cache can be designed as a hardware queue that ensures the usage ordering from top to bottom. The top is always the MRU (most recently used), then the next entry, the third entry, and the bottom entry is the LRU (least recently used). The LRU replacement algorithm throws away the LRU entry in the cache as described below.[38]

```
IF a match is found:
THEN Pull the matched entry out and push it on top;
ELSE Issue a central memory request to read the new entry;
  Wait until the new entry arrives, then push it on top;
ENDIF;
```

Due to the property of flip-flops, all the cache updates can be done in one clock cycle. In Figure 7.6c, pull out the matched entry E2 and push it on top. Thus, the entries E0 and E1 drop one level and E1 fills the hole once occupied by E2. Later, if a new entry E4 needs to be inserted, push it on top and the bottom entry E3 falls out as shown in Figure 7.6d.

If the fully associative cache has four entries, we need four hardware comparators as shown in Figure 7.7. Given a 33-bit block address as the key, the 8B block can be fetched in one clock. Henceforth, the address is compared with the key of every entry at the same time. The MAR (memory address register) contains the 33-bit address as the search key, and each comparator has 33 EOR gates. If a match is found, the comparator fires a signal to select the data field of the matched entry. Only one comparator can fire, so the data outputs of the E (enable) gate are OR-tied to the MDR (memory data register), which is connected to the 64-bit cache data bus. Because the full address is the key, the cache word length is 97 (64 + 33) bits. A fully associative cache has a speed advantage, but its smaller size is its limitation. Note that in a virtual memory system, the processor uses a fully associative TLB (translation look-aside buffer) that contains the MRU translation records. The speed of the TLB must be fast, as discussed later.

There are two other cache designs: direct mapping and set associative. The L1 instruction cache may use direct mapping, while the L1 data cache uses set associative search. In either case, the data operand in the cache may vary from one byte to four bytes. The L2 cache is usually off-chip and it is an optional design in the system. Usually, the L2 cache uses set associative search and its entry is an 8B block containing both instructions and data. To speed up instruction retrieval, we may also have the instruction queue that is designed as a parallel shift register. The instruction queue contains the IR as a subset left justified. That is, if the queue contains an instruction, it can be clocked into the IR in 2 ns or less. Therefore, the instruction queue can be thought of as the L0 instruction cache as introduced next.

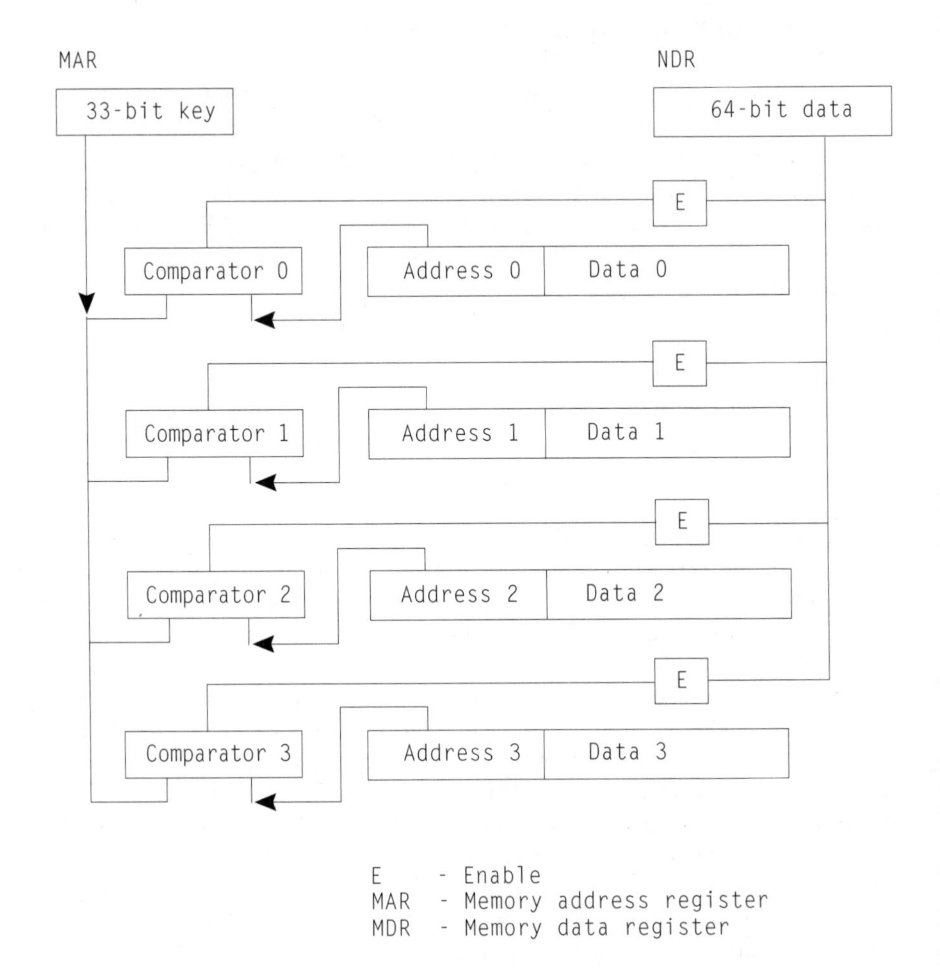

Figure 7.7 Four comparators operate in parallel in a fully associative cache.

7.3.4 Instruction Queue

If the instructions can have two different lengths, 16 bits or 32 bits, the alignment problem is intricate. This is especially true if an instruction is allowed to cross the boundary of a quadword. To alleviate the problem, an instruction queue can supplement the instruction cache. That is, if a cache miss is encountered, instructions are retrieved from memory into the L1 instruction cache and then loaded into the instruction queue.

For example, the instruction queue is an 8W parallel shift register such that 16 or 32 bits can be shifted in one clock. As shown in Figure 7.8a, the queue is empty due to a branch condition so its hole size is 8W. The hole size is kept in a four-bit counter, so 1000 in binary means that the queue is 100% empty and 0000 means that the

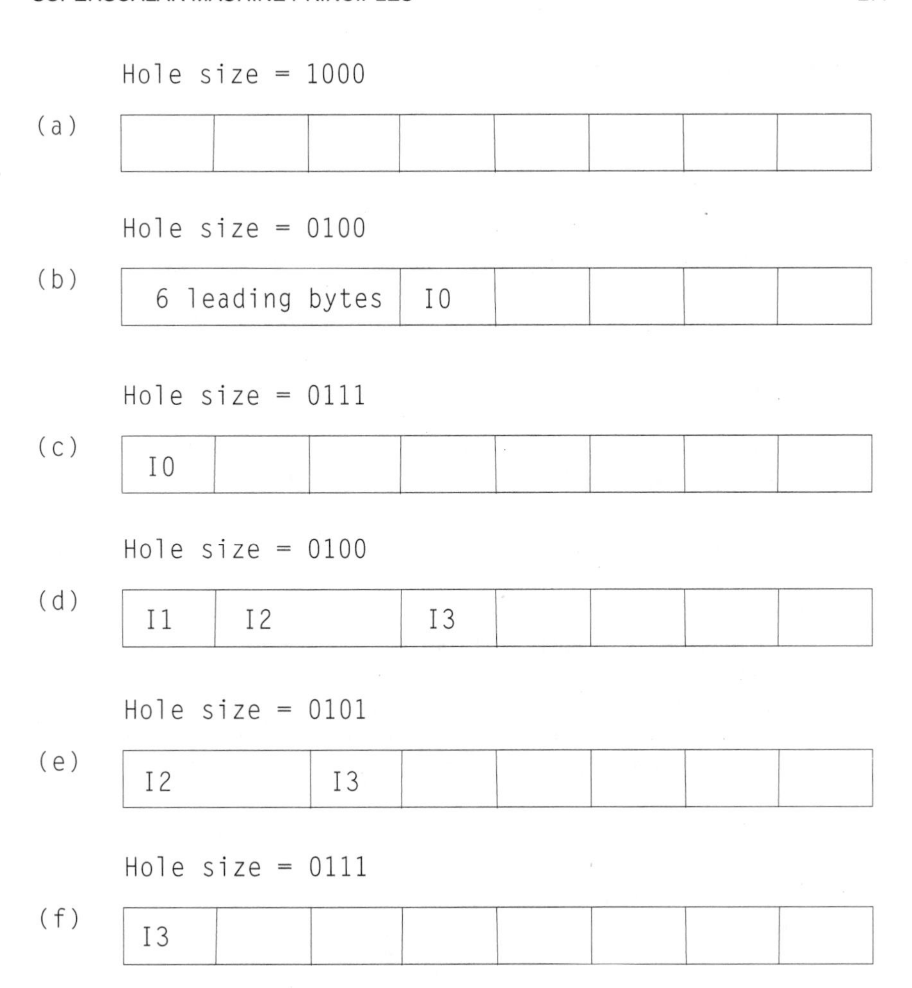

Figure 7.8 The holes in an instruction queue: (a) after a branch, (b) after loading an instruction block, (c) after shifting the leading bytes, (d) after shifting I0, (e) after shifting I1, and (f) after shifting I2.

queue is 100% full. A number in between means that the instruction queue is partially full. The leftmost field in the queue is the IR where the decoding circuit is hardwired. Any instruction must be left justified in the queue before decoding can start. After decoding and shifting the queue, the hole is increased by the instruction length of one or two words. The hole size also tells the starting position in the queue where the next instruction block is being loaded. After loading a 4W block, the hole size is decreased by four.

7.3.4.1 Instruction Queue Management:

```
IF a branch condition is encountered,
THEN
Flush the queue and set the hole size to 8W;
retrieve and wait for the 4W instruction block from the instruction cache;
After loading the block into the hole left justified, decrement the hole size
by 4;
    IF the branch address starts in the middle of block,
    THEN
    Shift left the queue x bytes where x is the lower 3-bit integer
    in PC (i.e., instruction address register);
    Increase the hole size accordingly;
    ENDIF;    {Now, the instruction is left justified in IR.}
ENDIF;
IF the queue contains a complete instruction,
THEN
After decoding, shift left the queue and increase the hole size accordingly;
    IF the hole size is greater than or equal to 4 in words,
    THEN
        IF the next 4W instruction block is ready in the instruction queue,
        THEN load the block into the queue and decrement the hole
        size by 4; ENDIF;
    ENDIF;
ELSE
Stall decoding;
Wait for the next 4W instruction block to be loaded;
ENDIF;
```

Let us trace the instruction queue and its hole size. To begin with, a branch condition generates an 8W hole, as shown in Figure 7.8a. After a 4W instruction block is loaded into the hole left justified, the hole size is reduced to 4W, as shown in Figure 7.8b. For example, assume that I0 (instruction 0) is retrieved at the branch address that has a lower three-bit 110. As I0 must reside in the IR before decoding can start, one clock is wasted to shift left the queue by 3W (i.e. six bytes) as shown in Figure 7.8c. Accordingly, the hole size is increased by three to become seven. After decoding I0, the queue is shifted to the left, so the hole size is eight. However, at the same time the next 4W instruction block is loaded, as shown in Figure 7.8d and the hole size is four. Now, we have I1 (instruction 1), I2, and I3 in the queue. After decoding I1, shift left the queue by one word so the hole size is five as shown in Figure 7.8e. After decoding I2 that is two words long, the hole size becomes seven as shown in Figure 7.8f. In sum, the instruction cache always anticipates the next block in order to fill the hole. Decoding continues as long as the instruction queue has a complete instruction. As soon as the hole is greater or equal to 4W, the cache tends to load the next 4W instruction block left justified and reduce the hole size.

7.3.5 Instruction Cache

The L1 instruction cache may have an access time of two clocks with a cache line of eight bytes. The CPU looks into the instruction cache to fill up the instruction queue. The instruction cache always looks ahead to retrieve a memory block of 16 bytes upon a miss. If the target instruction at a backward loop branch resides in the instruction cache, the target instruction can be obtained in two clocks. How is an instruction cache designed? First, the number of entries in the cache is determined. A search technique is selected. If each cache entry contains an 8 B data field, then 64 entries would provide a 512 B block of instructions (deemed adequate for most applications). Because the instruction stream is obtained from one segment at a time, The direct mapping search method is more attractive.

7.3.5.1 Direct Mapping

If the physical block address has 33 bits, the upper 27 bits are stored as the key, and the lower six bits are decoded to directly select a cache entry. This entry contains only a partial address because its lower six bits are implied to obtain the block. That is, the six lower bits constitute an index to the instruction cache. Because the six lower bits are decoded to select the entry, the cache requires only one comparator. After the comparison, if a match is found, it is a hit; otherwise it is a miss. The format of a directly mapped entry is shown in Figure 7.9a. The entry of 91 (64 + 27) bits includes a 27-bit key and a 64-bit data.

In Figure 7.9b, the cache has 64 entries, E0, E1, ..., E63. The search operation is done in two steps. First, the lower six bits in a given block address are decoded to select the entry into the MDR. Next, the upper 27 bits in the MDR are compared with the key field in the MAR. If it is a hit, the 64-bit data field is enabled as output as clocked into the instruction queue. If it is a miss, the cache then retrieves two 64-bit instruction blocks from central memory at the given address. The two blocks are then stored into two consecutive entries in the cache. The index to the first entry is determined by the lower six bits in the block address. After adding one to it, we obtain the index to the second entry.

Because the instruction cache only supports read operations, this does not mean that instructions in a program can not be modified. In fact, any instruction can be modified just like data if it is loaded into a data register first. This leads to our discussion of a data cache which handles both load and store operations. In other words, if the data block in the cache is written, its memory image must be changed accordingly.

7.3.6 Data Cache

An L1 data cache which can use a two-way set associative search technique resides on the processor chip. If a data cache entry has eight bytes, a 16 KB cache may be divided into 1 K sets and each set has two entries grouped as a pair. The indi-

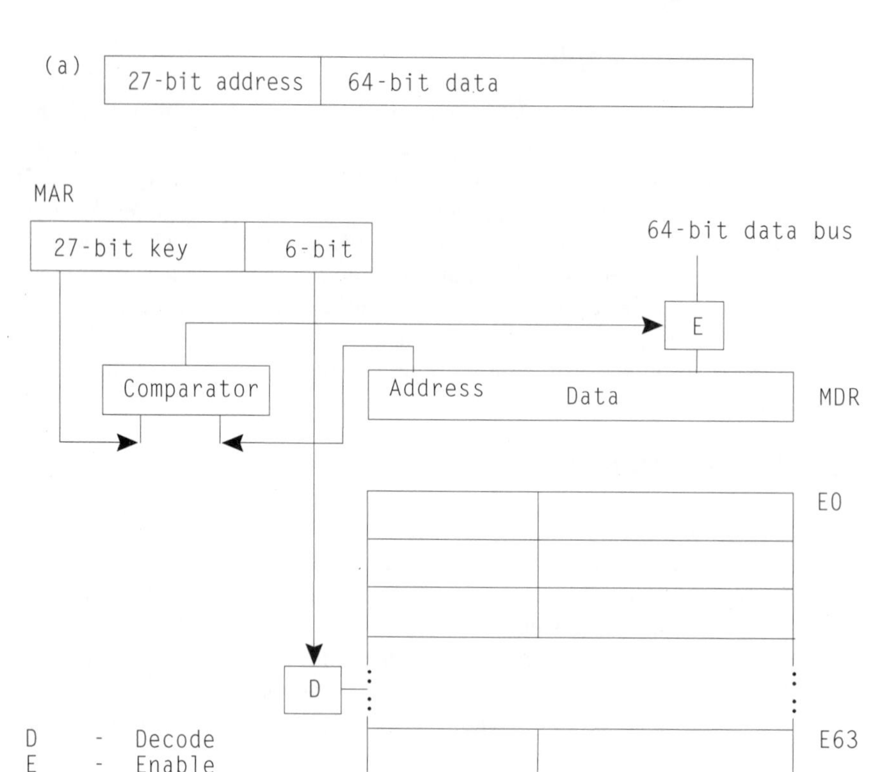

Figure 7.9 Direct mapping cache: (a) the 91-bit entry and (b) one 27-bit comparator for 64 entries.

vidual data operand may be eight bits, 16 bits, 32 bits, or 64 bits. Because the data cache is on-chip, it is labeled as the L1 data cache. Assuming that a word is aligned in word boundary, a double word in double word boundary, and a quad word in quad word boundary, the data cache must handle the alignment problem when accessing the operand. That is to say, the central memory can access a 64-bit block but the L1 data cache can handle an operand of different size. In order to access an 8 B block in central memory, only the upper 33 bits of the 36-bit byte address are needed on the address bus. In order to access an operand in the L1 data cache, the full 36-bit byte address must be given, so the upper 33 bits are used to access cache entry, and the three least significant bits are further decoded to select the byte level in the block. In other words, the lower three bits of the 36-bit byte address and the generic data type jointly specify the offset and size of the operand in the cache entry.

7.3.6.1 2-Way Set Associative

Typically, a data cache is larger than an instruction cache, so its design is somewhat different. In a program, data are usually obtained from two segments, static and dynamic. A static segment contains read only data while a dynamic segment contains writable data. It is possible for two pieces of data to both be recently used, and to have the same lower 10 bits in their block addresses. Both data should be in the cache, so we design a two-way associative set. Because each set contains two entries, the two-way set associative cache uses two comparators. Similarly, in a four-way set associative cache there are four entries in a set, so it needs four comparators. In concept, the set associative cache is a compromise between direct mapping and fully associative. The fully associative cache has one set with multiple entries, and the direct mapping cache has many sets and each set has only one entry.

In a two-way set associative cache, the 16 KB memory is divided into 1 K sets, and each set has two entries. As shown in Figure 7.10a, each entry has 88 bits: a 23-bit key, a 64-bit data, and the dirty bit. As shown in Figure 7.10b, we have 1024 sets, denoted as S0, S1, ..., and S1023 and two 23-bit comparators. The MAR contains a full 36-bit byte address divided into a 23-bit key, a 10-bit set address, and a three-bit byte address. Because a set has two entries, the middle 10 bits in a 33-bit block addresses can be indexed into the set of two entries at the same time. That is to say, the middle 10 bits are fed to D (decoder) to select the set that is loaded into the MDR of 176 (88 * 2) bits. Next, the 23-bit key in the block address is compared with each of the two keys in the MDR at the same time, a process called set associative search. If any one of the comparators fires, the data field of the selected entry is enabled to the output logic box. In sum, the set associative search is performed in two steps. First, the lower bits in a block address are decoded to select a set of multiple entries. Second, the upper bits in a block address are compared with the key in each entry to determine a hit or miss.

Let us try to read a byte from the data cache. In the MAR, the lower three bits are used to locate the byte in the data field if it is ready. The middle 10 bits are decoded to select a set of two entries in the MDR. Then, the upper 23 bits in the MAR are compared with the two keys in the MDR in parallel. If a match is found, the fire signal enables the selected data field to the enable gate. The lower three bits in the MAR are then decoded to select the byte in the data field of the matched entry. After a write operation the dirty bit in the entry is set, so a store operation is in order. In general, there are two methods to write a dirty entry into memory as described below.

7.3.6.2 Write-Through

The write through method means whenever the entry is written in, the dirty bit is set and the cache immediately writes the entry into memory. Unfortunately, if the same word is written 10 times by an instruction in a loop, the write through opera-

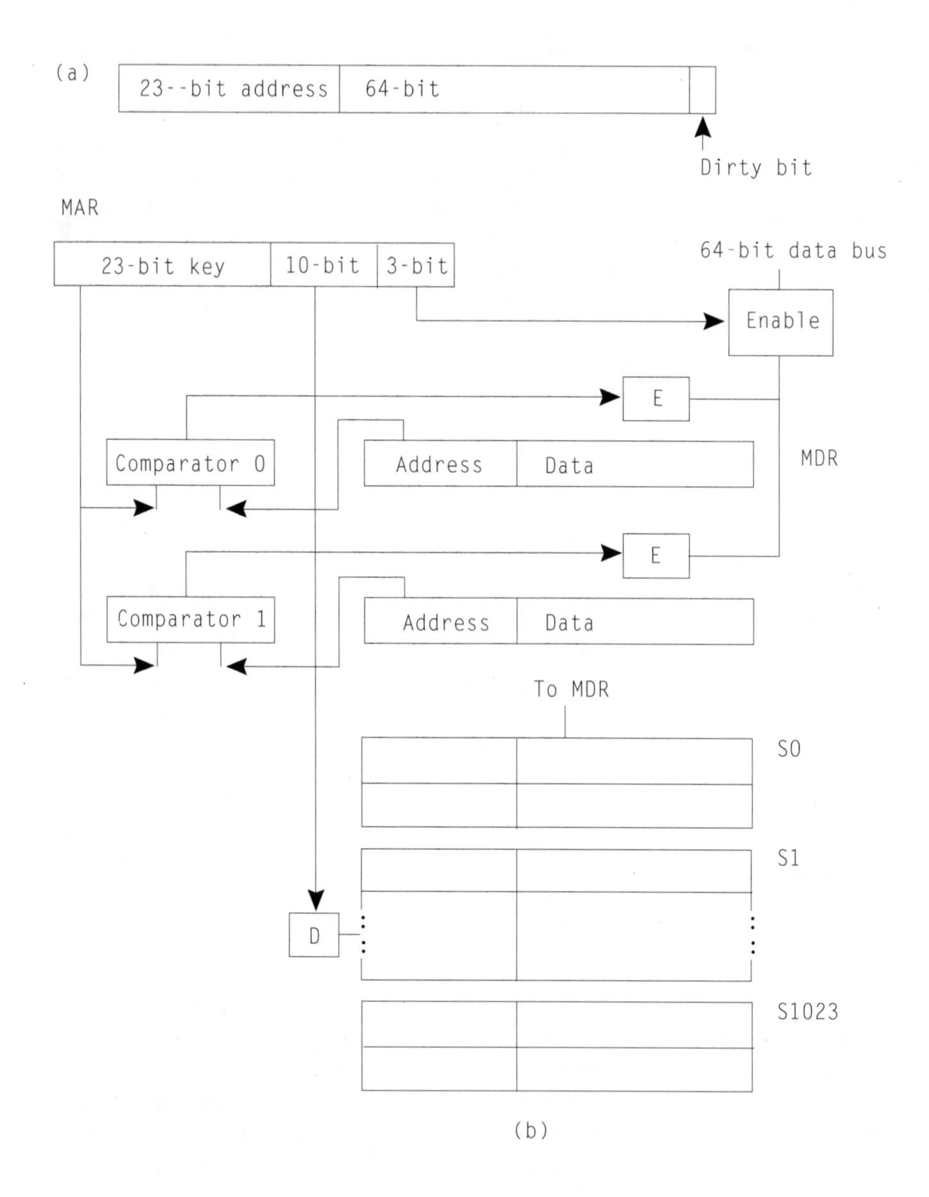

Figure 7.10 Two-way set associative cache: (a) the 88-bit entry and (b) two comparators for 1 K slot pairs.

tion is performed a total of 10 times. Such a design is simple but initiates too many write operations. A lot of memory bandwidth is wasted. As an alternative, the store operation can be deferred until the entry is to be replaced by a new one.

7.3.6.3 Write Back

The write back method is clever in that when the entry is written in, the dirty bit is set but the cache defers the write operation until the entry is ready to be replaced by a new one. That is to say, the data cache does not write the entry into memory each time it is modified. The data cache uses extra logic to defer the write operation, i.e., the data cache writes back the dirty entry only when it is ready to be replaced. Thus, the cache may write the central memory 10 times in write through mode, but only once if operating in write back mode. That is, a data cache can be turned on (enabled) or off (disabled). If the cache is on, it can operate in either write through or write back mode.

Finally, we study how to replace an entry in a two-way set associative cache. With two entries, each set can be designed as a queue using the LRU replacement algorithm. As shown in Figure 7.11a, the set has two entries, E0 and E1.

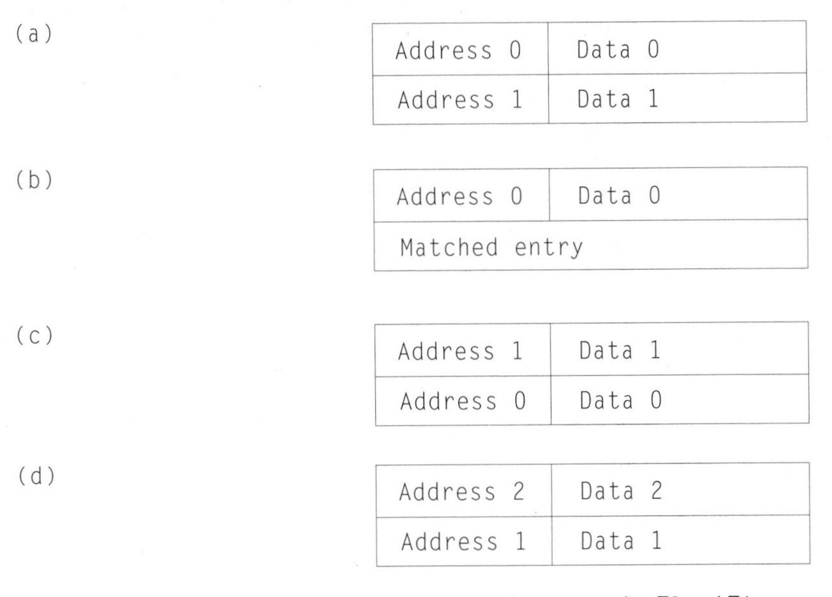

Figure 7.11 Replacement in a two-way set: (a) the set has two entries E0 and E1, (b) the bottom entry is matched, (c) after interchange, and (d) push the new entry E2 on top.

7.3.6.4 Replacement Algorithm in a Two-Way Set Associative Cache

```
CASE of the matched entry;
Top:  Take no action;
Bottom:  Interchange the top and the bottom;
  Write the interchanged pair to cache;
None:     Retrieve and wait for the new entry from central memory;
  Push it on top and the bottom falls out;
ENDCASE;
```

Of the two entries in the set, the top is the MRU. Thus, if the top entry E0 is matched, take no action. If the bottom entry E1 is matched as shown in Figure 7.11b, interchange E0 and E1 as shown in Figure 7.11c. In addition, write the interchanged pair to the cache. If no match is found, retrieve and wait for the new entry E2 from central memory. When E2 arrives, push it on top as shown in Figure 7.11d.

If the central memory is too slow, an off-chip L2 cache that contains both instructions and data may be considered. However, this design is merely an added option in order to speed up memory access. By doing so, control lines are needed between the L2 cache and the memory on the host system bus.

7.4 PENTIUM DECOUPLED PIPE

The Pentium processor is designed as a decoupled pipe whose block diagram is simplified in Figure 7.12.[28] The off-chip L2 cache is an intermediate storage between the central memory and the CPU to supply instructions and data. On the chip, the L1 ICache (instruction cache) supplies instructions, and the L1 DCache (data cache) supplies data. There are four major hardware units to constitute a 12-stage superpipeline. All stages are decoupled in order to achieve a higher clock rate. Features include branch prediction, speculative instruction retrieval and execution, and data flow analysis. The overlapped operations ensure high performance as described below.

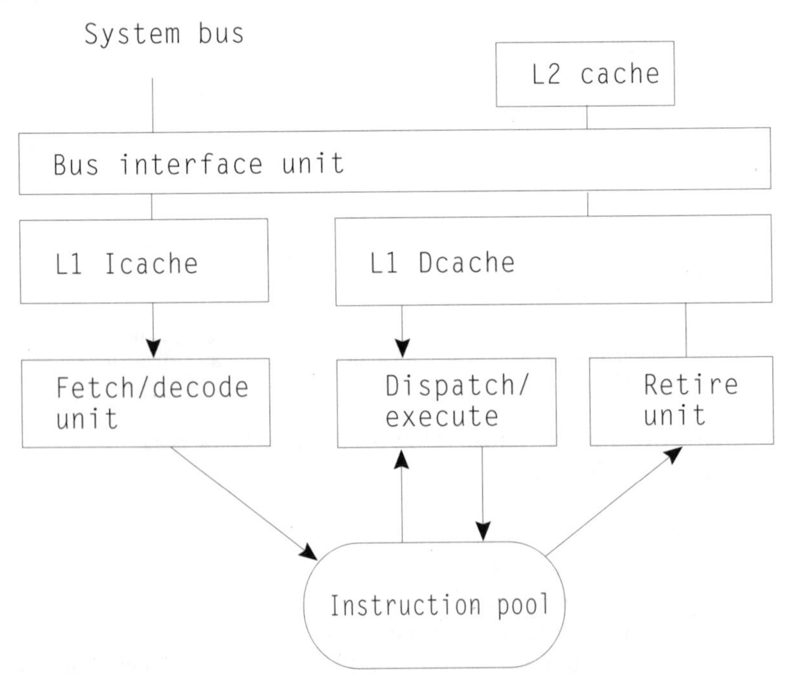

Figure 7.12 Block diagram of a Pentium CPU.

7.4.1 Hardware Units

The four major hardware units are BIU (bus interface unit), FDU (fetch and decode unit), DEU (dispatch and execute unit), and RU (retire unit). The BIU is an external unit while the other three are internal units. The BIU provides load/store interface between the L1 instruction or data caches and the L2 cache. In other words, the BIU connects three internal units or engines to the outside world.

If an internal unit operates in sequence, it is an in-order unit. Otherwise, it is an out-order unit. Thus, each unit is a stage in a decoupled pipe using microcoded logic. The three internal units jointly work with an instruction pool that contains micro orders to be executed, or have been executed but not yet retired. Such a design allows an instruction to be started in any order, but it must always be completed in the original program order. Let us fetch and decode four instructions as follows:

```
R[1] <— M[R[0]]   ;I0 (instruction 0)
R[2] <— R[1] + R[2]       ;I1
R[5] <— R[5] + 1          ;I2
R[6] <— R[6] - R[3]       ;I3
```

The first instruction I0 loads R1 with a memory operand. If a L1 DCache miss is encountered, the DEU looks at its instruction pool for subsequent micro orders to execute rather than being stalled. Obviously, the second instruction I1 has R1 specified as a sopd (source operand), which is also the dopd (destination operand) from I0. Therefore, the data dependency problem prevents I1 from being executed. Nonetheless, I2 and I3 do not have such a problem so they can be executed out of order.

7.4.1.1 Bus Interface Unit

The BIU handles two types of memory access: load and store. After receiving a μop (micro operation or order), the unit carries out its execution. A load operation needs the memory address, operand size, and temporary result as encoded into a single μop. A store operation needs to specify the memory address, operand size, and source operand to be written. In contrast, a store operation is encoded into two μops. The first one generates the address, the second one generates data, and the two must recombine later for the store operation to complete.

7.4.1.2 Fetch and Decode Unit

The FDU is an in-order unit that provides two functions, fetch (retrieve) and decode. The fetch unit retrieves the instruction stream from the ICache. A register named BTB (branch target buffer) contains the predicted address of a conditional branch. Another subunit named NextIP provides the start ICache index, based on inputs from the current instruction, BTB, trap/interrupt status, or branch misprediction indications from the execution unit. The ICache fetches the 64-bit cache line and the next line according to NextIP and presents 16 aligned bytes to the decoder. The lower three bits in the full instruction byte address constitute a byte count to shift left the instruction. After that, it is justified for the decoder. Because the beginning and end of the instructions are marked, the three parallel instruction decoders (ID) can accept this stream and proceed to decode the instructions in parallel. Each decoder converts the instruction into μops that represent the data flow of the instruction. The μop is triadic in the sense that it contains two logical sources and one logical destination. Most instructions are converted directly into single μops, some are converted into 2 – 4 μops, yet the complex ones require microcode. In the FDU, a store named MIS (microcode instruction sequence) provides the microcode, i.e., a sequence of normal μops.

The μops are queued, and sent to the RAT (register alias table) unit, where the logical register references are converted into physical register reference (i.e., decoded form). The allocator stage in the RAT unit adds status information to the μops and enters them into the instruction pool. The status bits may indicate to be executed, executed but not yet retired, etc. Note that a branch μop is tagged with the fall-through address if the prediction is wrong and it also contains the predicted destination address. The instruction pool is an array of content addressable memory called the ROB (reorder buffer). Note that some μops in the pool may be speculative, therefore their executions are discarded if not needed later.

7.4.1.3 Dispatch and Execute Unit

The DEU is an out-order unit that accepts the data flow stream and schedules the execution of μops. The dispatch unit selects μops from the instruction pool to the EU (execution unit) upon their status. If the status indicates that the μop has all of its data ready, then the dispatch unit checks if the execution resource needed by the μop is also available. Each resource is a functional unit, and the resources include fixed point, floating point, address compute for load, address compute for store, and jump. That is to say, in order to dispatch a μop its data and EU must be ready. Conceptually, the dispatch unit keeps a busy flag register for all the resources including the data registers. A bit zero indicates that the data or resource is free. The hardwired logic masks the busy flag register with the operation control and sources in the μop. If the result is zero, the two conditions are met, so the unit removes that μop from the pool to the resource where execution takes place. The DEU also provides registers for temporary execution results that may be speculative so the results of the μop are returned to the instruction pool for later examination.

Recall when a branch type instruction is decoded, the decode unit sets the BTB to the predicted branch address. Thus, the ICache can retrieve instructions at the new address. If during execution, the branch address coincides with this predicted address, then all the instructions retrieved after the branch address are valid in the pool and they will be duly retired. If the branch address does not coincide with this predicted address, e.g. no more branch to exit a loop, then the jump execution unit

(JEU) does two things. First, it sets the BTB to the fall-through address so the fetch unit can restart the pipeline from this new instruction address. Second, it changes the status of those μops right after the predicted branch address to invalid so they can be discarded later by the retire unit.

7.4.1.4 Retire Unit

The RU is an in-order unit that knows exactly when and how to commit (i.e., retire) the temporary, speculative results to a permanent architectural machine state. The retire unit looks for those μops that have been executed and can be removed from the pool. The unit first reads the instruction pool and finds the potential candidates for retirement. It then checks the status of the μops in the instruction pool, decides which μops are completed, and imposes the original program order on them. It must also do this in the face of interrupts, traps, and mispredictions. This is why the μop should contain the full physical address of the instruction being executed. The RU is capable of retiring up to three μops in one clock.

For programming convenience, the Pentium CPU supports a few vector instructions, such as push/pop all registers, move a block of memory, search a block of memory, etc.

7.4.2 Push All vs. Pop All

On a Pentium, the pusha (push all) instruction shares the same one-byte opcode with the pushad (push all double) instruction. Each instruction can push eight registers of 16 bits or 32 bits, as the operand size is specified in a hardware register by the OS. Similarly, the popa (pop all) and popad (pop all double) instructions also share the same opcode to pop all eight registers from the TOS, as shown below.

Code image	Statement	Description
60	pusha	Push eight 16-bit registers.
60	pushad	Push double eight 32-bit registers.
61	popa	Pop eight 16-bit registers.
61	popad	Pop double eight 32-bit registers.

The pusha/pushad and popa/popad operations are shown below:

Mnemonic	Description
pusha/pushad	;IF the operand size is 16 bits, ;THEN Temp <— sp ;Push ax, cx, dx, bx ;Push temp, bp, si, di ;ELSE Temp <— esp ;Push eax, ecx, edx, ebx ;Push temp, ebp, esi, edi ;ENDIF

```
popa/popad        ;IF the operand size is 16 bits,
                  ;THEN Pop di, si, bp
                  ;sp <— sp + 2 {skip next 2 bytes.}
                  ;Pop bx, dx, cx, ax
                  ;ELSE Pop edi, esi, ebp
                  ;sp <— sp + 4 {skip next 4 bytes.}
                  ;Pop ebx, edx, ecx, eax
                  ;ENDIF
```

Note that the pop sequence is the reverse of push, and one opcode is adequate for either push or pop regardless of the operand size. The reason for this is that the CPU can only operate in 16-bit or 32-bit mode as determined by the D bit (b22) in the segment descriptor to specify the operand size. If the D bit is one, it means one 32-bit operand, otherwise we have a 16-bit operand. That is to say, one bit is specified in the CPU as an extended opcode to determine the operand size.

7.4.3 String Instructions

There are five string instructions: lods (load string), stos (store string), movs (move string), cmps (compare string), and scas (scan string). Each instruction has a six-bit opcode followed by a D (destination) bit, and a W (width) bit. Since the operand can be one byte or one word (two bytes), the W bit in the main opcode makes the difference. Also, the D bit makes the difference between load and store. Note that each instruction contains no addresses because all the operandi are implied as listed in Table 7.1.

Table 7.1 String Instructions

Code image	Mnemonic	Description
AC	lodsb	Load string byte al <— M[ds:si] {DS is the base.}
AD	lodsw	Load string word ax <— M[ds:si]
AA	stosb	Store string byte M[es:di] <— al {ES is the base.}
AB	stosw	Store string word M[es:di] <— ax
A4	movsb	Move string byte M[es:di] <— M[ds:si] <7:0>
A5	movsw	Move string word M[es:di] <— M[ds:si]

Code image	Mnemonic	Description
A6	cmpsb	Compare string byte Compare M[ds:si]:M[es:di] <7:0> and set CC.
A7	cmpsw	Compare string word Compare M[ds:si]:M[es:di] and set CC.
AE	scasb	Scan string byte Compare al:M[es:di] <7:0> and set CC.
AF	scasw	Scan string word Compare ax:M[es:di] and set CC.

The lods instruction loads an operand from memory to the accumulator while the stos instruction stores an operand from the accumulator to memory. The movs instruction moves an operand from memory to memory. The scas instruction compares the accumulator with memory and set CC in the F register. Similarly, the cmps instruction compares memory with memory and set CC accordingly.

A string instruction is often issued in conjunction with a prefix, namely REP (repeat), REPE (repeat while equal), or REPNE (repeat while not equal) as listed in Table 7.2.

Table 7.2 Repeat Prefixes

Code image	Mnemonic	Description
F3	REP	Repeat
F3	REPE	Repeat While Equal
F2	REPNE	Repeat While Not Equal

If a one-byte prefix (i.e., opcode) is specified in the front, the two-byte instruction becomes a hardware loop. The implied registers must be properly set, and the CX register should contain a loop count. While the REP prefixes work with lods, stos, or movs, the REPE or REPNE prefix works with a compare type, such as scas,or cmps. Each prefix acts like a DO-WHILE construct as shown below:

```
DOWHILE (CX .NE. 0) .OR. (compare condition);
CX <— CX - 1;
Do the operation;   {Note: compare operation sets CC.}
Modify SI;
Modify DI;
ENDDO;
```

If either of two cases is true, the loop continues, otherwise the loop terminates. As the first case, CX is not zero. In the second case, the compare condition is true. After the loop terminates, either CX is decreased to zero or the condition changes to false after the comparison. In fact, the REP prefix means repeat while not end-of-string and shares the same opcode as REPE. Furthermore, after the operation, the index registers are increased or decreased by the operand size based on the DF (direction flag) in the F register. For example, if the operand size is one byte and DF is zero, the index registers SI and DI are each increased by one after the operation.

Note that the REP or REPE prefix has an LSB (least significant bit) equal to 1 and the REPNE prefix has an LSB of zero. If the LSB is denoted as z (pronounced zee), after the operation, the compare condition in the DO-WHILE is as follows:

compare condition = ZF .EQ. z

The compare condition is true if the z bit is the same as ZF. If the z bit is different from ZF, then the compare condition becomes false. Note that it is not necessary for a programmer to initialize ZF before the repeat loop. A coding example to move a 256B block from one memory location to another location is shown below.

```
; _____ Recall for string operations, the destination operand
;          uses ES as the default segment register. The implied
;          SI is a source index, DI is a destination index, CX
;          contains a count, and the direction flag is 0 to move
;          up so both SI and DI are increased by 1 after each
;          move. The CX register always counts down and the REP
;          MOVSB instruction moves a byte block as follows:
;          DO-WHILE (CX .NE. 0);
;          CX <— CX - 1
;          M[es:di] <— M[ds:si] <7:0>  {CC is not changed.}
;          SI <— SI + 1
;          DI <— DI + 1
;          ENDDO;
; ______________________________________________________________

           . . .
           mov   ax, ds
           mov   es, ax
           mov   si, offset blk1
           mov   di, offset blk2
           cld                    ;clear DF to move up
           mov   cx, 100h         ;block count is 256 in bytes.
      rep movsb
```

A scas instruction internally subtracts the destination string indicated by the DI from the implied accumulator, and sets the . Similarly, the cmps instruction internally subtracts the destination string indicated by the DI from the source string indicated by the SI and sets the CC. Because the destination string uses the ES as the base, the ES is set to the DS in this example. Either string, destination or source, remains unchanged after an internal subtract. Using the prefix REPE or REPNE with scas or cmps, we can search a string, and a coding example is shown below.

```
. . .
str1   db   `THIS IS TEXT'
. . .
mov    ax, ds
mov    es, ax
mov    di, offset str1
mov    al, 'E'
mov    cx, 12       ;byte count
mov    bx, cx       ;save a copy
repne  scasb
; ——— DO-WHILE (CX .NE. 0) .OR. (ZF .EQ. 0)
;      CX <— CX - 1
;      Compare M[ds:si]:M[es:di] <7:0> to set CC.
;      SI <— SI + 1
;      DI <— DI + 1
;      ENDDO
jne    notFound  ;Note 1
sub    bx, cx    ;Note 2
. . .
```

The jne instruction is tricky because the ZF in F is tested. If control is passed to notFound, it means no match. If control is passed to the next sub instruction, a match is found. Since CX is decreased by one each time in the loop, it is possible to have CX equal to zero so the last char is the matched character. In this example, CX contains two after exiting. After the sub instruction, BX contains 10, which means the 10th character in the string is the matched 'E'.

7.4.4 Multimedia Instructions

The Pentium processor has eight 64-bit MMX registers, denoted as MM0, MM1, ..., and MM7 to support multimedia applications as shown in Figure 13a.[50] As shown from Figure 7.13b-e, an MMX register may contain eight bytes, four words, two double words, or one quad word. Based on the MMX technology, the instructions are designed for some of the unary and binary operations. The operand size can be two, four, or eight bytes, and a binary MMX instruction operates between two MMX registers. Even though the number of parallel operations is limited, the design has the flavor of a SIMD processor.

Figure 7.13 Multimedia registers: (a) 64-bit MM0 to MM7, (b) an MMX register contains eight bytes, (c) four words, (d) two double words, and (e) one quadword.

7.4.4.1 Wrap-around Mode vs. Saturated Mode

If the display supports a resolution of 1024 by 1024 pixels, the number of pixels is one mega. In fact, each pixel is composed of three colors, R (red), G (green), and B (blue), and each color is an unsigned byte representing intensity. Wrap-around mode means regular twos complement arithmetic. That is, after adding two unsigned integers F000 and 3000 in hex, the result is 2000 as shown in Figure 7.14a. Because a carry is generated, the sum result has a smaller magnitude. If intensity is of concern, the result becomes lighter instead of darker. To solve the problem, saturated mode arithmetic is developed for both add and subtract operations. That is, if a carry is generated after add, the result is all ones to be the largest unsigned integer (i.e., FFFF), as shown in Figure 7.14b. If a borrow is generated after subtract, the result is zero as shown in Figure 7.14c.

Other MMX operations include multiply, shift, and logical. A move instruction can load a 64-bit block from memory to a register and store in the reverse direction. Moreover, two instructions can pack or unpack one register into another register.

```
(a)      F000   in hex           1111 0000 0000 0000   in binary
     +)  3000                    0011 0000 0000 0000
        -----                    -------------------
         2000                    0010 0000 0000 0000

(b)      F000   in hex           1111 0000 0000 0000   in binary
     +)  3000                    0011 0000 0000 0000
        -----                    -------------------
         FFFF                    1111 1111 1111 1111

(c)      3000   in hex           0011 0000 0000 0000   in binary
     -)  F000                    1111 0000 0000 0000
        -----                    -------------------
         0000                    0000 0000 0000 0000
```

Figure 7.14 Operation modes: (a) wrapped-around after add, (b) saturated after add, and (c) saturated after subtract.

Before packing a word to byte operation, each of the source words has a hole byte in the front, as shown in Figure 7.15a. After packing, the four holes are eliminated, and the four respective low-order bytes are right justified in the destination register, as shown in Figure 7.15b. The unpack operation does the reverse. Figure 7.15b also shows the four packed bytes, B0, B1, B2, B3 before unpacking. After unpacking, the destination register contains the unpacked bytes, as shown in Figure 7.15a.

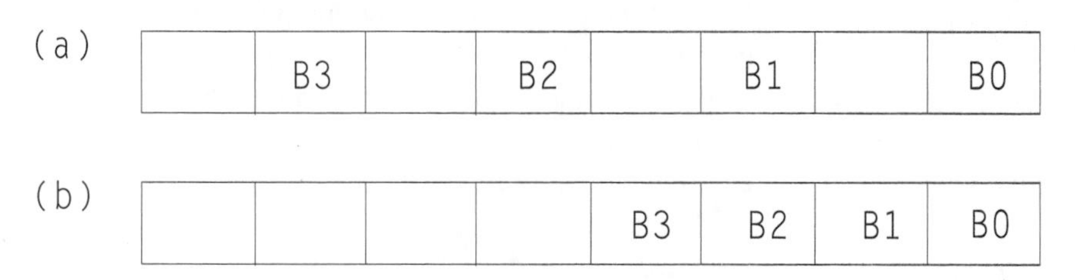

Figure 7.15 Pack and unpack operations: (a) four unpacked bytes, B0, B1, B2, B3, in an MMX register before pack or after unpack and (b) four packed bytes after pack or before unpack.

The summary of MMX instructions is listed in Table 7.3 and the first letter P in the mnemonic stands for parallel.

Table 7.3 Summary of the MMX Instruction Set

Mnemonic	Description
Padd[b|w|d]	Add byte, word, double word in wrap around or saturated mode.
Psub[b|w|d]	Subtract byte, word, double word in wrap around or saturated mode.
Pmullw	Multiply low of 4 signed words. The low-order 16 bits are in the destination register.
Pmulhw	Multiply high of 4 signed words. The high-order 16 bits are in the destination register.
Pmaddwd	Multiply and add 4 signed words. Adjacent pairs of 32-bit product are added and the result is a double word.
Pcmpeq[b|w|d]	Compare equal. If true, then the result is mask of ones otherwise zeroes.
Pcmpgt[b|w|d]	Compare greater than. If true, the result is mask of ones otherwise zeroes.
Psra[w|d]	Shift right arithmetic with shift count in register or immediate.
Psll[w|d|q]	Shift left logical with shift count in register or immediate.
Psrl[w|d|q]	Shift right logical with shift count in register or immediate.
Punpackl[bw|wd|dq]	Unpack low byte to word, word to double, or double to quad.
Punpackh[bw|wd|dq]	Unpack high byte to word, word to double, or double to quad.
Packss[wb|dw]	Pack word to byte, or double to word in saturated mode.
Pand	And two 64-bit registers
Pandn	Nand
Por	Or
Pxor	Exclusive or
Mov[d|q]	Move between memory and register

7.5 VIRTUAL MEMORY

A program address is a logical address pointing to an instruction or data in a program. With indexing, a program address becomes a program effective address abbreviated as EA, which is the sum of an address coded in the instruction and any index registers if specified. Without virtual memory, the program address is a physical or real address. With virtual memory, the program address is a virtual address. To access memory, address mapping, also called DAT (dynamic address translation) is required to convert a VA (virtual address) to a PA (physical address).

The VAS (virtual addressing space) is defined to be the logical programming space as seen by a user. The PAS (physical addressing space) is the real memory size. That is, the number of address pins on the processor chip sets the limit. If a system supports a VAS of 16 MB, a user believes that he has 16 MB, but in reality, the memory size may be only 8 MB. In theory, a VAS may be greater than, equal to, or less than a PAS. There are many virtual memory systems with the same goal of solving the relocation, protection, and overlay problems.

7.5.1 Segmentation

A segment is a contiguous block of code with a variable size. More often than not, a segment is logically intact and relocatable. A virtual memory may support one single segment or multiple segments with or without paging.

7.5.1.1 Single Segment Systems

With only one segment, the computed EA is a relative offset to the beginning of the segment. The PC determines the size of virtual memory. A single segment system without paging uses one RA (relocation address) register containing the physical address of a segment.[5] That is, the base address of a segment is in a hardware register. Before a memory cycle, the VA or computed EA is added to the RA register to generate the physical address. Simply put, the adder maps a VA to a PA, as shown below:

$$PA = RA + VA$$

There are two types of RA registers: implicit and explicit. An implicit base means that the RA register address is not coded in the instruction. An explicit base means that the register address of RA is specified in an instruction. In practice, the RA register is associated with a limit or FL (field length) that contains the segment length. If EA is greater than FL, it is out of bounds and an interrupt (address exception) is triggered.

In fact, many computers support only one single segment. However, the segment can be implemented by one-level paging, two-level paging, or three-level paging. As an example, the IBM 360/370/390 system uses a two-level paging system for one segment. Because the computed EA is 32 bits and its IAR is also 32 bits, the maximum segment size is 4 GB. Later, we will discuss paging, a technique to solve the fragmentation problems.

7.5.1.2 Multiple-Segment Systems

If a virtual memory has multiple segments, we obtain a multiple segment system. In such a system, each segment resides in a contiguous memory block indicated by an RA register. If a virtual memory has 4 GB, its 32-bit VA can be divided into an eight-bit S (segment) and a 24-bit D (displacement) as shown in Figure 7.16a. That is, the system has up to 256 segments and each segment. has up to 16 MB.

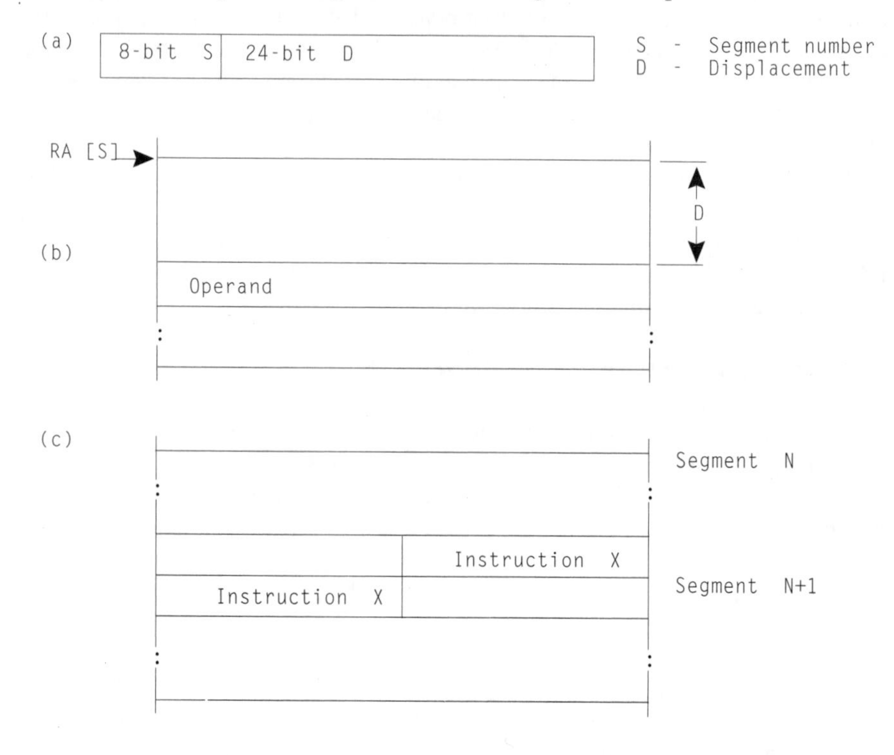

Figure 7.16 Linear segment system: (a) the virtual address is an ordered pair (S, D), (b) the operand in a segment, and (c) the instruction X crosses segment boundary.

The eight-bit S represents an RA register address, and the 24-bit D is an index to the segment of 16 MB, as shown in Figure 7.16b. Note that the D field in a virtual address is different from the Disp coded in an instruction. The D field in a VA is an index to a segment, while the disp in an instruction is relative to a base address in the program. When forming the virtual address (S, D), S is an independent segment number not subject to change and D is the sum of Disp and an index register, if specified. That is, the computed D is a segment displacement, so segment boundary is not crossed.[7]

Mathematically, the computed VA is an order pair of S and D. The PA is the sum of D and an RA register addressed by S, as shown below:

$$VA = (S, D)$$
$$PA = RA[S] + D$$

A total of 256 RA registers can be implemented by hardware. If a system supports a large S, the RA registers can be implemented by software. Nevertheless, an add operation is required to compute the sum of the base address from a table entry in memory and the displacement. If extra bits denoted by S are provided in an instruction to select an RA register, the computed EA is independent of S. As a result, the system supports multiple independent segments, and mathematically we have

$$VA = (S, EA)$$
$$PA = RA[S] + EA$$

The VA is an ordered pair of S and EA. If the computed EA is 32 bits, each segment can go up to 4 GB. If S is small, the hardware RA registers can be used to store segment descriptors. That is, each RA register contains the physical base address, control bits, and limit of the segment, as shown in Figure 7.17. If a segment is allocated in blocks, the base address is less than 32 bits. To make it simple, an R bit can be used to indicate that the segment is read only. That is, if the R bit is one, a write operation would trigger an interrupt (protection exception). The limit means segment length or size in bytes. At run time, any computed EA greater than the limit would trigger an interrupt (address exception).

Base address	Limit	Control

Figure 7.17 The format of a segment descriptor.

In a pure segment system, a large segment resides in a large memory block. As program terminates, if a new program can not fit in an old block, holes are developed in memory. That is to say, the memory becomes fragmented as explained below.

7.5.1.3 Fragmentation Problem

If segment base registers are used, holes or free space may be developed in memory from time to time. This is known as the fragmentation problem. Because all the user tasks do not terminate in sequence, when one user task terminates, it is possible that another task can not fit in an old segment once occupied by the previous run. As shown in Figure 7.18a, three holes are generated after segment zero, two, and four are purged. If the new task can not fit in any of the holes, the OS may take action to compact many small holes into a big hole. That is, the OS moves code blocks, segment partitions one, three, and five to a contiguous area in low memory so a big hole is left in high memory, as shown in Figure 7.18b. The system chore is quite easy if the code is relocatable. Compaction is an important system concept. The disk space may also become fragmented if files are deleted and created too many times. Usually, a sequential file need not occupy a contiguous block on disk. However, if all the files need to occupy a contiguous block for performance reasons, a disk compaction is in order. That is, the OS moves and packs all the files to a contiguous disk area in low address so a big hole is left in high address.

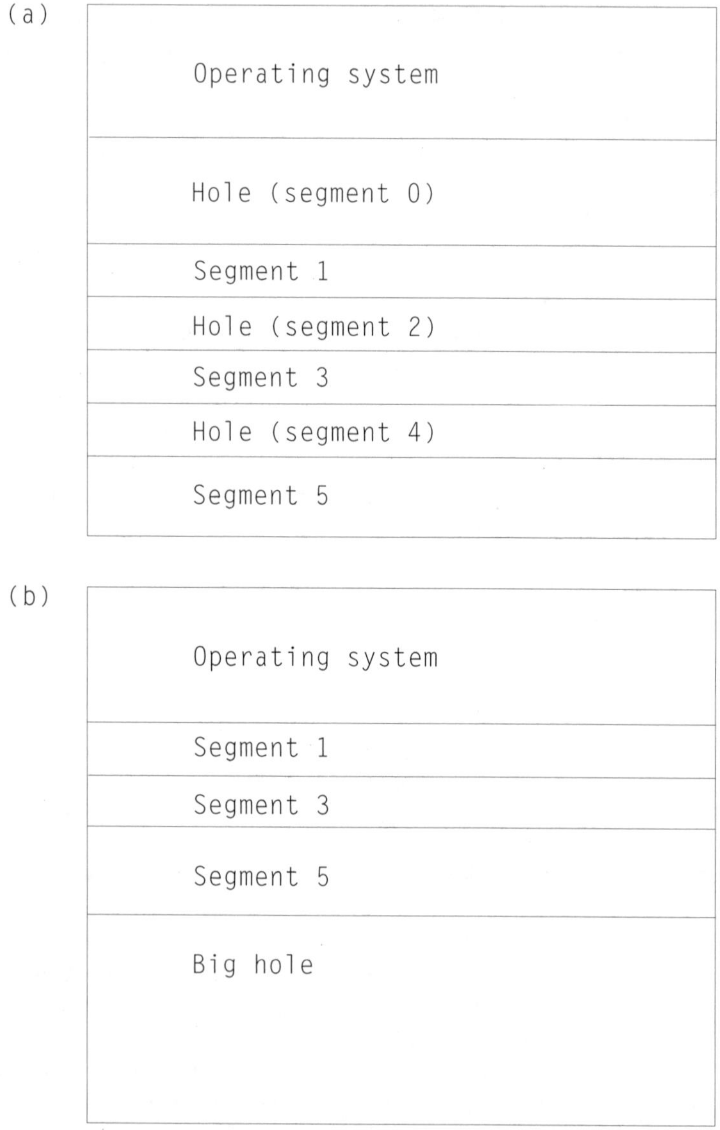

(a)

Operating system
Hole (segment 0)
Segment 1
Hole (segment 2)
Segment 3
Hole (segment 4)
Segment 5

(b)

Operating system
Segment 1
Segment 3
Segment 5
Big hole

Figure 7.18 Compaction: (a) three holes in memory before compaction and (b) a big hole in high memory after compaction.

Paging is another alternative as discussed in the next section. A single segment may be implemented by way of paging. If the VA has 32 bits, the segment size is 4 GB. The lower bits in the VA constitute a page number, and the upper bits constitute a page number that can be divided into one or two fields. If the upper bits in the VA represent a large page number in one field, we obtain a one-level paging system. If it is divided into two fields, we obtain a two-level paging system. That is to say, the VA is a triple: the first level page number, the second level page number, and the displacement. The terminologies may be different as the first level page number is also known as a segment number by IBM or a page directory number by Intel.

7.6 PAGING SYSTEM

A page is a contiguous block of code of fixed size designed to solve fragmentation problems. Various single segment models are implemented by one-level paging, two-level paging, or three-level paging. A VA is computed as the sum of the address in an instruction and any index registers if specified. Mathematically, the VA is a (P, D) pair in a one-level paging system.[34] Assume that the 32-bit VA is divided into 12-bit P and 20-bit D, as shown in Figure 7.19a. Because the computed P may contain bit pattern, all the pages are linear in a segment. From the given VA, we know that the VAS is 4 GB with 4 K pages, and each page, is 1 MB. In concept, each P selects an RA register that contains a page descriptor as shown in Figure 7.19b. The physical base address is a PF (page frame) or PFN (page frame number) provided that the PAS is divided into many PFs. Each page frame is 1 MB so a page is loaded into a free page frame just like a picture is placed in an empty picture frame. The pages, P0, P1, P2, etc are sequential in the VAS as shown in Figure 7.19c. Because a page has a fixed size, it is possible to split an instruction or data between two consecutive pages. Physically, the pages scatter all over in memory as shown in Figure 7.19d.

If the RA registers are hardware, we need a total of 4K such registers. Because of fixed page size, after mapping PF is concatenated with D to form PA as shown below:

$$
\begin{aligned}
\text{VA} &= (\text{P}, \text{D}) \\
\text{PA} &= (\text{PF}, \text{D}) \\
&= \text{PF}^\wedge\text{D}
\end{aligned}
$$

In a one-level paging system, the VA is an ordered pair of P and D. After mapping, PF is concatenated with D (PF^D) to form the physical address in bytes without going through an add operation. Note that P, PN (page number), and virtual page number all mean a virtual memory block address. Similarly, PF, PFN (page frame number), and physical block number mean a physical memory block address.

Even though the 4 K RA registers can be implemented by hardware, it is even better if the registers are implemented by software. That is, a page table (PT) is prepared in memory that contains page descriptors. One hardware (page table base reg-

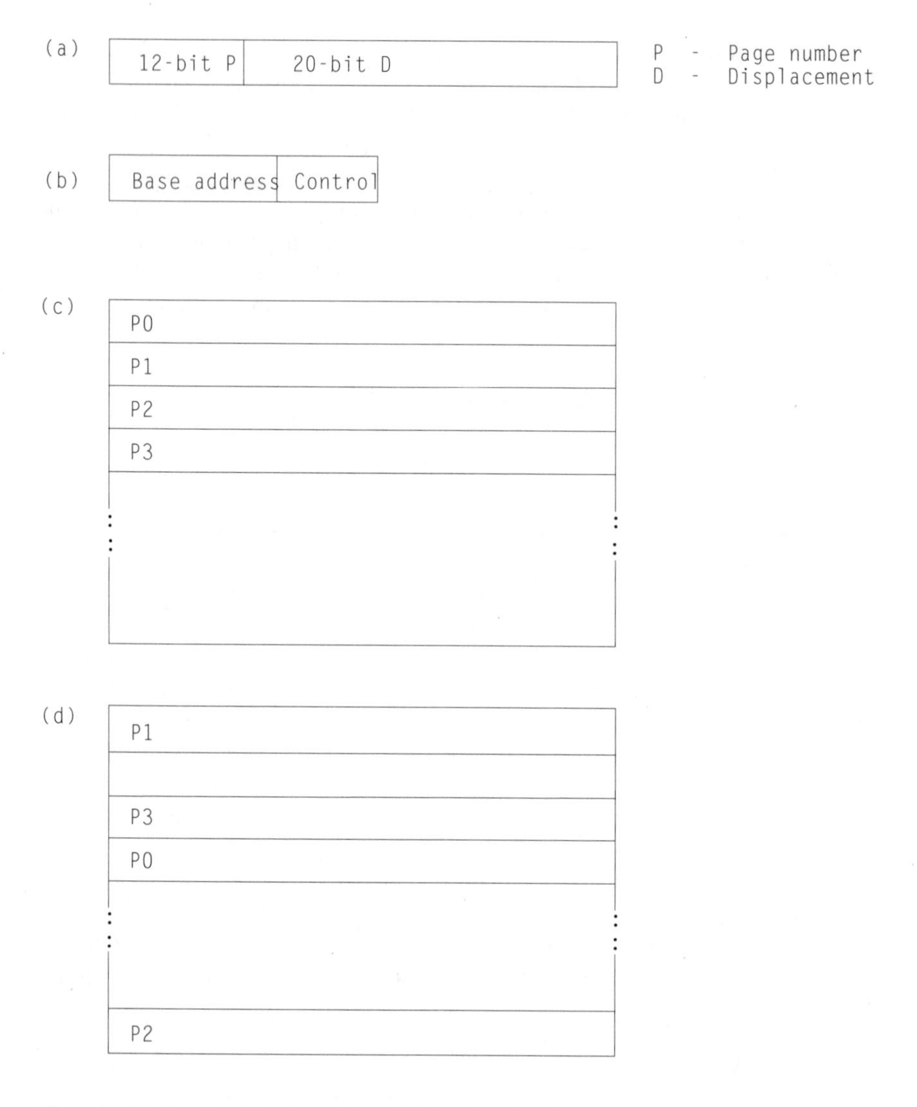

Figure 7.19 One-level paging system: (a) the virtual address is an order pair (P, D), (b) the page descriptor, (c) the pages are sequential in VAS, and (d) the pages may be scattered in PAS.

ister) (PTBR) contains a physical base address pointing to the page table. The page number, P, in conjunction with the address in the PTBR, is used to retrieve a PF in the page table entry. Note that the entry is just like a hardware register except it takes one extra memory cycle to fetch the PF field in it. The PT is a local data base, owned by the user task but managed by the paging supervisor.

It should be mentioned that the SR (status register) has a bit named V=R (virtual equal to real). If this bit is set, the CPU bypasses dynamic address translation, so the computed program address is a real address. In fact, when the nucleus or a real-time program is running, the dynamic address translation (DAT) is bypassed for performance reasons. If the PT is in memory, the one extra cycle slows down the CPU. A clever scheme presented in the next section potentially solved this problem.

7.6.1 Translation Look-Aside Buffer

A translation look-aside buffer (TLB) is a fully associative cache that contains the most recent translation records; its entry is shown in Figure 7.20. Given the 12-bit P as the key, if a match is found in the TLB, the PF is retrieved in one clock. The bit length of the PF determines the physical memory size. There are three possible designs, as listed below:

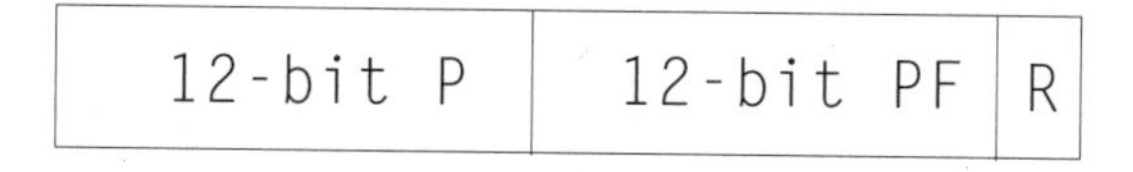

Figure 7.20 Translation look-aside buffer entry.

1. If the number of bits in PF < 12, then VAS > PAS.
2. If the number of bits in PF = 12, then VAS = PAS.
3. If the number of bits in PF > 12, then VAS < PAS.

The TLB speeds up address translation and the hardware unit DAT, i.e., MMU (memory management unit)[26,28] contains the TLB plus logic for address mapping. In addition to the physical base address, the PTBR may also contain access right and limit, i.e., the largest page number allowed. Each PT entry is a page descriptor that contains the PF and control. In the control field, the P (presence) bit tells whether the page is in memory or not. After retrieving a PT entry, if the P bit is one, the PF is obtained to complete mapping. A P bit of zero, triggers interrupt, i.e., page fault. The R (read only) bit, if set to one, specifies that the page can not be written in. Other pages may allow write access; when the OS removes such a page from memory, it must be written back to disk.

During address mapping, first use P as the key to search the TLB. If a match is found, mapping is done. Otherwise, P is used as an index to search the PT. The physical address of the PT entry is formed by hardware without going through add. If the PT contains 4 K entries and each entry is 2 B, the block base address in PTBR is appended with P and an implied zero. That is, the byte displacement in the block is P times two as shown below:

$$\text{Physical address of the PT entry} = \text{PTBR}\wedge(P * 2)$$

While fetching the block base address in the PTBR, the page number P is compared with the limit. If P is out of the limit, an interrupt (address exception) is triggered. Otherwise, the PT entry is retrieved and it has a PF for the page. Note that an extra memory cycle is used to read the PT so a TLB is necessary to improve performance. As a simple example, each TLB entry contains P, PF, and the R bit. During an update, the R bit is copied from the PT for write protection. Given P as the key, all the TLB entries are searched in parallel. If a matched is found, the PF is fetched and mapping is done. If a match is not found, P is used again as an index to search the PT. If there is no address or protection exception and the page is in memory, the PF is fetched to complete mapping. Meanwhile, the new entry is the most recent translation record so it is pushed onto the TLB based on the LRU (least recently used) algorithm. Since each user task has its own VAS, before a new task is scheduled to execute, the OS issues a privileged instruction to purge the TLB.

7.6.2 Paging Supervisor

The paging supervisor is part of the OS nucleus that receives control after a page fault. This OS routine manages two sets of addresses: virtual and physical. Only the paging supervisor developers know the existence of physical addresses. Other groups, including the job scheduler and programming tools, do not see the physical addresses. Demand paging is designed such that a page is brought into memory literally on demand at run time. A large amount of system overhead is spent to utilize the memory space more efficiently. The P (presence) bit is in PT, and if P is zero, the hardware, while searching the PT, triggers page fault. Consequently, the paging supervisor takes over control and loads the missing page from disk into memory.

7.6.2.1 External Page Table

The PT (page table) entry mainly contains the base address of the page, and the EPT (external page table) entry mainly contains its disk addresses. Both tables are local to the user task. The PT of a user task is shown in Figure 7.21a and its associated EPT is shown in Figure 7.21b. In a demand paging system, each PT entry contains a valid PF if the page is in memory. In contrast, the corresponding entry in the EPT always contains its disk address and access right. It is possible to eliminate the disk address field in an EPT entry if P can be converted into a disk address directly. That is, every page in the VAS must occupy a block on disk. Therefore, if the VAS is very large, there will be many hole blocks on disk if some of the pages are not in use. The EPT is part of a load module but the PT is not.

7.6.2.2 Page Frame Table

After a page fault occurs, the paging supervisor first allocates a free page frame for the page. The page frame table (PFT) is a global data base, as shown in Figure 7.21c. After boot, the table is initialized so that each of its entries contains information about a page frame that is swappable. Because the page frames in low memory

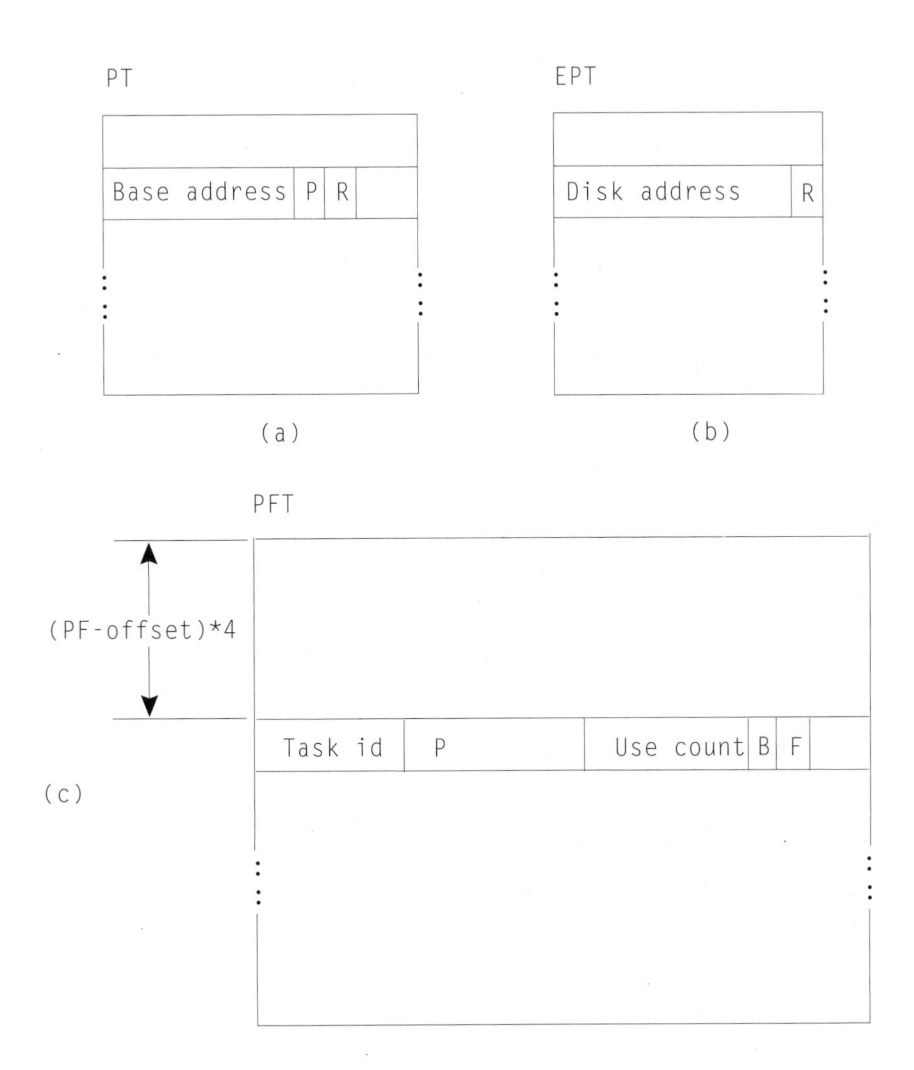

Figure 7.21 Data structures in a paging system: (a) page table, (b) external page table, and (c) page frame table.

are not swappable, they have no representations in the PFT. The search index N is equal to (PF-offset) * 4, where offset is the number of non-swappable frames in low memory and 4 is the entry size in bytes. Because the PF number can be derived from the search index, one design is to not store it in the 32-bit entry, as described below:

- Task id — the 8-bit local (internal) number for the task
- P — the 12-bit private page number of the occupied page frame
- UC — the 8-bit use count of a shared page (number of current users)

- B — the busy bit means the page frame is occupied
- F — the fixed bit means the page can not be swapped out
- 2 reserved bits

7.6.2.3 Next Search Method

The routine uses the N index to search the 'next' free frame in the PFT as follows:

```
REPEAT;
  IF the entry pointed by N is free,
  THEN report the free frame and set the Done flag;
  ELSE increase N by 4, the size of the entry;
    IF N is equal to the size of PFT,
    THEN wrap around so N points to the beginning of PFT; ENDIF;
    IF N is equal to its original value before the search,
    THEN report no free frame in PFT and set the Done flag;
    ENDIF; ENDIF;
UNTIL Done.
```

7.6.2.4 Dynamic Address Translation

The DAT algorithm is based on three steps. The first two steps are done by hardware. The third step is done by software in a demand paging system that uses the next search method to find a free page frame.

```
1. Search TLB;
   IF a match is found,
   THEN fetch PF and exit; ENDIF;
2. Search PT;
   IF the P (Presence) bit is 1,
   THEN
   Fetch PF;
   Push new entry on TLB based on LRU algorithm and exit;
   ELSE trigger page fault; ENDIF;        {Demand paging system.}
3. Paging supervisor takes over control;
   Locate the disk address from EPT;
   Next search PFT;
   IF a free page frame is found,
   THEN goto to loadPage; ENDIF;
   Throw away an old page;
          {A preferred sequence may be the read-only page
```

```
                first, dirty page, and shared page.}
        Reset the P bit in its PT to 0;
                {The old page is no longer in memory.}
        IF the old page is written in,
        THEN issue a request to write it back on disk; ENDIF;
loadPage: Issue a disk read request to load the new page into
        the page frame;
        Wait for the arrival of the new page;
        Set the task id, B (Busy) bit, P (Page) no. in PFT;
        Set the P (Presence) bit and its PF (Page Frame) no. in its
        PT;
        Pass control to Dispatcher.
```

If the interrupted task has the highest priority, it will be dispatched next. After restoring its environment and purging the TLB, the dispatcher passes control to the task. Thus, the instruction that triggered the page fault executes a second time. If the search of the TLB fails, the PT is searched. If the entry is there, the PF is fetched and the applications exits. At the same time, the new translation record is pushed on the TLB. Suppose the address that caused the page fault is an instruction address. After increasing the PC, if its P (page number) remains unchanged, then the TLB contains the translation record. Hence, during the next cycle, the PF of the instruction is fetched from the TLB and the application exits.

In a two-level paging system, the VA is a triple (P1, P2, D), where P1 is the first level page number, P2 is the second level page number, and D is the displacement in page. The basic concept is similar, as discussed later.

7.6.3 Relationship between the MMU and L1 Caches

The L1 instruction and data caches can be implemented along with paging. Both the L1 caches and the MMU are transparent to the users. Where should the L1 caches be positioned logically, before or after the MMU? Both the L1 caches should reside after MMU for two reasons. First, if a physical address is used as the key to search the cache, there is no need to flush the cache after control is switched to a new task. Second, upon a TLB miss, the DAT tends to fetch the PT entry from the L1 data cache. If the PT entry is already in the data cache, it can be fetched in less time.

7.6.4 Paging Related Problems

7.6.4.1 Physical Addresses in the I/O Controller

If a system supports virtual memory, the CPU must pass the physical address of a memory buffer to the I/O controller because the controller has no address mapping hardware. This means extra system overhead is needed to translate a VA to a PA before issuing a direct memory access I/O.

7.6.4.2 Page Fix

Not all the pages in memory can be swapped. There is an F bit in the PFT, indicating that the page can not be swapped to disk. An example is an I/O buffer. If the memory buffer is swapped out and the I/O operation continues, the bits stored in the page will no longer be correct. To solve this problem, the OS sets the F bit in the PFT, so the page is fixed or locked in memory.

7.6.4.3 Thrashing

Thrashing in a paging system occurs when the physical memory size is not adequate. As a consequence, page fault occurs too frequently and the system does non-productive work by swapping pages in and out. When that happens, the paging disk rattles. As a solution, the operator simply kills some of the user tasks (i.e., jobs) in memory and therefore relieves memory requirement. As a rule, the ratio between the VAS and PAS should not be greater than two. In other words, a small real memory just can not support a large VAS under heavy load.

7.7 SEGMENTATION WITH PAGING

If the VAS is divided into independent segments and each segment is further divided into pages, we have segmentation with paging for implementing a very large VAS. Let us use the Pentium processor to illustrate the concept. When operating in protected mode, each extended address register or PC is 32 bits, so the EA is 32 bits to address a 4 GB segment. Adding the EA to the relocation address results in a 32-bit linear address. The paging feature is optional. Without paging, the linear address is a physical address and the entire segment is either in memory or not. With paging, the generated linear address, is translated into a physical address but the entire segment may be partly in memory and partly in disk storage. According to the spec, the processor chip has 33 address pins to access a block, and the data bus is 64 bits (8 B). Because the full PA in bytes can be 36 bits, the PAS can be as large as 64 GB.

7.7.1 Supporting Registers

A few control registers are specifically designed to facilitate paging. In particular, CR0 (control register zero) contains control bits to set the mode and status of CPU operations. B31 in CR0 is the PG (paging) bit. If PG is one, then paging is enabled after segmentation. Otherwise, paging is bypassed. If paging is enabled, then CR3 is the page directory base register (PDBR). The upper bits in the linear address represent the page directory number. Therefore, the page directory table is equivalent to a segment table in a linear segment system.

Note that Pentium supports multiple segments, and each segment is independent. That is, a program address is added to a relocation address to form the linear address in a segment. Four base registers are implemented on top. The global descriptor table register (GDTR) is shown in Figure 7.22a and it points to the GDT in Figure 7.22b. The interrupt descriptor table register (IDTR) is shown in Figure 7.22c and it points to IDT in Figure 7.22d. The local descriptor table register (LDTR) as shown in Figure 7.22e points to LDT in Figure 7.22f. Finally, the task register (TR) in Figure 7.22g points to task segment status (TSS) as shown in Figure 7.22h. Each register at least contains a 64-bit segment descriptor. All the descriptors are similar with some minor differences. That is to say, each descriptor contains the basic segment attributes, such as base address, limit, and access/control bits. The GDTR contains the segment descriptor for the GDT (global descriptor table), and its full name should be global segment descriptor table base register. Note that GDT is a segment but it has no entry in itself, so the GDTR does not have a selector in front. The LDTR contains the segment descriptor for the LDT (local descriptor table), and its full name should be Local Segment Descriptor Table Base Register. However, the LDT has an entry in the GDT, so the LDTR has a 16-bit system segment register in front to select an entry from either descriptor table, GDT or LDT. The IDTR contains the segment descriptor for the IDT (interrupt descriptor table) and it has no selector either. When an interrupt occurs, the interrupt vector is used as an index to access a gate descriptor in the IDT. The gate descriptor specifies the starting address of an interrupt handler. The TR contains the TSS (task state segment) descriptor retrieved from the GDT and it has a selector. The TSS descriptor contains status information of the running task. For example, a B (busy) bit of one indicates that the task is running or ready to run. If the busy bit is zero, the task is in wait, i.e. blocked state. The TSS descriptor has the base and the limit for the task state segment that mainly contains the running environment, (i.e., register save area), of the task.

Several instructions are provided to load and store the four descriptor registers. A segment selector, selector in short, is used to access a segment descriptor in the GDT or LDT, as shown in Figure 7.23a. In the selector, the upper 13-bit field contains an index to the entry in a segment descriptor table, global or local. That is, if the TI (table indicator) bit is 0, the entry in the LDT is accessed, otherwise the GDT is accessed. The two-bit field, b1 and b0, denotes a request privilege level (RPL) with 00 to have the highest privilege. During an intersegment reference, if the calling segment has a lower privilege, then an interrupt (i.e., protection exception) is triggered. The logical layout of a 64-bit segment descriptor has the base, limit, and access/control bits as shown in Figure 7.23b.

The 32-bit base is the physical address of a segment, the 20-bit limit indicates the segment size in 4 KB blocks, and the remaining 12-bit field provides both access control and status information of the segment. In particular, the data bit (physically, b22) indicates the operand size: one means 32 bits and zero means 16 bits. The P

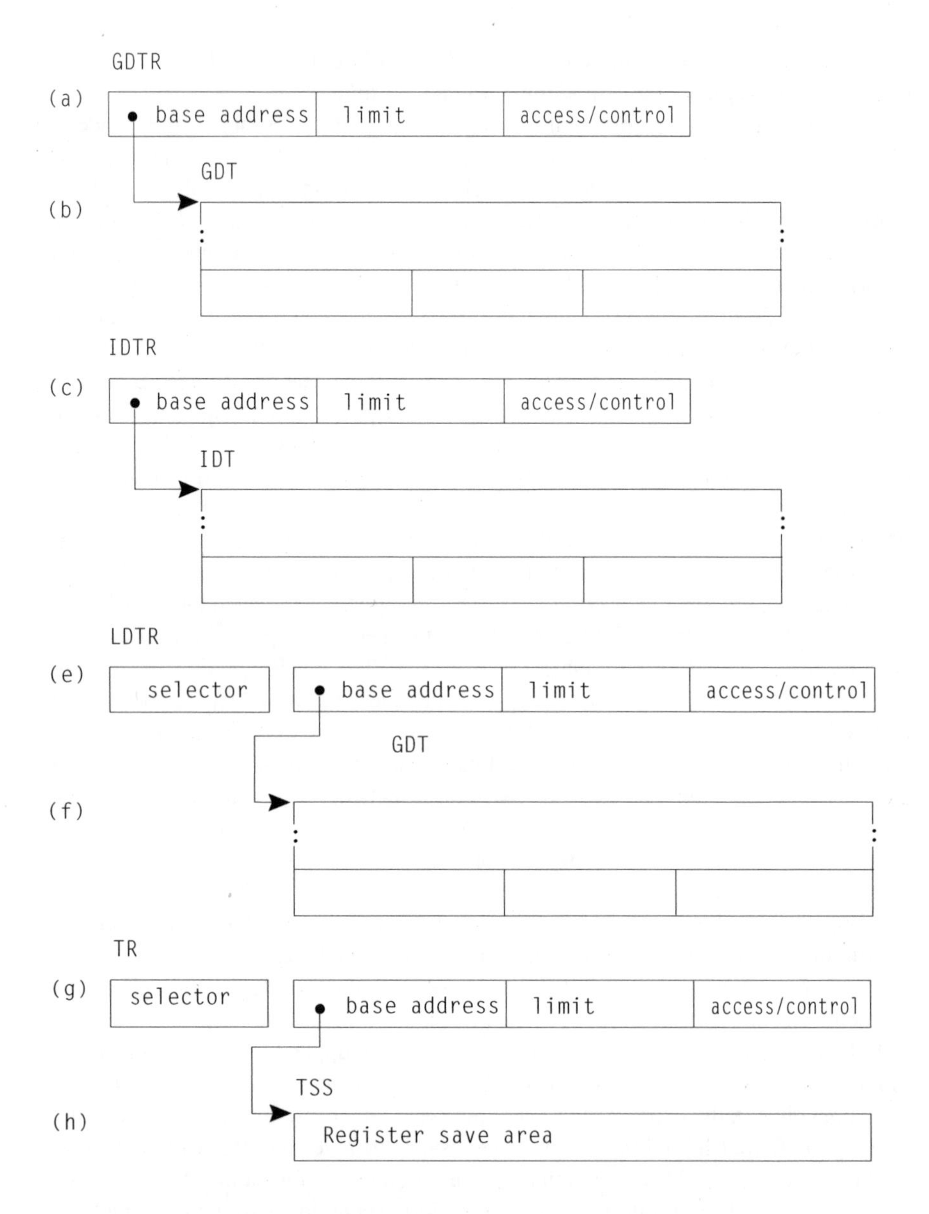

Figure 7.22 Top-level base registers and tables: (a) GDTR, (b) GDT, (c) IDTR, (d) IDT, (e) LDTR, (f) LDT, (g) TR, and (h) TSS.

(presence) bit (physically, b15) indicates whether the segment is in memory: one means in and zero means not in. During a segment reference, if b15 is zero, a segment fault is triggered to implement a demand segment system. Other bits may indicate access right, such as Read/Write, Read Only, privilege level, etc.

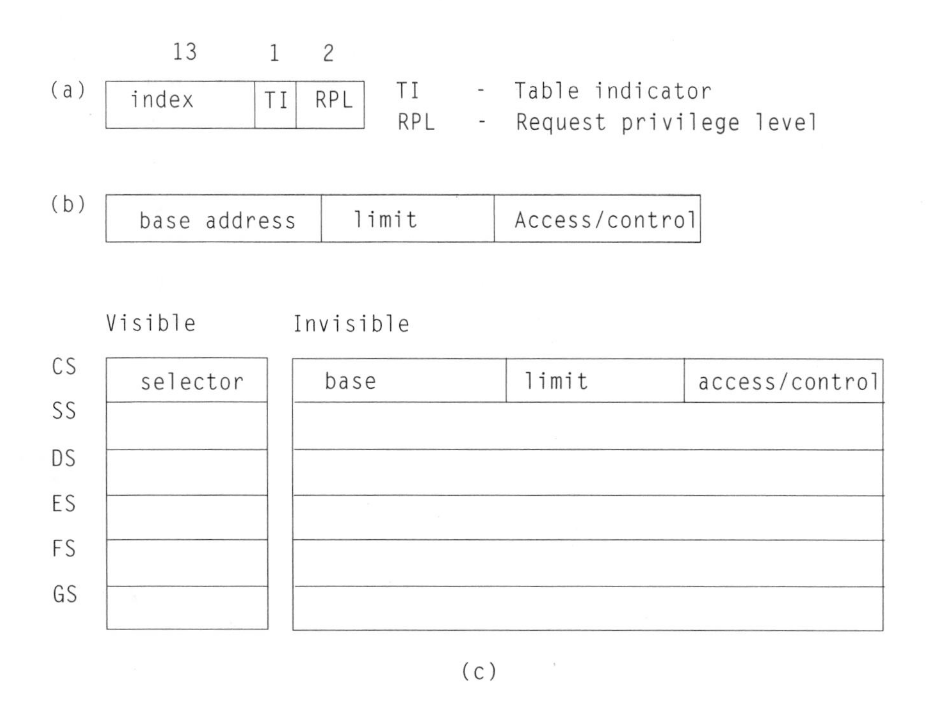

Figure 7.23 Registers: (a) the 16-bit selector, (b) the logical layout of a 64-bit segment descriptor in GDT or LDT, (c) the six segment registers.

The six segment registers are shown in Figure 7.23c and each register has two parts: visible and invisible. The visible part is a 16-bit segment selector, and the invisible part is its associated 64-bit segment descriptor. Usually, a mov (move) instruction is issued to change a segment register, namely the index, TI, and RPL. Based on TI, the index is used to fetch the entry in the descriptor table, GDT or LDT, and load the invisible part accordingly. In general, the OS uses the GDT and a user task uses the LDT.

7.7.2 Page Translation

On a Pentium, a program or logical address is always added to the base in one of the six segment descriptors in order to form the 32-bit linear address. Each segment, CS, SS, DS, ES, FS, and GS, has a visible segment selector and an invisible segment descriptor. If paging is enabled, this linear address is further mapped into a physical address through two memory references. The linear address in the segment is divided into a 10-bit PD (page directory number), a 10-bit P (page number), and a 12-bit D (displacement), as shown in Figure 7.24a. The PDT (page directory table) and PT (page table) have 1 K entries each. Because each table entry is 4 B, the size of the PDT, PT, or page is the same, that is 4 KB. The PD in the linear address is an index to an entry in the PDT. The base address of the page directory table is a 22-bit PF (page frame) or physical block address stored in CR3, aka PDBR (page directory base register) as shown in Figure 7.24b. Concatenating the PF fetched from the PDBR with the PD, we obtain the physical address of the PD entry without going through an add operation.

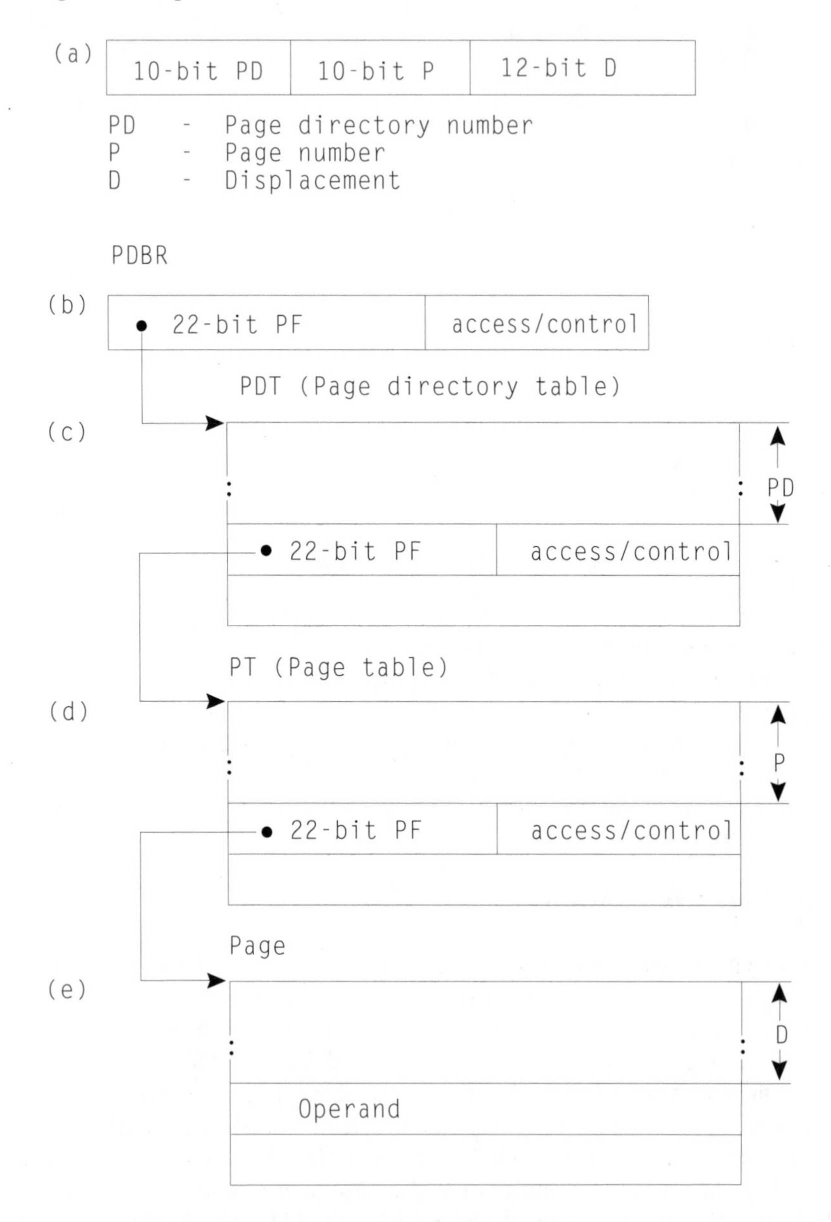

Figure 7.24 Data structures for address translation: (a) 32-bit linear address, (b) page directory base register, (c) page directory entry, (d) page table entry, and (e) operand in page.

As shown in Figure 7.24c, the fetched PDT entry contains a 22-bit PF pointing to the PT plus control bits. Concatenating the PF fetched from the PDT with P, we obtain the physical address of the entry in the PT. The fetched PT entry contains a 22-bit PF (page frame number) indicating the page plus some bits for access, control, and status, as shown in 7.24d. If the P (presence) bit is zero, the page is not in memory and so triggers a page fault. As a result, the paging supervisor takes over and loads the page from the disk to memory. Meanwhile, all the related tables are updated. If the P bit is one, the page is in memory, so the fetched PF in the PT entry is concatenated with D to form the physical byte address of the operand, as shown in Figure 7.24e. Obviously, the two extra memory cycles mean a penalty in speed. To alleviate the problem, a TLB is designed to contain the most recently used translation records, and the key is the 20-bit PD^P (PD concatenated with P). If the TLB search is successful, then address mapping is completed without having to search the PDT and PT.

7.7.3 Concluding Remarks

A time-sharing system may support a VAS that is smaller than its PAS in order to solve the relocation problem. If the relocation address is stored in a base register, mapping can be done fast enough so as not to impose any speed penalty. Paging tends to solve the fragmentation problem. As pages are swapped in and out between memory and and disk at high speed, the CPU can run more user programs that are I/O bound.

A very large VAS solves the overlay problem, but the price paid is stiff. As memory is getting cheaper and cheaper, fragmentation is no longer an issue. Thus, the debate is whether paging should be implemented at all. In a single segment system with paging, more programs are executed so the throughput is increased by 1.5 percent. However, this is still application dependent, and there have been cases where the vendor was told to strip the paging hardware because of the decrease in performance. For example, when compute bound jobs or real-time jobs are running, dynamic address translation is turned off. At the system level, a large VAS with paging requires an EPT in the load module. In the nucleus, the memory space occupied by the global PFT and local PTs is just too large.

7.8 SUMMARY POINTS

1. Based on instruction and data flows, all the machines can be generally classified into SISD, SIMD, and MIMD.
2. An SISD machine is a scalar or superscalar processor in that each instruction, arithmetic or logical, acts on one piece of data.
3. A superscalar processor allows many instructions executed in parallel, but each instruction still operates on one piece of data.
4. The four-level storage structure consists of a register set, cache, central memory, and disk.

5. In a memory system with addresses interleaved in two ways, the two physical banks can be busy at the same time.
6. A two-port memory may receive requests from both the CPU and the I/O processor at the same time. Its arbitration logic grants the cycle to the I/O if the addressed bank is free.
7. To achieve overlapped operations in a CPU, we have data dependency and control dependency problems.
8. The CPI of a machine is defined as the average number of clocks required to execute an instruction.
9. The MIPS is the clock rate in Mhz/CPI.
10. The MFLOPS is the clock rate in Mhz/Clocks per FLOP.
11. A cache is an associative memory, and each entry mainly contains two fields: key and data.
12. There are three cache memory search techniques: fully associative, direct mapping, and set associative.
13. A direct mapping cache uses one comparator, while a two-way set associative cache uses two comparators.
14. An instruction queue coupled with an instruction cache reduces the instruction retrieval (fetch) and decode time.
15. A superscalar machine may support a few vector instructions, such as push, pop, move, search, etc.
16. The VAS is the logical programming space as seen by a user, while the PAS is the real memory size.
17. A VAS may be greater than, equal to, or less than a PAS.
18. A DAT box or MMU contains a TLB and logic to search through tables in memory.
19. A segment is a contiguous code block of variable size, while a page is a contiguous code block of fixed size.
20. In a multi-segment system, the computed EA can not cross segment boundaries.
21. Pages are linear in a segment to solve fragmentation problems.
22. The paging supervisor, part of the OS nucleus, receives control after a page fault interrupt.
23. A TLB is a fully associative cache that contains the MRU translation records.
24. The L1 caches should be placed after the MMU.
25. An entry in the EPT mainly contains the disk address and access right of the page.
26. The PFT is the global data base in the OS nucleus, and each PFT entry describes the status of a page frame.
27. A paging system has many related problems, e.g., physical address in an I/O controller, page fixed, and thrashing.
28. Segmentation with paging means that the VAS is divided into segments, and each segment is further divided into pages.

PROBLEMS

1. Based on instruction flow and data flows, briefly describe the three classes of computer systems.
2. What is a superscalar processor?
3. What is a four-level storage structure?
4. Describe the concept of memory interleaving.
5. Describe the data dependency and control dependency problems in a pipelined CPU.
6. Define the CPI (clocks per instruction) of a machine.
7. Describe the difference between MIPS and MFLOPS.
8. As shown in Figure 7.4c, the logic box (on top) can prepare signed multiplicands. That is, the sign is extended all the way to the left. To multiply signed -1 by signed -1, the product is +1 as follows.

```
     1  1  1  1  1  1  1  1     First multiplicand            (-1)
     1  1  1  1  1  1  1        Second multiplicand           (-2)
     1  1  1  1  1  1           Third multiplicand            (-4)
  +  0  0  0  0  1              Twos complement of the Fourth
                                multiplicand                  (+8)
  ----------------------------
     0  0  0  0  0  0  0  1     Eight-bit signed product      (+1)
     P7 P6 P5 P4 P3 P2 P1 P0
```

   The tricky part is to generate the twos complement of the fourth multiplicand in two steps. First, feed the ones complement of the fourth multiplicand to the six-bit CSA. Second, during the final arithmetic add cycle, remember to turn on the Cin (carry in) bit of the CPA. If so, what are the bits in SV1, CV1, SV2, CV2, and Cin?
9. In a 32-bit multiplication tree, how many CSAs are cascaded before the final 33-bit CPA?
10. Name the three search techniques used in cache memory design.
11. Describe the difference between an instruction queue and an instruction cache.
12. Is it true that a VAS may be greater than, equal to, or less than a PAS?
13. What is the difference between a linear segment system and an independent segment system?
14. What is the key difference between a segment and a page?
15. During address mapping, describe the two hardware steps in a paging system.
16. What is contained in a TLB (translation look-aside buffer)?
17. Describe the functions performed by the paging supervisor.

18. Describe the difference between a PT and its EPT. Under what condition can the disk address in the EPT be eliminated?
19. Why do we set the V=R bit in the PSW to bypass the DAT?
20. What are the three related problems caused by a paging system?

CHAPTER 8

Vector and Multiple-Processor Machines

8.1 VECTOR PROCESSORS

A SIMD machine provides a general purpose set of instructions to operate on arrays, namely, vectors. As an example, one add vector instruction can add two arrays and store the result in a third array. That is, each corresponding word in the first and second array are added and stored in the corresponding word in the third array. This also means that after a single instruction is fetched and decoded, its EU (execution unit) provides control signals to fetch many operandi and execute them in a loop. As a consequence, the overhead of instruction retrievals and decodes are reduced. As a vector means an array in programming, the terms vector processor, array processor, and SIMD machine are all synonymous. A vector processor provides general purpose instructions, such as integer arithmetic, floating arithmetic, logical, shift, etc. on vectors. Each instruction contains an opcode, the size of the vector, and addresses of vectors. A SIMD or vector machine may have its data stream transmitted in serial or in parallel. A parallel data machine uses more hardware logic than a serial data machine.

8.1.1 Serial Data Transfer

The execution unit is called the processing element (PE) where the operations are performed. If one PE is connected to one processing element memory (PEM), we have a SIMD machine with serial data transfer as shown in Figure 8.1a. That is, after decoding a vector instruction in the CU (control unit), an operand stream is fetched and executed serially in a hardware loop. That is, serial data are transferred on the data bus between the PE and PEM on a continuous basis until the execution is completed. In a serial data SIMD machine, there is one PE and one PEM. However, one instruction retrieval is followed by many operand fetches.

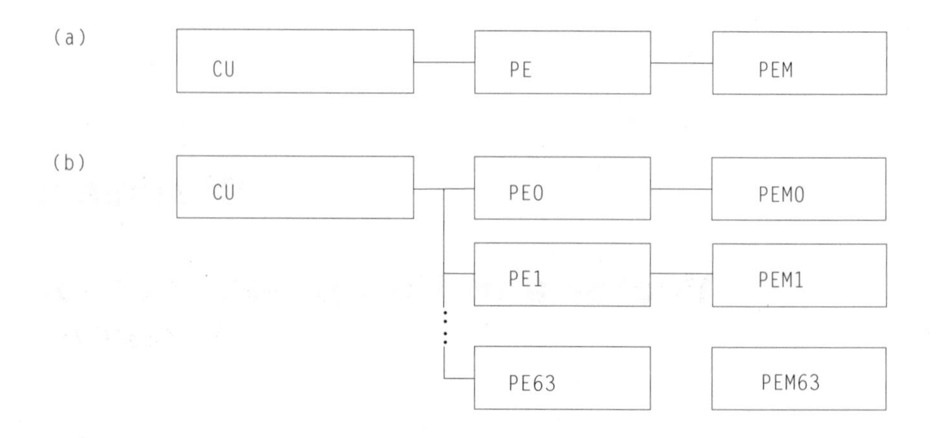

Figure 8.1 SIMD machines: (a) serial data transfer and (b) parallel data transfer.

8.1.2 Parallel Data Transfer

If multiple PEs are tied to the CU and each PE is connected to a PEM, we have a parallel data machine, as shown in Figure 8.1b. For example, the ILLIAC IV (Illinois Advanced Computer IV) has 64 PEs denoted as PE0, PE1, ... , PE63. Each PE is connected to its own PEM denoted as PEM0, PEM1, ..., PEM63. Operations are performed in parallel because the data transfer is done in parallel. On a SISD machine, if we multiply two arrays A and B and place the result in array C, we write a loop as shown below:

```
DO i = 1 to 64, step 1
C[i] = A[i] * B[i];
ENDDO;
```

In the loop, each entry in array A is multiplied by its corresponding entry in array B, and the product is stored in the corresponding entry in array C. The same operation is repeated a total of 64 times. But on an ILLIAC IV, this loop is translated into one vector instruction, and when executing this instruction, there are 64 multiply operations in parallel. One can argue that the market of vector machines is never good for two reasons. First, if the vector instruction is frequently used, it can be implemented on a superscalar machine as an enhancement. Second, many processors can be interconnected, so a cluster of machines can do team work, as introduced next.

By definition, a MIMD machine consists of many processors that are interconnected. Because each processor can execute its own set of instructions on an independent basis, we see that multiple instructions and multiple data are fetched at the same time. In other words, each processor can act on its own data fetched from memory. In a broad sense, the MIMD class consists of all the multiple-processor machines, namely the multistation systems, multiprocessing systems, and computer networks. Due to its sophisticated software structure of network I/O, a computer network stands out as its own class. There are many issues in designing a multiple computer systems. We start from the basics and move to more advanced topics in the following sections.

8.2 INTERPROCESSOR COMMUNICATIONS

A hardware connection must exist between two processors. If the connection is wireless, it is not visible. Thus, a connection means that one side has a transmitter and the other side has a receiver. As far as hardware is concerned, there are three ways to facilitate interprocessor communications as follows:

- Interrupt
- Shared Memory
- Via Channel-to-Channel I/O

8.2.1 Via Interrupt

As shown in Figure 8.2a, an interrupt bus exists between two CPUs. Usually, CPU 1 is the main processor and CPU 2 is an I/O processor. When CPU 2 finishes its I/O, it generates a hardware signal to interrupt CPU 1. That is, CPU 2 can force CPU 1 to relinquish control of its current program in execution. As a result, an interrupt service routine (ISR) on CPU 1 takes over control and processes this event. If a microcomputer has many I/O controllers, each controller is classified as a processing unit. The controller sends an interrupt request signal after an I/O completion to the CPU. After accepting the request, the CPU sends an acknowledge signal to the controller. Since each of the I/O controllers can issue an interrupt request at the same time, an arbitration scheme is needed for the CPU to accept one interrupt. Thus, a request-acknowledge mechanism using daisy chain is different from centralized control that uses separate control lines.

8.2.1.1 Daisy Chain vs. Centralized Control

As shown in Figure 8.2b, the interrupt bus connects the main CPU to three I/O controllers in a daisy chain. The interrupt bus provides request, acknowledge, and data signals. The interrupt request signal from each controller is or-tied to a common line, which means many controllers can request interrupt at the same time. The interrupt grant signal from the CPU ripples through a daisy chain. Therefore, each controller has an input port to receive the grant signal and an output port to transmit

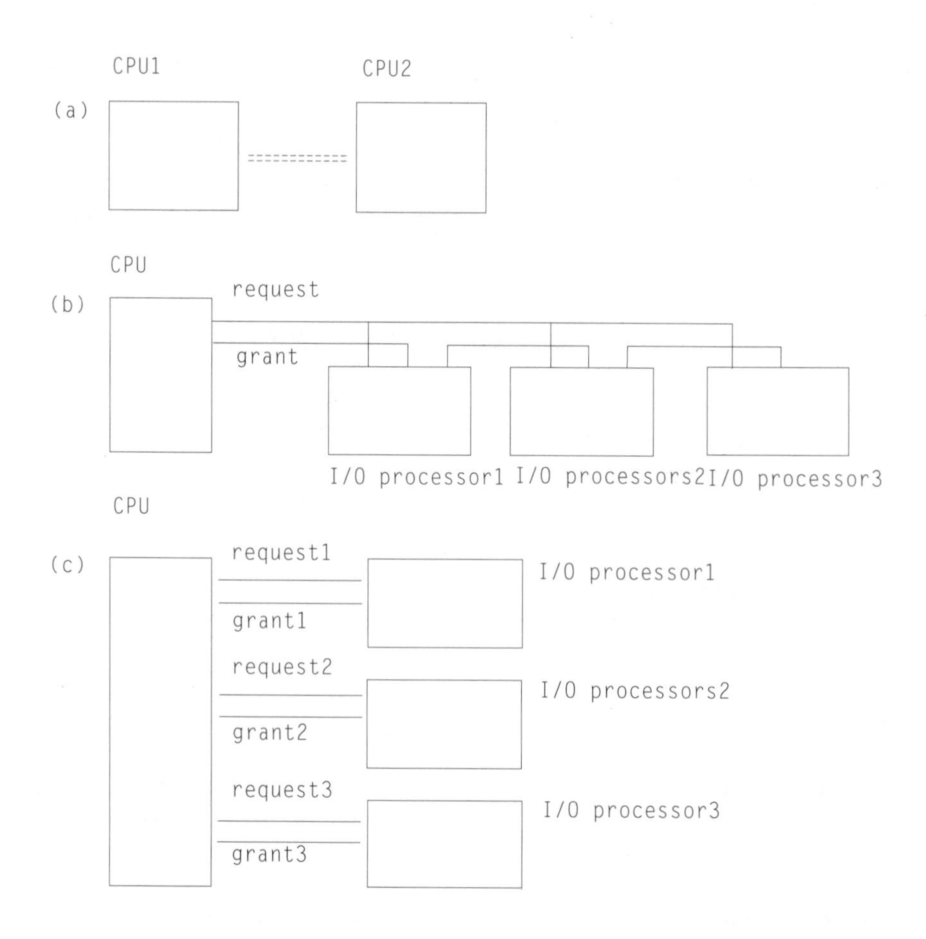

Figure 8.2 Interrupt buses: (a) an interrupt bus between two CPUs, (b) the request/grant bus as a daisy chain, and (c) the centralized request/grant pairs.

the grant signal. The first controller receives the grant signal directly from the CPU and the last controller in the chain can only receive the signal and can not transmit it.

The daisy chain concept works as follows. The CPU does not know whose interrupt request this is. If the CPU is in a state to respond to the interrupt, the CPU transmits a grant (acknowledge) signal. After receiving a grant signal, the first controller takes action based on whether it has requested an interrupt. If it has, the I/O controller does not pass the grant signal to the next processor. Thus, the controller, after being acknowledged, may place an interrupt vector address on the interrupt data lines. If the controller has no interrupt request, it passes the grant signal to the next controller in the chain. That is to say, any controller with an interrupt request waits in the chain until its turn comes to receive a grant signal. The assigned priority is determined by the position of each processor in the chain. The leftmost processor has the highest priority and the rightmost processor has the lowest priority. In fact, the priority is fixed by the position of each processor. This scheme works fine because only one controller can respond at one time.

The acknowledge control mechanism may be centralized, as shown in Figure 8.2c. Each controller has an interrupt request line and an interrupt grant line. In concept, each controller may send an interrupt signal to the CPU. If the CPU is in a state to respond to an interrupt, it transmits the grant signal to the controller with the highest priority on a separate wire. That is to say, the CPU acknowledges an interrupt based on priority as physically assigned by its line position. In a more sophisticated system, the bus priority can be dynamically changed by a software routine to suit application requirements. This arbitration idea can be applied to memory contention so many CPUs may issue requests to access a shared memory at the same time.

We now understand that after an I/O completion, the I/O processor can interrupt the CPU. But, the CPU can also interrupt an I/O processor. In practice, after the I/O routine prepares I/O messages in shared memory, it issues a privileged instruction or writes to a special location in shared memory to interrupt the I/O processor. An example of executing a privileged instruction is shown below:

```
SIO   x'284'   ;Start I/O device
```

where SIO means Start I/O, and the I/O address of a particular device is 284 in hex. As a result of its execution, the I/O processor gets interrupted. Consequently, its service routine fetches the I/O message, interprets, and starts I/O. Each I/O message contains the operation code, memory block address, and block count in bytes. Note that on some mainframes, the I/O processor runs the OS so executing an exchange jump instruction interrupts the main CPU as shown below:[5,65]

```
EXN         ;Exchange Jump
```

where EXN stands for exchange jump which is privileged. In fact, all the instructions on the I/O processor are privileged because they can not be accessed by a user. The execution of EXN causes the main CPU hardware to perform context switching that is, save the register set and load a new set from memory whose address is coded in a PPU register so the CPU executes a new program from an address in its program counter. The SIO or EXN instruction is different from a system call because the instruction interrupts the other processor, not itself.

At the system level, the interrupt hardware mechanism means a one-bit message sent from one processor to the other processor. A string of bits can be passed from one processor to another processor by placing the message in shared memory or transmitting it over an I/O channel.

8.2.2 Via Shared Memory

If two processors are located in one room, the message can be placed in a shared memory, as shown in Figure 8.3. The memory has two connections; one goes to the

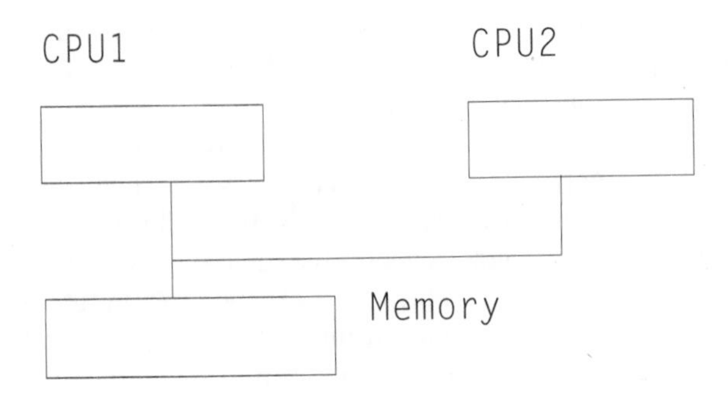

Figure 8.3 The shared memory system.

CPU and the other goes to the I/O processor. It is also possible that both processors are main CPUs. If a system has shared memory, it is classified as a closely coupled system. There is a subtle difference between interprocessor communication and intertask communication. The former is a hardware connection and the latter is a software capability. A task is a program in memory that has all the resources allocated by the OS. We define intertask communication as a system capability that one task can send a message to another task on a different processor. A hardware connection scheme must exist between the two processors, and special network I/O software must be written to support intertask communications over the network.

Note that the shared memory concept can be extended to a DMA controller, which can be classified as an I/O processor. Each hardware register in the controller is an IOR (I/O register) that is connected to a separate I/O bus or mapped into a memory word. In either case, the IOR is shared by both processors. That is, the main CPU writes into the IOR for the I/O processor to read, interpret, and execute. Upon I/O completion, the controller writes into the status register for the CPU to read and interpret. Thus, the IORs act like shared memory because both processors have access to them.

8.2.2.1 Cache Coherency

Assume the design of a dual processor system that uses shared memory to communicate, with each CPU having its own data cache, as shown in Figure 8.4.

Because both CPUs can access the shared memory, the rule is established that only one CPU can write a block at one time. This works fine if there is no data cache. With a data cache enabled, it is possible for CPU 1 to write into the shared memory block while CPU 2 still has a stale copy of the old data. This is defined as the data cache coherency problem. That is, the copy in the data cache is not consistent with the copy in memory. One common solution is to use the snooping hardware technique, as described below:[63]

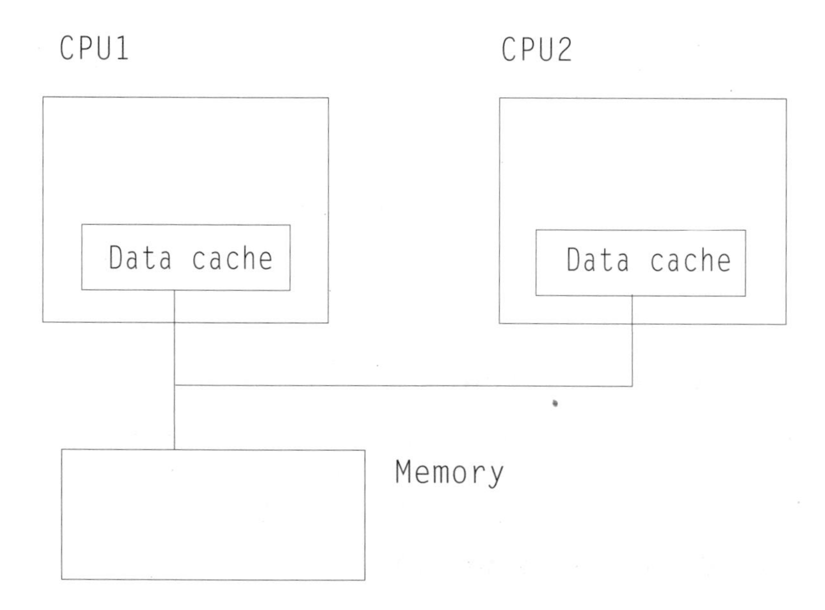

Figure 8.4 Data caches and shared memory in a dual processor system.

8.2.2.2 Snooping Technique

There are special command lines between the two CPUs. One of them is the snoop command line and others may be response lines. Before CPU 1 writes the memory block, it sets its data cache to write-through mode so data is written into memory immediately. While CPU 1 writes the shared block, it also asserts the snoop line to signal CPU 2. Because the snoop line is asserted, CPU 2 uses the address on the shared bus and reads its cache lines. If there is a hit, it invalidates the matched entry. After that it asserts a response to signal CPU 1 that action has been taken.

However, the coherency problem can be solved if the data cache entry can be invalidated by an instruction. Suppose the central memory has two ports, as shown in Figure 8.5. One port is used by the CPU, while the other port is used by an I/O processor. As one processor places a message in shared memory, the other processor fetches and interprets. The problem arises when the main CPU has a data cache but the I/O processor does not. The I/O processor needs no data cache because its data rate is slow. The main CPU has the cache coherency problem because after the I/O processor writes into memory, the data is not reflected in its cache. In other words, the data cache in the main CPU may contain an old copy of the data. Take the read operation for example; after the I/O processor moves data from an I/O device to memory, it places a channel status word at a fixed location in shared memory to indicate that the data block is ready. Nevertheless, the status word written by the I/O processor is not stored in the data cache as seen by the main CPU. As one solution, the main CPU may

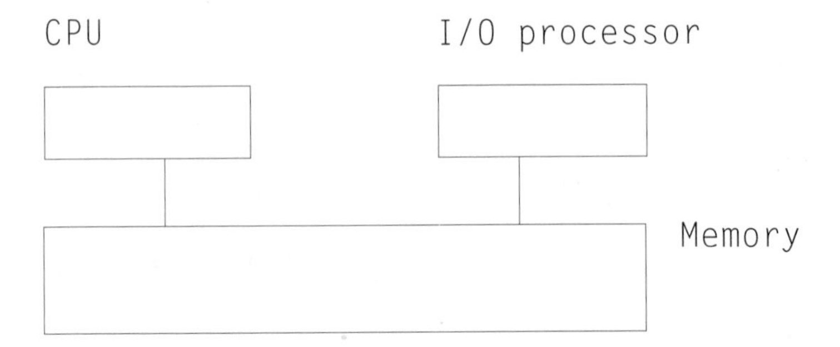

Figure 8.5 The shared central memory has two ports.

issue an instruction to invalidate the data cache by the address before it reads the I/O status word from shared memory. Another solution, at the cost of system performance, is to disable or turned off the data cache altogether.

8.2.3 Via Channel-to-Channel I/O

The third hardware connection scheme is channel-to-channel I/O, as shown in Figure 8.6. As far as hardware is concerned, one processor transmits a message via its I/O channel and the message is received by the remote processor via its I/O channel. Thus, the software on one host processor issues a write request to its channel to transmit bits on the line while the remote routine issues a read request to its channel to receive bits. Each I/O channel or processor may have its own communication controller.

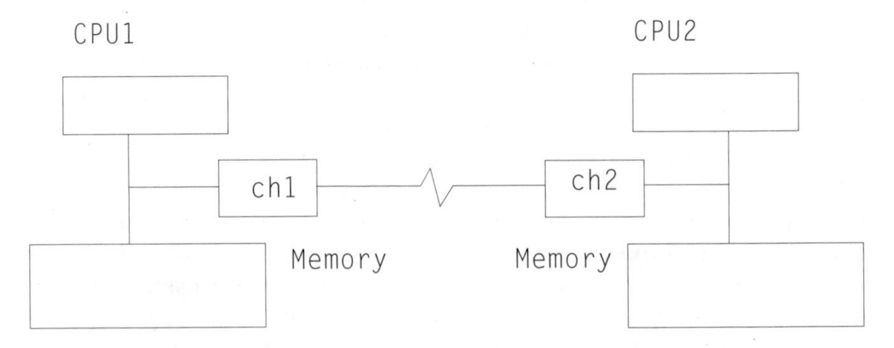

Figure 8.6 Channel-to-channel I/O connection.

8.3 INTERPROCESSOR MESSAGES

An I/O message is passed from the main CPU to the I/O processor and vice versa. Usually, the I/O message can be placed in shared memory for the other processor to fetch, interpret, and execute, or the messages can be transmitted on the line via its

I/O channel between two processors. The format of a message is application dependent so it is a software design issue. The messages for a disk controller, display controller, or communication processor are described below.

8.3.1 Channel Command Word vs. Channel Status Word

On an IBM 370 mainframe, the CPU places a channel command word (CCW) in shared memory for the I/O processor to fetch, interpret, and execute.[26] In return, the I/O processor places a channel status word (CSW) containing the status of an I/O operation. If we want the freedom to place a CCW anywhere in shared memory, we need to have a command address word (CAW) that must have a fixed address in low memory. That is to say, the CAW points to the CCW list, as shown in Figure 8.7a. Because the I/O instructions and the low memory are protected, a user must issue a system call to request an I/O service from the OS. Usually, the I/O routine on the CPU prepares the CCW list and issues an SIO instruction. After that, the I/O processor gets interrupted so its ISR takes over control. First, it fetches the CAW at the fixed address.

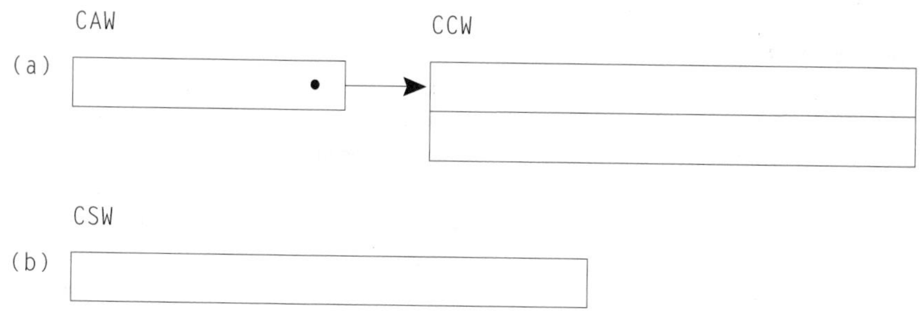

Figure 8.7 Processor – channel communication messages: (a) CAW points to CCW and (b) CSW.

Next, from the pointer in the CAW, it fetches the CCW, interprets, and executes. A typical CCW contains an opcode, flag bits, a 24-bit byte count, and a 32-bit physical memory address. After executing the CCW, the flag bits tell the I/O processor what to do next. Let us try to read a sector from a disk or tape into memory. The CCW contains opcode, flag bits, a memory address, and the block size in bytes. After executing the CCW, the ISR on the I/O processor examines the flag bits in the CCW and decides what action to follow. More often than not, the flag bits tell the I/O processor to interrupt the CPU.

The flag bits tell the I/O processor to fetch the next CCW in sequence and interpret it. As a matter of fact, it is possible to construct the CCW list in a loop so that the last CCW branches to the beginning of the loop. Such a design was often used to handle I/O operations of real-time data acquisition. As the main CPU continues to receive and process data, the I/O processor continues to supply data. After each real-time data frame arrives, the I/O processor signals interrupt to the main CPU to request attention.

After the I/O is completed, the I/O processor sends an interrupt request signal to the CPU. If the interrupt request is acknowledged, the I/O processor places a message in shared memory at a fixed low address. This reply message is known as the CSW, as shown in Figure 8.7b. A CSW contains the address of the last CCW being executed by the I/O processor, a channel status byte, a device status byte, and a residual block count. If the I/O completion is successful, the residual block count is zero. However, if the I/O operation is not successful, the residual block count contains a non-zero integer indicating the number of bytes yet to be transferred.

After the main CPU gets interrupted, its ISR fetches the CSW and examines the residual block count, the channel status byte, and the device status byte. The ISR posts the I/O event if it is a successful completion. Posting an I/O event means to wake up the task which originally issued the I/O request. In case of any error, the ISR on the main CPU will tell the I/O routine to retry. In computer network parlance, the CCW is a request message to the service provider, and the CSW is a confirm message to the service user. Both the CCW and CSW are I/O messages to describe an I/O operation.

8.3.2 Message for Display

The PC contains the display controller or processor whose job is to fetch bits in its screen memory, interpret, and execute. The screen memory for display is divided into bit planes that are mapped into main memory at a fixed high address. That is, the screen memory for display is shared so the CPU program prepares the display messages in it. Note that the display message is passive because the main CPU does not issue a signal to interrupt the display processor. Thus, the display controller, (i.e. processor) must routinely fetch the bits from the shared memory, interpret, and execute.

The display screen is a cathode ray tube whose glass surface area is divided into pixels (picture elements). Each pixel is a dot on the screen, and the total number of pixels on the screen surface defines its resolution. The display controller can operate in text or graphics mode, and each mode has its own display messages. In text mode, the screen is divided into 80 columns by 25 rows, so the total number of characters on screen is 2,000. For each character, the message is composed of two bytes. The low-order byte is the ASCII character, and the high-order byte is its color attribute, as shown in Figure 8.8a. The cathode ray tube has three basic color guns: R (red), G (green), and B (blue). The lower four bits in the attribute byte specify a foreground color. By mixing the basic colors with the I (intensity) bit, we can derive a specific color. The next three bits specify the background color and the last bit tells whether to blink the character or not.

In graphics mode, each pixel is a dot on the surface with its associated color attribute. A resolution can be high or low as determined by the number of colors allowed. Take the SVGA (super video graphics array) card for example; its resolution is 1024 by 768 with 256 colors. That is to say, the number of pixels on the screen is (1024 * 768), and each pixel is represented by a one-byte color attribute as shown in Figure 8.8b. The screen (shared) memory size on the card is 768 KB. On the other

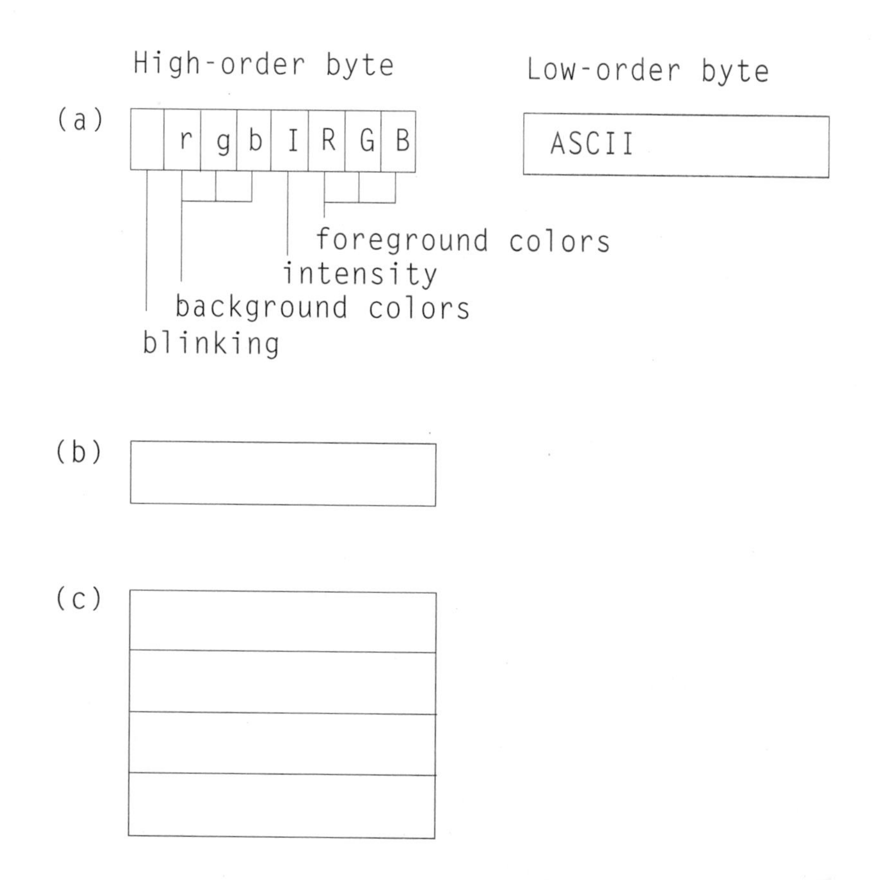

Figure 8.8 Messages for display processor: (a) the ASCII character and its attribute, (b) the one-byte color pixel, and (c) the four-byte color pixel.

hand, an advanced display may support a resolution of 1280 by 1024 pixels. If each pixel is represented by a four-byte attribute, the screen memory size is 5.12 MB for just one screen, as shown in Figure8.8c. In such a design, each of the first three bytes represents a basic color with 256 intensity levels, and the last byte is called the alpha (alphanumeric) plane for overlaying text on graphics. The program on the CPU must move bits in the screen memory fast enough. In other words, the CPU has quite a demanding graphics processing requirement.

8.3.3 Message for Communication Processor

There are two network models: VC (virtual circuit) and datagram. The message on a VC model is called a packet and the message on a datagram model is called a datagram. The difference between them lies in the fact that a packet follows a fixed route and a datagram does not. Each corporate computer site may have a two-processor pair:

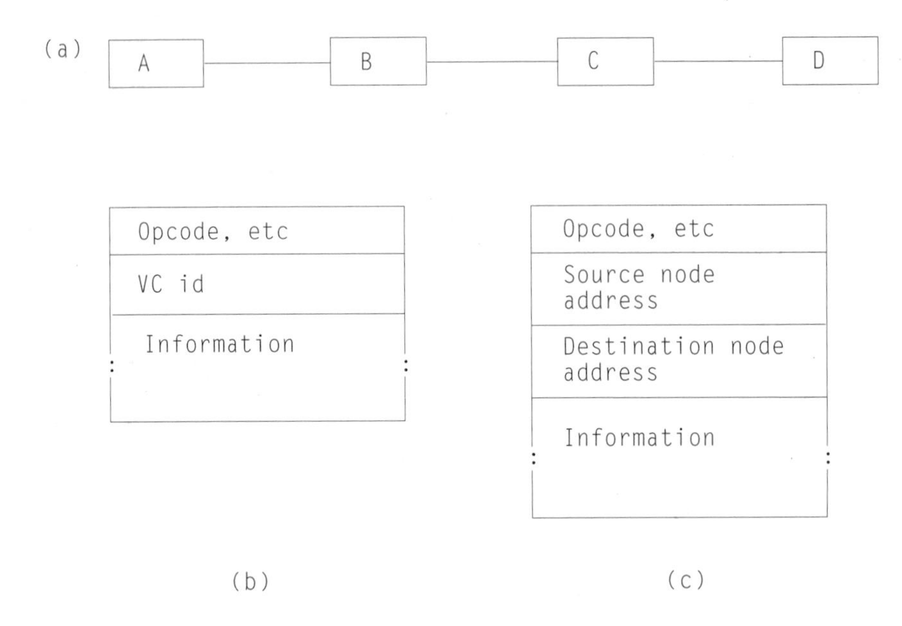

Figure 8.9 Network communications: (a) host A to host D connection, (b) the packet, and (c) the datagram.

the host and the communication processor. Typically, the host processor passes a message to its communication processor via shared memory. The communication processor passes the message to another processor in a network. Eventually, at the remote site, the message is received by the communication processor which, in turn, passes the message to its destination host. Figure 8.9a depicts a host to host VC connection with the communication processors omitted on purpose.

The host computer A in San Francisco is called the origination node that initiates the connection phase. The host computer D in New York is the destination node. There is a VC between the two hosts, and the two nodes B and C are intermediate nodes in the middle of the route. In our example, node B is in Denver and node C is in Chicago. Both B and C are communication processors with or without a host. The packet is the basic transmission unit as seen by the two hosts. The packet layout is shown in Figure 8.9b and its short header contains opcode, data type, and routing information, e.g., VC id. In contrast, a datagram model has no connection phase, so its route is not fixed. The datagram header is longer and contains opcode, source (sending) node address, destination (receiving) node address, etc., as shown in Figure 8.9c. In either model, the header of a packet or datagram provides routing information so that each node in the route knows how to relay the message from the sending host to the receiving host. [22,24]

8.4 MULTISTATION SYSTEM

A multistation system has one main CPU surrounded by many I/O processors. Each I/O processor is called an I/O station that moves data between memory and an I/O device. The I/O processor known as the PPU (peripheral processing unit), has a much slower speed than the main CPU. This concept was first employed by mainframes, so all processors can operate at the same time. Nowadays, even a personal computer has many I/O stations, such as keyboard, display, communications, etc. The OS usually runs on the main CPU if it is a scalar or superscalar processor. However, in a few exceptions, the OS runs on an I/O processor.[5] Note that the main CPU can be a vector processor. In such a design, the OS runs on one or several I/O processors. For example, the CPU of the ILLIAC IV is a powerful array processor; its OS runs on a team of I/O processors made of mainframes.

8.5 MULTIPROCESSING SYSTEM

A multiprocessing system consists of multiple processors interconnected and characterized by (1) a program running on one CPU can communicate with another program running on a different CPU; and (2) the CPUs are usually homogeneous, interchangeable, and adjacent.

In a multiprocessing system, each CPU may have its own memory in addition to shared memory. Each CPU performs a specific function provided that a supervisor routine runs on one CPU to dispatch workload to other CPUs. In a closely coupled system, the processors are adjacent so they use shared memory to pass messages. In a loosely coupled system, the processors can communicate via channel-to-channel I/O.

One design case is to designate one processor as the master (i.e., supervisor) and the rest as slaves. After receiving a message from the master, the slave interprets and carries out the execution. The master coordinates the work done by all the slaves. Each time as the master dispatches work load to the slave, it also monitors the completed results. If the master fails due to hardware glitches, it can not act as the monitor. Well designed software will ensure that the system operates with a smooth transition in passing power to the next node. Henceforth, the supervisor running on the master computer leaves messages in shared memory constantly to claim its position as the boss. The processor that is second in the command chain checks the message on a periodic basis to make sure that the master is in command. If at any time the master is not able to leave a message in shared memory, the next processor in command declares its position as the new boss and takes over the duties of the master.

In a broad sense, a multiprocessing system can be thought of as a simple computer network where intertask communications are provided so that all the CPUs work as a team to achieve a common goal. Some of the communication issues are addressed below.

8.5.1 Shared Memory Bus

As regards processor interconnection, all the pins on the chip constitute a host system or local bus to communicate with the outside world. The memory bus has many address, data, and control signals that may be shared by other CPUs or controllers. Because the number of pins on the chip keeps changing, the host bus structure also changes accordingly. For example, the 8088 chip has 40 pins and the 286 has 64 pins. The Pentium chip has 242 pins, including a 33-bit address bus, a 64-bit data bus, many pins for voltages, ground, control, etc. The new IA-64 chip has even more pins, including a 128-bit data bus so that many instructions or data can be accessed in a bundle. Because the I/O devices have their own characteristics, a common I/O bus standard is desired so that manufacturers can connect their I/O devices to the bus via a bus controller.[77] In practice, some standard I/O buses in the field are Multibus, Micro Channel, ISA (industry standard architecture), EISA (extended ISA), PCI (peripheral component interconnect), and Ethernet, to name a few. Generally, each I/O bus uses many copper wires to transmit control, address, and data signals. In concept, the address and some of the control lines are simplex (i.e., transmission is allowed in one direction). The data lines are mostly bi-directional, parallel, and half-duplex (i.e., transmission is allowed in both directions but only one at a time). In the following, we introduce two well-known buses: the 188-pin PCI and the two-wire Ethernet.

8.5.1.1 PCI Bus

In the 1990s, Intel patented the 188-pin PCI bus standard and placed the mechanical, electrical, and functional specifications in the public domain.[61] Consequently, manufacturers can connect their I/O devices or CPUs to the bus. The signal pins on a particular processor chip constitute its host system or local bus, which is proprietary. However, the host processor can be connected to a PCI bus via a host – PCI bridge, as shown in Figure 8.10. This bus bridge is the controller on a card that connects two buses. On the host side, we see the CPU, L2 cache, and central memory modules. The L2 cache is optional since it contains the mirror image of a portion of memory that includes instruction and data. Both the L2 cache and memory modules are tied to the host system bus. Thus, a block of instructions or data can be accessed from the L2 cache or memory per each request. The PCI side contains the graphics controller which drives a display. On either side of the bus bridge, the address and data wires are multi-dropped so they are shared by all the processors and devices. In other words, when a processor or controller places bits on the bus, other logical devices on the bus can read the bits at the same time. A bus bridge controller can make signal conversions and can arbitrate bus requests from multiple processors. That is, after each processor makes a request, it must receive a grant signal from the bridge so that it can become the bus master of the next cycle. Without receiving a bus grant signal, the requesting processor must patiently wait for its turn. If multiple requests exist at the same time, the bridge is the centralized arbiter and only one

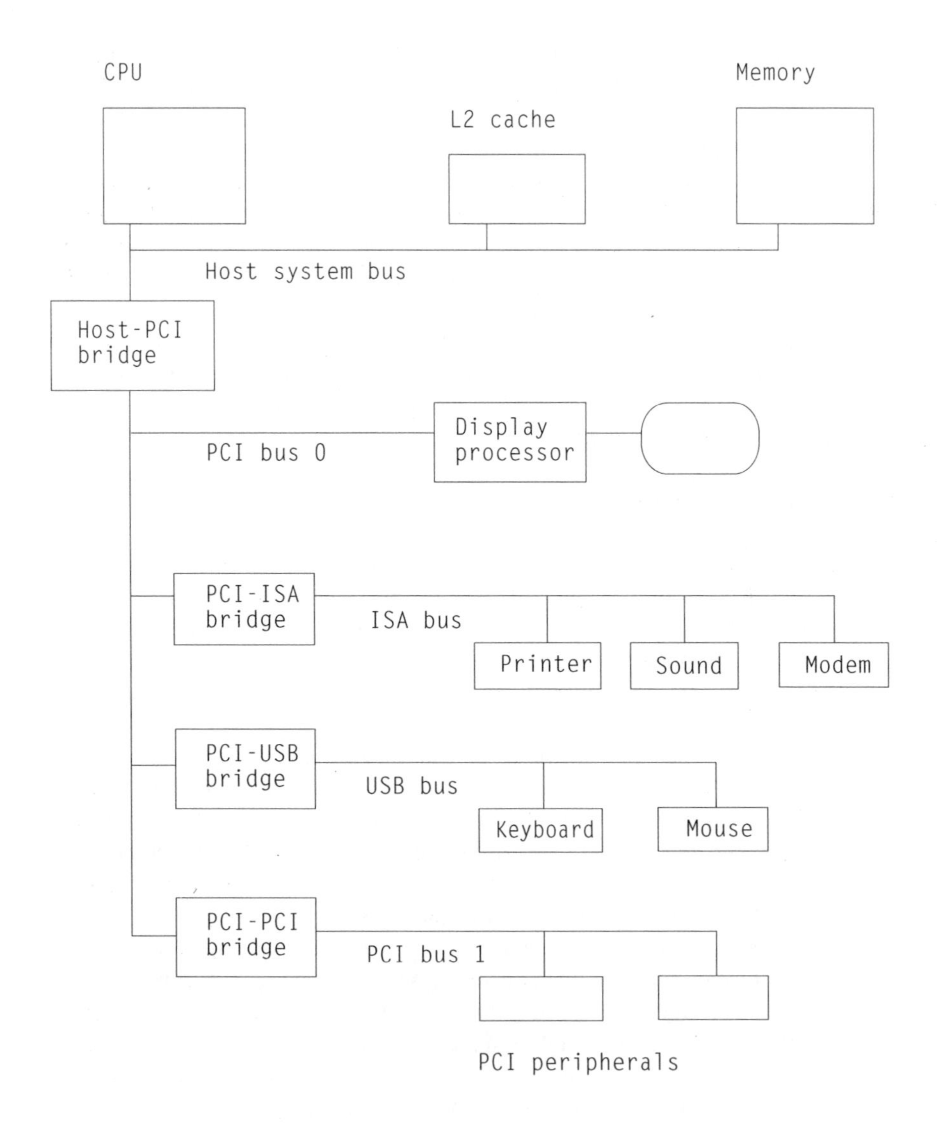

Figure 8.10 The host to PCI bridge.

processor with the highest priority can receive the grant signal. Conceptually, a bus master is active because it can request a bus cycle. In contrast, the L2 cache or memory is a passive bus target that can not initiate a cycle request.

Interestingly, on the PCI bus the address and data signals are transmitted on the same wires, i.e., the wires are time-shared to save pin numbers. To request a memory write, the bridge places an address on the lines, and after a while it places data on the same lines. To request a memory write, the bridge places and address on the lines, and after a while it reads the data bits from the same set of lines. In order to stay flexible, the PCI bus can handle 32-bit data transfers as well as 64-bit. The PCI bus cycle can go up to 100 Mhz, so if an eight-byte block is transferred per cycle, the bandwidth of the bus is 800 MBps (megabytes per second). In addition, the PCI bus can operate in either 32-bit or 64-bit address mode.

Since there are many standard buses, there are many different bridges. The PCI – PCI bridge connects one PCI bus, to another PCI bus, thus expanding the number of PCI peripheral devices. The PCI – ISA bridge connects a PCI bus to an ISA bus. Consequently, some I/O devices, such as printer, sound card, and modem on the ISA side can transmit 16 data bits in parallel. A third type of bridge, the PCI – USB bridge, connects the PCI bus to USB (universal serial bus) that supports one data signal line. On the USB side, the mouse and the keyboard use serial data transfers. An I/O device on one bus can be connected to another bus via a bus bridge without having to change its hardware interface.

8.5.1.2 Ethernet Bus

The term "wire technology" means that the transmission medium is a coaxial cable, twisted pair, or optical fiber. The first two schemes use copper to transmit electric signals and the third one uses fiber to transmit light rays. Using copper, the line can be multi-dropped, i.e., shared access by all the processors. The term "wireless technology" means no wire, so the transmission medium is air or a vacuum, that is not visible.

To make the I/O bus more flexible, we can reduce the number of wires to two, signal and ground, known as Ethernet. The address and data bits are all transmitted serially, as shown in Figure 8.11a. One coaxial cable can be used to transmit data at 10 Mbps (megabits per second). A different design, e.g., the fast Ethernet uses a group of four twisted pairs to achieve a combined data rate of 100 Mbps. There are also Ethernets using fiber optics to achieve data rates of gbps.

By way of an Ethernet, we can connect many local computers in a network. Each processor can place bits on the line as a data link (DL) frame. A data link (i.e., link) means a com port with its associated software. That is to say, the software drives the physical link in hardware. Each DL frame contains a header, the info (information) field, and a trailer as shown in Figure 8.11b. The header contains the 48-bit address of the sending node and the 48-bit address of the receiving node. The info field has a variable length of up to 1500 bytes. The trailer contains FCS-32 (32-bit frame check sequence) as an error detecting code (explained by Hsu).[24] All the node addresses are unique and they must be assigned by Xerox.

The rules of how to place bits on the bus by each processor constitute the Ethernet DL protocol. Ethernet data transmission is half-duplex, so each processor takes turns placing bits on the line. However, when one processor transmits, every node on the bus (including itself) can receive the transmitted bits at the same time. The data signal is Manchester code as a combination of high-low rectangular voltages. Because there is neither carrier nor modulation in the transmission, it is base-

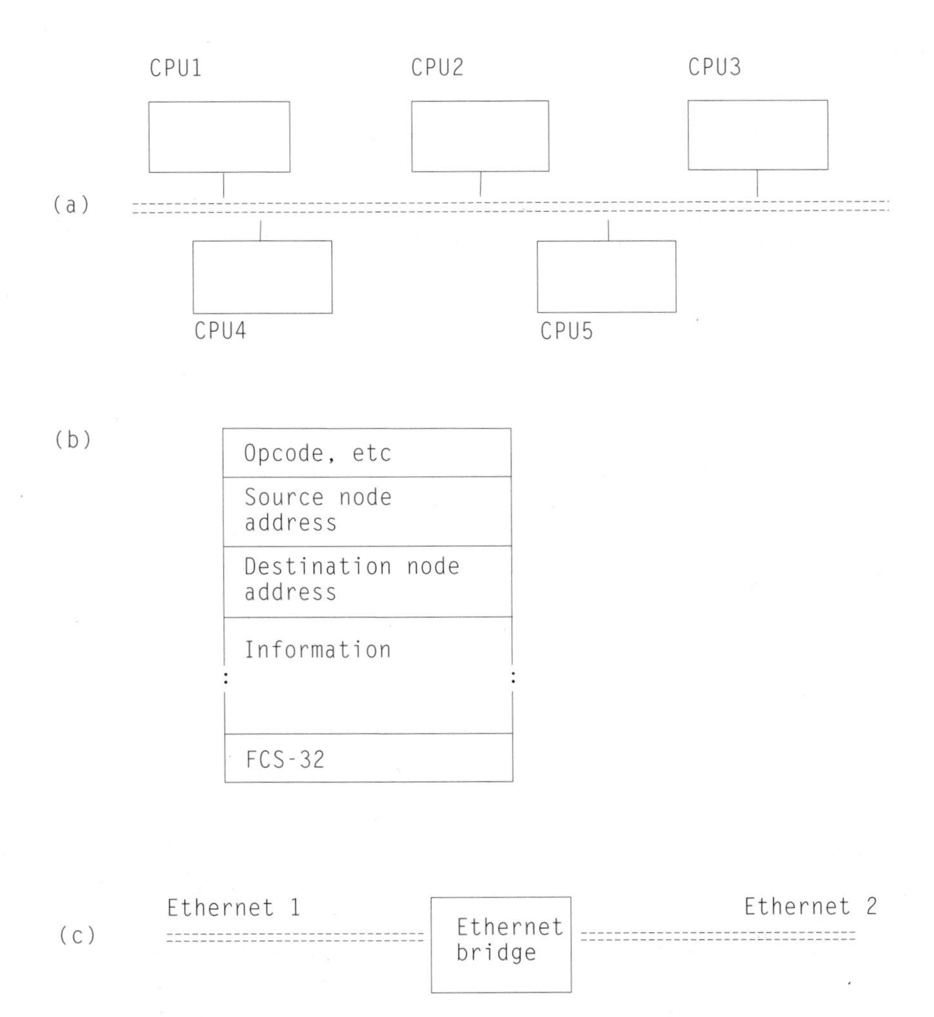

Figure 8.11 Ethernet connections: (a) multi-drop coaxial cable, (b) data link frame, and (c) Ethernet network bridge.

band. If two or more processors transmit at the same time, bits are corrupted on the line, which is defined as a collision. Whenever a collision occurs, the system I/O software retransmits the bits after waiting for a random period of time.

A network bridge is a switching device that connects two Ethernet networks as shown in Figure 8.11c. The network bridge has its own processor to run network system software. After receiving bits from one side of the bus, the software on the bridge examines the destination address in the DL frame header. If the address belongs to the same side, the bridge does nothing. If the address belongs to the other side, the bridge retransmits the bits on the other side. That is, the Ethernet bridge relays the bits to the other side if necessary.

8.5.2 Semaphore

If one processor updates a shared memory block while the other processor does the same, there may be a problem. As a simple example, if one processor adds one to a variable in memory and the other processor also adds one to the same variable, the end result should be the addition of two. However, if both processors fetch the same variable in two subsequent cycles, add one and store the result in memory, the final result is adding one instead of two. Of course, this is incorrect. Therefore, we must set up rules that allow only one processor to update a block in shared memory at one time. In other words, we need to guard the block of memory.

In concept, a binary semaphore is very much like a token in a plate. Whoever gets the token first has the right to write the memory block. After writing the block, the owning processor must put the token back on the plate. Thus, only one processor is allowed to write the memory block at one time. The block in shared memory usually contains a message queue that facilities communications between two system routines running on two processors.

A binary semaphore can be implemented as a memory word containing a one or zero. Using positive logic, a one means that the token is available while a zero means that the token is gone. The remaining problem is how to protect the semaphore, so no two processors can grab an available token at the same time. To accomplish this goal, we need hardware support. That is, reading the semaphore, testing its value, and setting it to a new value must be performed in one memory bus cycle that is indivisible. In other words, a read – modify – write hardware memory cycle must be atomic. That is, when the first CPU is testing and setting the semaphore, no other CPU can break into the middle of the cycle and do the same. In the seventies, core memories were popular and each tiny core was used to store one bit. Reading the core, we always switch its magnetic flux to zero. After reading, we need to restore the core's original magnetic flux by switching it to the other direction if the bit is one, that is, the memory always supports a full read – write cycle. A read – modify – write cycle means that the cycle time is prolonged so the processor can read the operand, modify it, and write it back in memory. If two or more CPUs want the token at the same time, only one CPU can grab the token based on its priority on the bus.

The 8086 processor uses half cycles to either read or write the semiconductor memory.[27] A special one-byte opcode known as the lock prefix must be placed in front of an instruction to lock the memory bus if desired. Locking the bus means that no other processor can access the memory at the same time. The lock prefix is a special opcode that tells the CPU hardware to lock the memory bus when the current instruction starts to execute.

A writable memory block is guarded by a semaphore, as shown in Figure 8.12a. Any task on a CPU must execute the code at the front before it can write into the block of shared memory. In Figure 8.12b, a counting semaphore is implemented in assembly language on a PC. The semaphore is initialized to one meaning only one token is available. Decrement the semaphore by one and set the condition code in the SR (status register) based on the result. There are two possible outcomes. If the

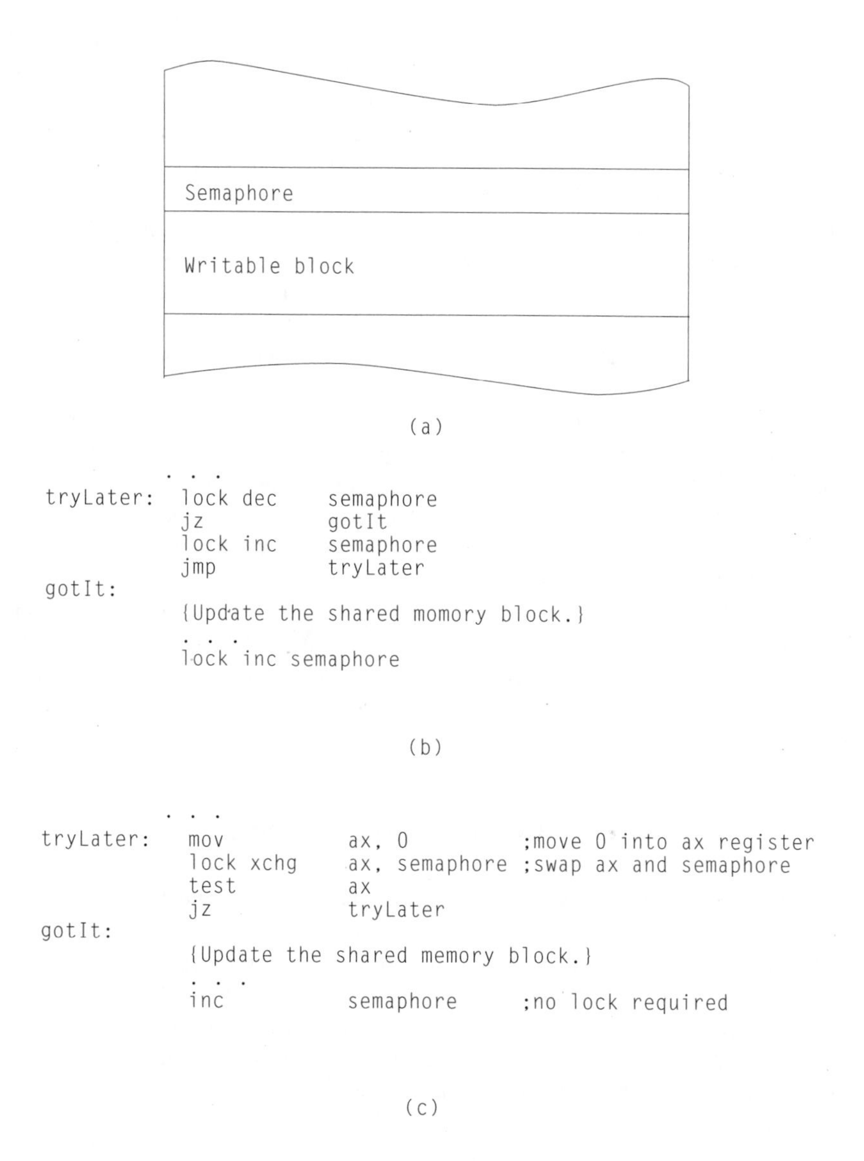

Semaphore
Writable block
(a)
. . .
tryLater: lock dec semaphore
jz gotIt
lock inc semaphore
jmp tryLater
gotIt:
{Update the shared momory block.}
. . .
lock inc semaphore
(b)
. . .
tryLater: mov ax, 0 ;move 0 into ax register
lock xchg ax, semaphore ;swap ax and semaphore
test ax
jz tryLater
gotIt:
{Update the shared memory block.}
. . .
inc semaphore ;no lock required
(c)

Figure 8.12 Semaphores: (a) semaphore and writable memory block, (b) counting semaphore, and (c) binary semaphore.

result is zero, control is passed to the label named gotIt, which means the CPU can proceed and update the shared memory block. The jz instruction means jump if result is zero. The second possibility is that the result is negative, which means another CPU has already grabbed the token, so the current CPU increments the semaphore to restore its original value and tries later. If the token is available, after grabbing it the task updates the message block. At the end of the update, the task must increment the semaphore by one to put the token back. If there are three CPUs, it is possible for one CPU to own the token while the other two CPUs in the loop are still competing for the token. Thus, it is possible to have the semaphore as -2 and then its absolute value indicates the number of tasks currently waiting on that semaphore. Consequently, when the owning task returns the token by incrementing its value by one, it is also possible for the semaphore to stay negative because the other two CPUs may have already decremented the semaphore in the same time period. Eventually, after incrementing more times, the semaphore will become one indicating that the token is available. This scheme works if the inc and dec instructions have a lock prefix.

An improved scheme is to use a binary semaphore whose value can be zero or one as shown in Figure 8.12c. The exchange instruction exchanges a register and a memory word containing the binary semaphore, and it needs a lock prefix. The xchg instruction replaces the decrease instruction to ensure that the token becomes available as soon as the holder task increases it, and no lock prefix is needed for the last increase instruction. In essence, the semaphore is a one-bit message to the receiving task. It is important to know that when the xchg instruction executes on a Pentium, the memory bus is automatically locked by hardware.

If we extend the multiple processor concept to multiple computers, we obtain a distributed system or a general purpose computer network. That is, each node may be located at a distance. In addition, each node can be made of different hardware running different OSs and all the host computers in a network have equal status regardless of their speeds.

8.6 COMPUTER NETWORKS

Physically, a computer network consists of many host computers interconnected via I/O channels. There are general purpose computer networks as well as distributed systems. A distributed system is a special purpose computer network that supports transaction processing, such as banking, credit checking, reservations, or monitoring a chemical plant or mine. The same functions can be achieved by a general purpose computer network as long as the right application software has been provided. The basic concept is that one processor may communicate with other processors via an I/O channel. A computer network has the following characteristics:

- Autonomous computers interconnected via I/O communication channels
- Intertask communications via I/O communication channels with routing intelligence
- Distributed hardware, software, and data base

Each node in the network is autonomous and the computer is usually heterogeneous with its own OS. A global network may contain several million or more computers. In a distributed system or in a general purpose computer network, note that the same set of network I/O routines are used to provide intertask communications. That is to say, a task on one CPU may communicate with another task running on a different CPU located several thousand miles away. Writing the network software is tedious, and challenging, yet manageable. The objectives of computer networks are listed below:

- Resource sharing
- Parallel computation
- Redundancy

The primary goal of designing a generalized computer network is to share resources: hardware, software, and data. Expensive shared I/O devices include color laser printers, high performance disks, and special devices for making films. Software routines implemented as remote procedure calls are also shared by many users at different locations. Data, such as disk file, stock market quotations,and e-mail are also considered resources.

Parallel computation can be achieved if all the processors work as a team to search, compute, etc. Redundancy means that the network system is fault tolerant to a certain degree. For instance, a small network may be ideal for flight control. Each decision must be voted by the majority of main processors. Any one failed processor will not affect the majority vote. Such a system is also designed for real-time transactions that must be fault tolerant. One failed hardware component should not bring down the entire system.

As far as technology is concerned, computer networks can be classified into LANs (local area networks) and WANs (wide area networks). A third category is MANs (metropolitan area networks), which are really LANs. For example, a MAN may be designed as a token ring LAN using the same hardware and software. The only difference is that a MAN covers an area as big as a city but the basic design remains the same. In contrast, the design of a LAN is quite different from a WAN. For example, the network I/O software routines, also called NAMs (network access method) are different in their NOSs (network operating system).[24]

8.6.1 Local Area Networks

A LAN contains all its computers within a local area, for example, a few square miles, as characterized by:

- Fast data rate, up to 100 Mbps or over
- Low error rate, 10^{-8} to 10^{-11} Epb (errors per bit)
- Simple routing
- Moderate distance between two nodes
- Mostly homogeneous computers
- Innovative technologies

Some LANs use copper wires, fiber-optics, or a combination of both. Because all the computers are close to each other, high speed data rates over 100 Mbps are not uncommon. As technology advances, Gbps (gigabit per sec) LANs are developed in laboratories. Because the distance between two nodes is usually less than 2.5 km, its error rate is low. With fiber optics, roughly one bit flips out of 100 gigabits transmitted. In a single LAN, each computer has one transmitter and one receiver. After one node transmits, every other node on the network sees the bits almost immediately. Consider a token-ring for example; when one node transmits, the bits pass through each node on the ring and return to the sender. That is to say, most of the nodes in a LAN have no routing intelligence.

However, a LAN can be connected to another network by a special switching processor, a bridge or router, to handle the switching function. In order to ease the maintenance problem, most companies prefer homogeneous nodes running the same set of software. But in a university environment, there may exist several different types of LANs that are connected to a backbone. Protocol conversions are often necessary to tie different LANs into a single computer network.

8.6.2 Metropolitan Area Networks

A MAN is a LAN that covers a larger physical area, e.g., the CATV (community antenna television) network. In terms of hardware or software design, a MAN is the same as a LAN. In the example of a token-ring LAN, the distance between two adjacent nodes can be 2 km. With 360 nodes maximum, the perimeter can be as long as 720 km, or 450 miles. A future proposal is to use the DQDB (distributed queue and dual bus) LAN to transmit TV signals. Such a design not only eliminates the collision problem of an Ethernet, but also negates the token passing problem on a token-ring, yet it covers an area as large as a city.

8.6.3 Wide Area Networks

The WAN technology was developed to connect two computers at a distance of, for example, 7000 miles. In a WAN, when the data bits arrive at the destination node, transmission stops right there. Even in one room, two PCs may be interconnected as a WAN. That is, if one PC is moved several thousand miles away and the other one stays in the room, the network applications still run on the network with no changes made to the hardware or software. Hence, WANs and LANs differ in technology but not in size. Well-known WANs are listed below.

- SNA (IBM System Network Architecture)
- Internet (Inter-networking)
- ISDN (Integrated Services Digital Network)

SNA is popular in industry for perform transaction processing or general purpose computing. The Internet is the oldest network that ties together all computers in educational institutions, government, organizations, and commercial companies all over the world. In other words, different computer networks are interconnected in a global network. The Internet was evolved from the Advanced Research Project Agency Network (ARPANET) in the early 1970s. Since then, ARPANET has changed its name a few times and is now called DDN (Defense Data Network), part of the Internet.

After understanding computer architecture from a system perspective, we can explore some contemporary processor designs in the next chapter.

8.7 SUMMARY POINTS

1. A vector machine may employ a design such that its data stream is transmitted in serial or in parallel.
2. In a broad sense, a MIMD machine consists of many processors operating in parallel.
3. A multistation system has one powerful CPU surrounded by many I/O processors.
4. Interprocessor communications are achieved via interrupt, shared memory, or channel-to-channel I/O.
5. The SIO (start I/O) or EXN (exchange jump) instruction is different from a system call because the instruction interrupts the other processor, not itself.
6. A multiprocessing system has multiple processors, and the processors are interchangeable and adjacent.
7. The memory bus has many address, data, and control wires that may be shared by other CPUs or controllers.
8. A bus bridge connects two buses and has logic to make signal conversions and to arbitrate simultaneous bus requests from multiple processors.
9. Reading a semaphore, testing its value, and setting it to a new value must be performed in one memory bus cycle in a multiprocessing system.
10. In a computer network, many autonomous computers are interconnected, and intertask communications are done via I/O communication channels.
11. The objectives of computer networks are resource sharing, parallel computation, and redundancy.
12. As far as technology is concerned, computer networks can be classified into LANs (local area networks) and WANs (wide area networks).
13. A network bridge is a switching device that connects two local area networks.
14. A distributed system is a special purpose computer network.

PROBLEMS

1. Explain the conceptual difference between serial data transfer and parallel data transfer in a SIMD machine.
2. What is a multistation system?
3. What is a multiprocessing system?
4. The key concept for the communication of two CPUs is passing messages. What are the three ways for one CPU to pass a message to another CPU?
5. On an IBM mainframe, explain the difference between the CAW (command address word), CCW (channel command word), and CSW (channel status word).
6. What is the difference between a system call and an SIO (start I/O) or EXN (exchange jump) instruction regarding interrupt?
7. In a multiprocessing OS, what is the hardware requirement to design a semaphore scheme so that only one CPU can enter a routine to modify a shared data base?
8. What is the functional difference between a bus bridge and an Ethernet network bridge?
9. What are the objectives of designing a computer network?
10. What is the key difference between a distributed system and a computer network?

CHAPTER 9

Processor Design Case Studies

9.1 COMPLEX INSTRUCTION SET COMPUTER

This chapter introduces many processors: IBM mainframe, PowerPC, Alpha, Itanium, and the reduced software solution computer. Issues include reliability, speed, system overhead, and cost. The chapter discusses CISC (complex instruction set computer), RISC (reduced instruction set computer), and VM (virtual machine). A software based VM can be implemented on any superscalar machine.

Generally speaking, a RISC machine uses less logic in a pipe that has an execution rate of one stage per clock. As technology advances, a CISC machine can be designed as a decoupled pipe and its stage can also execute in one clock. However, in regard to architecture, there are some fundamental differences. A CISC machine has the attributes listed below:

- Variable instruction length
- eight-bit opcode with two addresses
- Operations allowed between memory and register
- Many addressing modes
- Less extensive register set
- Hardwired or microprogrammed pipe

The two most famous CISC designs are the IBM 360/370/390 and Pentium. Both machines have two addresses in an add instruction. As far is external code is concerned, the former uses EBCDIC (extended binary coded decimal interchange code), while the latter uses ASCII. The IBM mainframe has enjoyed popularity in the corporate world for its reliability. On the other hand, the Pentium processor is popular on desktops for its market timing. Since the Pentium processor has been discussed in the previous chapters, it is now time to introduce the IBM mainframe.

9.1.1 IBM Mainframe

The IBM mainframe generation has evolved from the 360/370/390 and beyond. Each family or generation consists of many models, and each model supports the same instruction set architecture (ISA) with a different execution speed. The machine has 16 32-bit GPRs (general purpose register) denoted as R0 to R15 and four 64-bit floating point registers. Each GPR can be used as an index register, and R0 is assumed to be zero if it is used as an index. The instructions are grouped into five types: RR (register – register), RX (register – storage, indexed), RS (register – storage), SI (storage – immediate), and SS (storage – storage). Generally, each instruction has a one-byte opcode followed by two addresses. If the address points to a register, the field has four bits. If the address points to memory, the displacement is 12 bits augmented by one or two index registers. The format of a two-byte RR instruction is shown in Figure 9.1a.

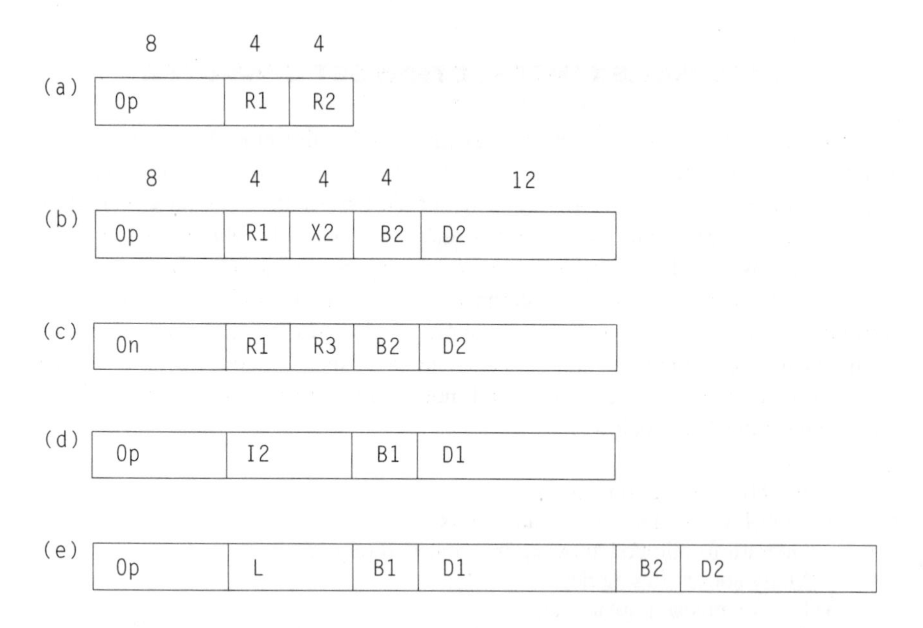

Figure 9.1 IBM 360/370/390 instructions: (a) RR, (b) RX, (c) RS, (d) SI, and (e) SS.

The four-bit R1 denotes source register 1, the first source that is also the destination. The four-bit R2 denotes source register 2, the second source. The digit 1 or 2 after the mnemonic represents the ordering of the operand. The next three types are RX, RS, and SI. Each instruction has four bytes with two addresses, and one of the two is a memory address. As shown in Figure 9.1b, the RX instruction uses two index registers, X2 and B2. That is to say, for source operand 2, X2 is the index register, B2 is the base, and D2 (displacement 2) is a 12-bit offset. The GPR zero can be used as a scratch pad (accumulator), but when it is used as an index, its content is assumed to be zero. In other words, if the X2 field is zero in the instruction, no indexing is specified. The short displacement field has caused some confusion about its addressing range. The index register is 32 bits long and address constant can be used. That is, by loading a 32-bit address constant in an index register, we obtain an addressing space of up to 4 GB (2^{32} B). As shown in Figure 9.1c, an RS instruction can load or store multiple registers, so it has the capability of limited vector operations. Because R3 specifies the upper bound of a register set and R1 specifies the lower bound, only one index register B2 is left as the base. Figure 9.1d shows the SI instruction. Note that the I2 field contains an immediate byte, and the destination operand is in memory. With 16 bits left, the index field is B1 and D1 is the 12-bit offset. The six-byte SS instruction is shown in Figure 9.1e. The instruction can move a block of 256 bytes from one memory location to another memory location. The one-byte L (length) specifies the block size in bytes, and for each address only one index register can be specified.

Instructions are provided to round, normalize, and perform floating point arithmetic for 32-bit, 64-bit, or 128-bit operands. Each machine may have a virtual memory using two-level paging. That is to say, the virtual memory has one single linear segment up to 4 GB. For example, the 32-bit virtual address of a second source operand is the sum of D2, R[X2], and R[B2]. The VA is an ordered triple (S, P, D), where S is called the segment number, P the page number, and D the displacement. The integer S is really a first level page index or page directory index to fetch a PT base address in ST (Segment Table). The integer P is really a second level page index to fetch a PFN in the PT.

Each machine supports an I/O structure of 16 I/O channels or more, and each channel can be tied to 256 I/O devices. Thus, up to 4096 I/O devices can access the central memory via the channels. Finally, the machine supports privileged I/O instructions and its OS is reliable inside out.

9.2 REDUCED INSTRUCTION SET COMPUTER

The VAX machine of the 1980s was a fast CISC machine with 304 instructions and 21 addressing modes. The idea was the more, the better. However, the complexity of a VAX processor triggered a revolution in computer design as the concept changed to less is more. Actually, the early CDC6600 and IBM801 were RISC machines. The concept did not receive much attention until Patterson and Hennessy published their research results.[18,47] A RISC machine is characterized by the following attributes:

- Uniform 32-bit instruction length.
- Six-bit opcode with three addresses.
- Load/Store architecture with register – register operations.
- Extensive register set, 32 or more.
- Hardwired pipe.

Logically, the instructions on a RISC machine have basically the same format. The RR instruction is shown in Figure 9.2a. The six-bit opcode is followed by five-bit sreg 1 (source register 1), five-bit sreg 2 (source register 2), 5-bit dreg (destination register), and 11-bit eop (extended op). The bit positions in an instruction may be arranged differently, but the semantics remain the same. For example, the eop may contain a function code or a five-bit shift count. The RX instruction contains a memory address as shown in Figure 9.2b. In addition to opcode and sreg 1, the sreg 2 field contains an index register address followed by the 16-bit disp (displacement). As an example, a load instruction has sreg 1 as the destination. The source operand 2, whose EA (effective address) is the sum of the index register and disp. is indexed in memory In a store instruction, sreg 1 represents the data to be stored at the effective address using indexing. The compare and conditional branch operations are combined into one instruction. That is to say, sreg 1 and sreg 2 are compared and branched if the condition is true. Usually, a relative addressing mode is used, so the branch address is the sum of the relative disp and program counter.

(a)

6	5	5	5	11
Op	Sreg1	Sreg1	Dreg	Eop

(b)

6	5	5	16
Op	Sreg1	Sreg2	Disp

Figure 9.2 Power PC instructions: (a) RR and (b) RX.

Because the instruction length is fixed, it is easy to retrieve an instruction from the instruction cache. Note that the fixed length instruction concept was tried on many second generation computers. For example, the second generation mainframes IBM 704 and PDP-10 all support fixed length instructions.[3] Such a design maps an instruction into a memory word of 36 bits. As a result of this design, some bits are not fully utilized, so the code size is larger. In practice, a typical RISC machine supports 32 general registers (GR) and 32 float registers (FR). A large register set has two advantages. First, during a subroutine call, the arguments are passed in registers instead of stack. As a stack is simulated in memory, the number of memory references is reduced. Likewise, after return, there is no need to pop the stack. Second, the registers can be used to store local and global variables in lieu of memory. Compilers can generate code to eliminate data dependencies. In consequence, parallel operations can be achieved in the execution unit. However, one obvious disadvantage is slow context switching during an interrupt because it takes longer to save and restore a larger set of registers.

RISC machines have not posed a threat to either mainframes or PCs. Power PCs are popular in embedded systems for many reasons. First, the RISC machine lacks a sophisticated I/O structure. Second, it is not protected without paging, which means overhead. It can not compete well in a multiprogramming system. Third, it lacks market timing or momentum. Nonetheless, discuss the design of the Power PC, Alpha, and Itanium in the following sections.

9.2.1 Power PC

The acronym POWER stands for performance optimized with enhanced RISC.[35,58] It represents a new technology, and the machine was intended to run the UNIX-like operating system for both scientific and commercial applications. The CPU has a hardwired pipe with 64 GRs and 64 FRs. An early model supported a register size of 32 bits, but a later model increased the register size to 64 bits. The chip has a 40-bit address bus and a 64-bit data bus to support PCI. Its pipelined CPU has many independent execution units, namely FXU (fixed point unit), FPU (floating point unit), LSU (load/store unit), BPU (branch processor unit), and CU (completion unit). The CU is similar to an RU (retire unit) that retires instructions in the program order. The system also supports an instruction queue, instruction cache, and data cache. The pipe is able to execute each stage in one clock.

In addition, the processor has a TLB (translation look-aside buffer) and an MMU (memory management unit) to support a two-level paging system. Just like an IBM 360/370 mainframe, the Power PC is a linear segment system. Thus, the VA is an ordered triple (S, P, D) where S is four bits, P is 16 bits, and D is 12 bits. As a result, the page is 4 KB, the segment is 256 MB, and the VAS is 4 GB. After mapping, the PAS is also 4 GB. The later model of the Power PC supports a virtual address of 64 bits, and the physical address is increased to 44 bits (16 TB).

9.2.2 Alpha Processor

The Alpha chip was the first 64-bit microprocessor on the market with an astounding clock rate of 417 Mhz.[60] It has many design features, as listed below:

- 32 GRs and 32 FRs, 64 bits each.
- No integer divide instructions.
- Conditional moves based on register content.
- Large VAS (55-bit).
- Large PAS (48-bit).
- Three-level linear paging.

The Alpha also uses fixed length 32-bit instructions, and each instruction has a six-bit opcode followed by three addresses. There are 32 GRs and 32 FRs. Each register is 32 bits on lower models and 64 bits on higher models with a 64-bit adder. Just like the CDC6600, the alpha has no integer divide instructions. To divide an integer, floating arithmetic is used instead in three steps. Data conversions are required before and after the divide operation. The machine supports conditional move instructions. That is, the instruction executes upon a condition, true or false, as set by a register. The goal is to implement an IF – THEN – ELSE construct without using branches. This clever idea is incorporated into the design of IA-64 (Intel architecture-64 bits) described later.

The alpha chip supports a three-level linear paging system because the VA is an ordered quadruple that supports a very large VAS. For example, the 55-bit VA is a program address, as shown in Figure 9.3a. The upper bits in front of D can be considered a large page number if a one-level paging system is implemented. Such a design would require a too large PT (page table). To solve that problem, the large page number is divided into three fields: P1, P2, and P3. The integer P1 is the first level page index; P2 is the second level page index; and P3 is the third level page index. The lower bits in the VA constitute D, the displacement, in a page. The TLB entry contains the large page number (P1, P2, P3), the 32-bit PF, plus control bits as shown in Figure 9.3b.

During address translation, the hardware searches the TLB first. If there is no match in the TLB, a three-level page table search method is used to find the PF (page frame number) for the page. That is, upon a TLB miss, the DAT hardware expends three extra memory cycles to poke through three different page tables in order to find the PF for the page. The page table entry at each level has 8 bytes (64 bits) divided into a 32-bit PF, 16 unused bits, and 16 control bits as shown in Figure 9.3c. The list structure of three-level page tables is shown in Figure 9.3d. A fixed length PF is retrieved from the PT each time. The hardware PTBR (PT base register) contains a PF that is the physical block address of the first level page table. Using P1 as an index to access the entry in the first level PT, we obtain the PF of the second level PT. Using P2 as an index to access the entry in the second level PT, we obtain the PF of the third level PT. Using P3 as an index to access the entry in the third level PT, we obtain the PF of the accessed page. The address translation is completed at run time, as shown in Figure 9.3e.

Note that the Alpha machine supports four different page sizes: 8 KB, 16 KB, 32 KB, and 64 KB. For each page size, the D (Displacement) field has different number of bits so the VAS is different. Regardless of the page size, two design rules remain true. First, the P1, P2, and P3 all have the same number of bits. If the D (i.e., page size) field is increased by one bit, the length of P1, P2, and P3 is also increased by one bit. Second, after each level of PT look-up, the fetched PF (page frame number) always has 32 bits, i.e., a fixed length. Both the VAS and PAS are determined by the page size as summarized in Table 9.1.

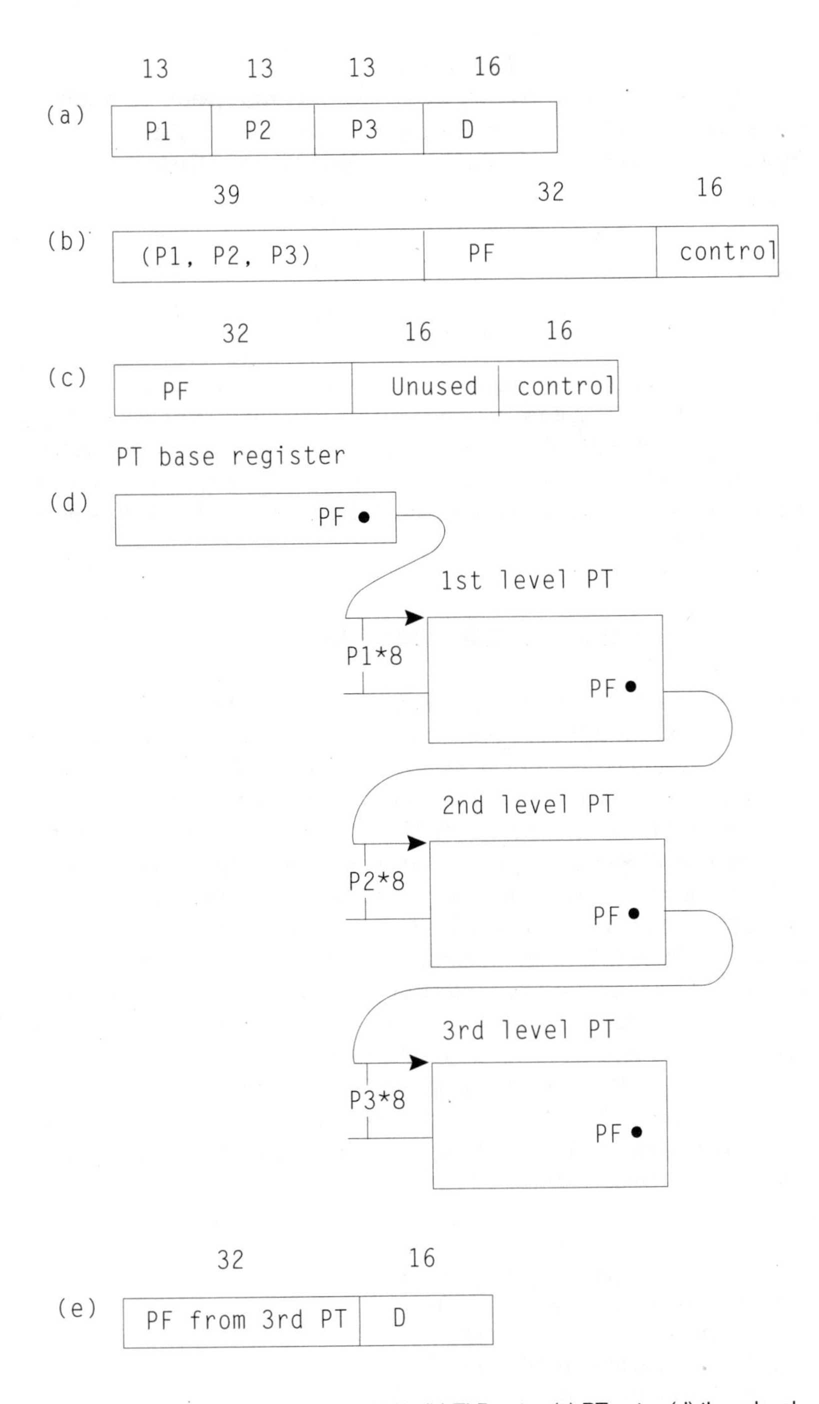

Figure 9.3 Alpha virtual memory: (a) VA, (b) TLB entry, (c) PT entry, (d) three-level paging, and (e) PA.

Table 9.1
Virtual Addressing Space vs. Physical Addressing Space on Alpha

Page Size (Bytes)	D (bits)	P1, P2, P3 (bits)	VA (bits)	PA (bits)
8 K	13	10	43	45
16 K	14	11	47	46
32 K	15	12	51	47
64 K	16	13	55	48

In concept, the Alpha program counter (PC) has 64 bits. Since each instruction is 32 bits long, the PC always has two trailing zero bits as implied. As the VA can go up to 55 bits, the virtual addressing space is 32 PB (peta byte) with seven leading zeroes in the PC. That is to say, the actual PC is shorter than 64 bits. In the field, the Alpha chip is considered a fast compute engine that uses a 750 Mhz clock as powered in many work stations.

9.3 ITANIUM PROCESSOR

The Itanium chip is the first IA-64 (Intel architecture-64 bits) processor based on the technology known as EPIC (explicit parallel instruction computing). The Itanium processor represents a family of chips. The low end will have tens of millions of transistors while the high end will have hundreds of millions. The design concept was derived from the research of VLIW (very large instruction word) machines.[6] The term VLIW means a long instruction word is retrieved in one memory cycle. The total number of bits in a long word also implies the number of data pins on the chip. As an example for multimedia applications, a 64-bit long instruction word may contain two 32-bit instructions, so both may be decoded and executed at the same time. Moreover, the VLIW processor may support two execution units but also supports two separate register files. Thus, after decoding, each execution unit operates on its own register file. As a result, we achieve two operations in parallel due to a superwide CPU. The concept of EPIC is similar because three instructions are always grouped in a 128-bit instruction bundle. If there are no dependencies, all three instructions are decoded and executed in parallel. The design features of the IA-64 are listed below:[11,30]

- 128 GRs (general registers), 64 bits each
- 128 FRs (floating point registers), 82 bits each
- 128 ARs (application registers), 64 bits each
- 64 PRs (predicate registers), 1 bit each
- 8 BRs (branch registers), 64 bits each
- 128-bit instruction bundle
- 41-bit instructions

- An RR instruction with opcode, predicate, and three addresses
- 64-bit data bus to be extended to 128 bits in the future
- Backward-compatible with IA-32

9.3.1 Operating Environments

The IA-64 architecture provides binary compatibility with the IA-32 instruction set. The Itanium processor supports two operating system environments, the IA-64 OS environment and the IA-32 OS environment as shown in Table 9.2.

Table 9.2 IA-64 Processor Operating Environments

System Environment	Application Environment	Description
IA-64	IA-32 protected mode	IA-32 protected mode applications
	IA-32 real mode	IA-32 real mode applications
	Virtual IA-32	Virtual 86 applications
	IA-64 instruction set	IA-64 applications
IA-32	IA-32 instruction set	IA-32 protected mode, IA-32 real mode, and virtual 8086 mode applications

In the IA-64 OS environment, IA-64 processors can run IA-32 applications under a native IA-64 OS that supports execution of all the IA-32 applications. In the IA-32 OS environment, IA-64 processors can run IA-32 application binaries under an IA-32 legacy operating system provided that the platform and firmware support exists in the system.

In addition, the IA-64 architecture provides the capability to support mixed IA-32 and IA-64 code execution. Within the IA-64 OS environment, the processor can execute either IA-32 or IA-64 instructions at any time. In order to provide hardware assist, the processor implements three special instructions. By executing any of these instructions, the CPU switches the hardware execution environment, as explained below:

1. jmpe (IA-32 instruction): jump to an IA-64 target instruction to change the instruction set from IA-32 to IA-64.
2. br.ia (IA-64 instruction): jump to an IA-32 target instruction to change the instruction set from IA-64 to IA-32.
3. rti (IA-64 instruction): the mnemonic means return from interrupt. The result of executing this instruction returns to an IA-32 or IA-64 instruction that was interrupted.

9.3.2 Instruction Bundle

As demonstrated in Figure 9.4a, each bundle consists of three 41-bit instructions and one five-bit template. The leftmost instruction is denoted as slot zero, the next instruction as slot one and the rightmost instruction slot two. The template contains information to tell the pipe about resource dependencies. That is to say, the three instructions in a bundle can execute in parallel if the template is 00 (00000 in binary), which means no stop. A stop is a non-zero bit pattern that asks one or more specified execution units to wait. The template is encoded because of many combinations of execution units that need to wait in the pipe. Different execution units are ALU integer, non-ALU integer, memory, float, branch, etc. As a simple example, if the template contains 01 (00001 in binary), the non-ALU resource for instruction in slot two is not available at this time so execution must be stalled until the resource becomes available. If the template is three (00011 in binary), the non-ALU resource for each of the two instructions in slot two and one are not available. As a result, both of their executions must be stalled. However, if the compiler finds a group of 16 instructions that have no mutual dependencies, they can be packaged into six different bundles. Each of the first five bundles contains three instructions with no stop. The sixth bundle has one instruction from the group and two other instructions with stop. The template in each of the first five bundles is 00, so the instructions are executed at the same time, bundle by bundle. Two points are observed. First, because the bundled instructions need not be in their original program order, they may represent different paths of a branch condition such as IF – THEN – ELSE. Second, because both dependent and independent instructions can be placed in a bundle, there is less chance for the compiler to insert nops into empty slots in the bundle.

Conceptually, the logical format of a 41-bit RR instruction is shown in Figure 9.4b. The opcode, including eop, is 14 bits, the predicate is six bits, and each of the three register addresses has seven bits. Usually, sreg 1 stands for source register 1, sreg 2 means source register 2; and dreg means destination register. The predicate register is one bit indicating true or false. At run time, the pipe first scans the templates, then picks out the instructions with no dependencies, and dispatches them to the execution units in parallel. The pipe then schedules instructions that are resource dependent according to their requirements. When the instruction is about to retire, the predicate condition is already known so the RU (retire unit) can make the machine state permanent according to the predicate.

9.3.3 Predication

The goal of predication is to eliminate conditional branches and execute as many operations as possible. As a solution, the pipe actually performs two sets of operations. After the condition is tested, the pipe discards one set of operations without retiring to permanent state. Thus, the processor performs the operations regardless of the condition of true or false. As a consequence, even though half of the operations are thrown away, the pipe can still increase throughput. The first example of

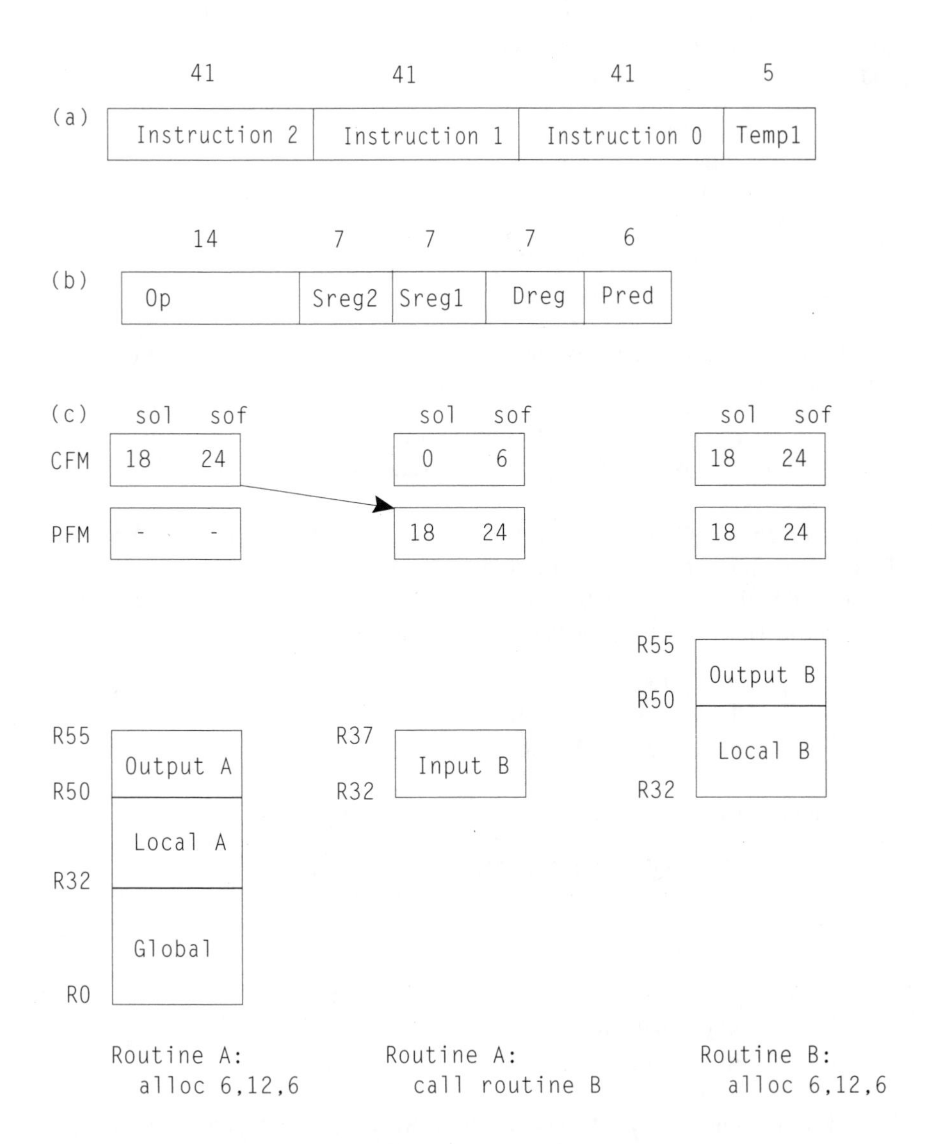

Figure 9.4 IA-64: (a) 128-bit instruction bundle and (b) 40-bit RR instruction.

this, shown below, uses conditional execution which is a precursor of predication. Its sole purpose is to implement an IF – THEN – ELSE construct without branches, as shown below:

> IF R1 .EQ. 0,
> THEN R3 <— R4;
> ELSE R4 <— R5; ENDIF;

On a conventional machine, we generate two labels, L1 and L2, as branch addresses, and the assembly code is as follows.

```
        CMP     R1, 0
        JNE     L1
        MOV     R3, R4
        JMP     L2
L1:     MOV     R4, R5
L2:     ...
```

On an Itanium, we issue two move instructions based on a test condition to get rid of the branches.

```
CMOVZ  R3, R4, R1  ;If R1 is 0, R3 <— R4
CMOVN  R4, R5, R1  ;If R1 is not 0, R4 <— R5
```

The mnemonic CMOVZ stands for compare move if zero, and CMOVN stands for compare move if not zero. The zero condition is tested in the third operand R1. A more complicated IF – THEN-ELSE may execute a block of code of any type of operations. Thus, we rely upon using a predicate, as shown below:

```
IF R1 .EQ. R2,
THEN R3 <— R4 + R5;
ELSE R3 <— R4 - R5; ENDIF;
```

We have the hypothetical Itanium assembly code as follows:

```
CMPEQ      R1, R2, P0       ;If R1 .EQ. R2,
                            ;then P0 <— 1
                            ;    P1 <— 0
                            ;else P0 <— 0
                            ;    P1 <— 1
<P0> ADD  R3, R4, R5
<P1> SUB  R3, R4, R5
```

The basic concept is that two predicates work as a pair. An even predicate teams up with an odd predicate in the next order. That is to say, P0 teams with P1, P2 teams with P3, etc. The conditional compare sets a pair of two predicates based on the tested condition. If the condition is true, the even predicate is set to true, but at the same time the odd predicate is reset to false. The reverse is done if the tested condition is false. The instructions in the THEN clause, as well as those in the ELSE clause, are conditionally executed based on the predicate pair. That is to say, by the time an instruction retires the predicate condition is already known and the instructions with false predicates are simply discarded without making the machine state permanent. Because there are 32 pairs of predicate registers, the IF – THEN – ELSE structure can be nested. For each pair of THEN – ELSE clauses, a pair of predicates is specified.

9.3.4 Speculative Loading

The speculative loading concept means that the compiler physically moves a load instruction from after the branch condition to before the branch. This concept is known as hoisting, a way of physically moving a load instruction to an earlier point in the instruction stream before a branch. Assume that the logic of a program is listed below:

```
R1 <— R1 + 1;
IF R1 .GT. R2,
THEN R3 <— M[R1 - R2];
   M[R3] <— M[R3] + 1; ENDIF:
```

The compiler realizes the memory operand M[R1 - R2] is data independent after increasing R1. Therefore, it moves the load instruction before the IF test and marks it as load.speculative (ld.s). If the load instruction generates an interrupt, page fault, or address exception, the interrupt is deferred by marking a token bit associated with the destination register. This interrupt will be honored later by a check.speculative (check.s) instruction that is inserted at the same position as the original load. Symbolically, we show the logic of the resulting code as follows:

```
R1 <— R1 + 1;
ld.s R8 <— M[R1 - R2];
IF R1 .GT. R2,
THEN check.s R8;
   R3 <— R8;
   M[R3] <— M[R3] + 1; ENDIF:
```

When the check.s instruction executes, it does nothing if the token bit of R8 is not set. If it detects an address exception, control is passed to fixed code in the OS that terminates the user task. If it detects a page fault, the OS (paging supervisor) brings in the page. After that, the execution of load is carried out and the execution of the instruction is resumed after check.s. That is to say, the interrupt is deferred but honored later.

9.3.5 Register Stack

Subroutine calls occur quite often in a program. After the calling routine issues a call, control is passed to the routine being called. The calling routine is referred to as the caller, and the routine being called is known as the callee. As a common practice, the callee saves all registers before they can be modified. The callee also restores all registers from memory before return. The register saving/restoring means overhead. To eliminate this overhead, the Itanium implements its register file as a stack. That is to say, a routine can issue an instruction to allocate a frame of registers for its local variables.

The register file is divided into two subunits: static and stacked. The static subunit contains the lower 32 GRs, denoted as R0 to R31, so its boundary is fixed. In programming, this subunit contains the global variables that are visible to all procedures. In contrast, the stacked subunit is designed as a register stack that contains dynamic data. The stacked subunit can further so divided into many register frames and each frame contains the local variables seen by each procedure. The register stack is a data structure rather than a working stack as in a stack machine.

Suppose routine A is the caller and routine B is the callee. Because routine A is the main routine, it has the first register frame allocated, whose size can vary from 0 to 96. Routine B is at the second level, so it can only allocate a register frame whose size is limited by the unused portion in the stacked subunit. For either routine A or B, its register frame always begins with R32 by renaming the general register addresses as a side effect of a call or return. Some of the key concepts are listed below:

1. Each of the first 32 GR addresses in an instruction is a physical register address.
2. Each of the remaining 96 GR addresses in an instruction is a virtual register address that is mapped into a physical register address at run time.
3. When executing a call, the CPU does two extra things. First, it saves the CFM (current frame marker) of the caller into the PFM (previous frame marker). Then, the CPU sets the CFM containing the frame status of callee. The CFM contains physical register bases, the sol (size of local), sof (size of frame), etc. The PFM is a field in AR64 (application register 64) referred to as PFS (previous function status).
4. When executing a return, the CPU restores the CFM from the PFM in AR64.

It is advised that each routine allocates registers to store its own local variables if there is unused or free space in the stacked subunit. If the number of registers allocated is less than the number of local variables, the routine should be responsible for storing and loading registers during computation. That is, the routine uses its registers in a time-shared manner. The caller allocates registers to store passing arguments as its output, i.e., the input of the callee. The output or input is for communications between two adjacent routines: the caller and its callee.

In practice, the first instruction in a routine is an alloc (allocate) that has three immediate operandi: size of input, size of local variables, and size of output. Adding the size of input and the size of local variables, we obtain the sol (size of local). Adding all three immediates, we obtain the sof (size of frame). Therefore, we divide each register frame into two parts. The first part is for storing all the local variables with input overlaid in the front, and the second part is for output that is overlaid with the locals of the callee as input.

The CFM (current frame marker) contains register base addresses (general, float, and predicate), the sol, and sof. During a call, the CPU does two extra things. First, the CFM is copied into PFM (previous frame marker). Second, the CPU modifies CFM so its sol is always 0 and its sof is equal to the size of caller's output. Once after receiving control, the callee issues an alloc instruction so the CFM reflects its new register frame status.

Figure 9.4c, shows the register frame status of routine A and B. In the left column, routine A issues alloc 6, 12, 6. As a result of its execution, the CFM contains a sol of 18 (6 + 12) and an sof of 24 (6 + 12 + 6). The lower 32 general registers constitute the global area. Because routine A is the main routine, its first register in the frame is a physical R32. The middle column shows the frame status when routine A calls routine B. After executing the call, the CFM contains the new sol zero and the new sof six which is the difference between the sof and sol of routine A. The right column shows the frame status when routine B issues alloc 6, 12, 6. As a result, the CFM contains an sol of 18 and an sof of 24. However, routine B also sees the first register in its register frame as R32, a virtual register address. At run time, the virtual R32 is mapped into the physical R50 because the base 18 (sol of routine A) is added to the virtual register address. In other words, the sum of 18 and 32 is 50, which is the physical register address. After routine B issues a return, the original CFM is restored from the PFM. Interestingly, routine B may call another routine or even itself.

IA-64 is a powerful engine for compute bound jobs for three reasons. First, it has extensive parallel operations. Second, it has fewer branches. Third, it has no register saving and restoring during subroutine linking. Nevertheless, if applications are I/O bound such as Web searches, games, and word processing, the IA-64 may not greatly impact the speed. There are other debatable issues. First, branch conditions always exist in a computer, e.g., looping, call, return, and interrupt. If we somehow design a CPU in such a way that the same time is spent to execute a branch; conditional or unconditional, we simply flush the instruction pipe. Second, during context switching the three extensive register sets cause too much overhead. In fact, some applications do not need a large register set. In addition, if the L1 cache size is reduced, it can match the speed of registers because both are on-chip. Third, a large code size slows down loading and transmission. Hence, we explore a new processor design from a different angle.

9.4 REDUCED SOFTWARE SOLUTION COMPUTER

The acronym RssC (pronounced arsk) stands for reduced software solution computer, which has an enhanced architecture derived from mainframes.[1,5,78] The machine supports four explicit segments without paging. The size of each segment can be up to 1 GB and the processor has multiple execution units. It is designed for real-time applications that include DSP (digital signal processor) applications.[36,81]

The machine supports floating point arithmetic via a microcode sequence. For scientific applications, a vector math unit can be added. In addition, many clusters can be interconnected, and each cluster has many CPUs sharing one memory. Such a system qualifies as a supercomputer system. The application program running on one CPU can place messages in shared memory for the other CPUs in the same cluster to interpret and execute.

The RssC machine supports 32 GRs (general registers) and 16 of them can be used as index registers. Some registers are designated for special functions, e.g., some are reserved for branch or floating point operations. Because each functional unit has a different speed, an instruction can be dispatched out of order. In other words, the CPU is a decoupled pipe. The machine supports an instruction queue in addition to an instruction cache so the instruction retrieval time can be reduced.

The instruction length can be one, two, or four bytes, so the program counter is truly a 32-bit counter. After decoding, the program counter is increased by the instruction length in bytes. An instruction usually has a six-bit primary opcode followed by two addresses. However, an instruction may have only one operand or no operand. The RR type instruction has two register addresses. The RX type has one register address and one memory address plus indexing. The register is the first source as well as the destination in a two-address machine. The 32 GRs are used to support scalar operations; but for load or store operation, one instruction can handle a vector of eight registers.

9.4.1 Compact Code

In Table 9.3, compares the opcode length, instruction length, number of addresses, number of registers, and their lengths for the popular computers.

Table 9.3 Comparisons of CISC, RISC, and RssC Machines

Name	Opcode Bits	Instruction Bytes	Addresses	Registers/Bits
IBM370	8	2,4,6	2	16/32
RISC	6	4	3	32/32 or 64
Pentium	8	1-13	2	8/32
IA-64	14	41 bits	3	128/64
Bytecode	8	1-5	0	0
RssC	6	1,2,4	2	32/32

In general, the code size of a RISC machine is bigger than that of a CISC machine.[82] The Pentium (IA-32) instruction length ranges from 1 to 13 bytes in byte increments. Similarly, the bytecode ranges from 1 byte to 5 bytes in byte increments. In either case, the shortest instruction has only one-byte opcode. The IBM mainframe also has a variable instruction length of 2, 4, 6 bytes, so its IAR is not a counter. Most of the RISC machines use one uniform instruction length, e.g., 32 bits. Because there are ways to expedite instruction execution, the RssC machine supports variable length instructions, namely 1, 2, or 4 bytes, for the sake of reducing the code size. The machine also supports vector instructions to a limited degree. The processor eliminates paging hardware and software altogether, so it is a good candidate for SOC (system on chip).

9.4.2 Register Set

The 32 GRs are denoted as R0, R1, ..., and R31 in Figure 9.5a. Each register has 32 bits, and the generic data types are byte, 16-bit word, 32-bit double word, and 64-bit quad word. Among the GRs, the first 16 can be used as index registers.

The PC has 32 bits as shown in Figure 9.5b. The SR (status register) also has 32 bits, as shown in Figure 9.5c and its lower eight bits constitute the SRB (status register byte) that can be programmed by an application programmer. The one-byte field includes carry flag, overflow flag, direction flag, branch flag, data cache enabled, data cache write-through, overflow enable, and operand size (16 or 32 bits). The upper 24 bits of the SR can only be modified by a privilege instruction. The field mainly contains bits for system control, such as supervisor state, hardware wait state, interrupt enables, and interrupt code. The PC contains a 32-bit byte address so each task can take up to 4GB virtual addressing space. Each task may have four explicit segments, namely SC (static code), DC (dynamic code), EC (extra code), and GC (global code) as shown in Figure 9.5d. For each segment, there is an RA (relocation address) register containing the segment descriptor. Each RA has the first half containing a 32-bit segment base address in paragraphs of 16 bytes and the second half containing one bit for read-only access and a 30-bit limit in bytes. Therefore, each segment is limited to a size of one gigabyte. Because the segment base address has four trailing zeroes implied, the physical address space is 64 GB, meaning a 36-bit PAS. In assembly code, a segment declaration is shown below:

```
scode    segment  static, read-only
           {Assembly statements.}
         ends
dcode    segment  dynamic
           {Assembly statements.}
         ends
```

The sample program has two code segments: static and dynamic. The symbol defined in column 1 represents the base address of the segment. If a segment is read-only, any attempt to write the segment will trigger interrupt. The segment declaration is a pseudo op with attributes specified in the operand field. The last statement is ends (end segment), also a pseudo op. Between two pseudo ops there may be a group of assembly statements.

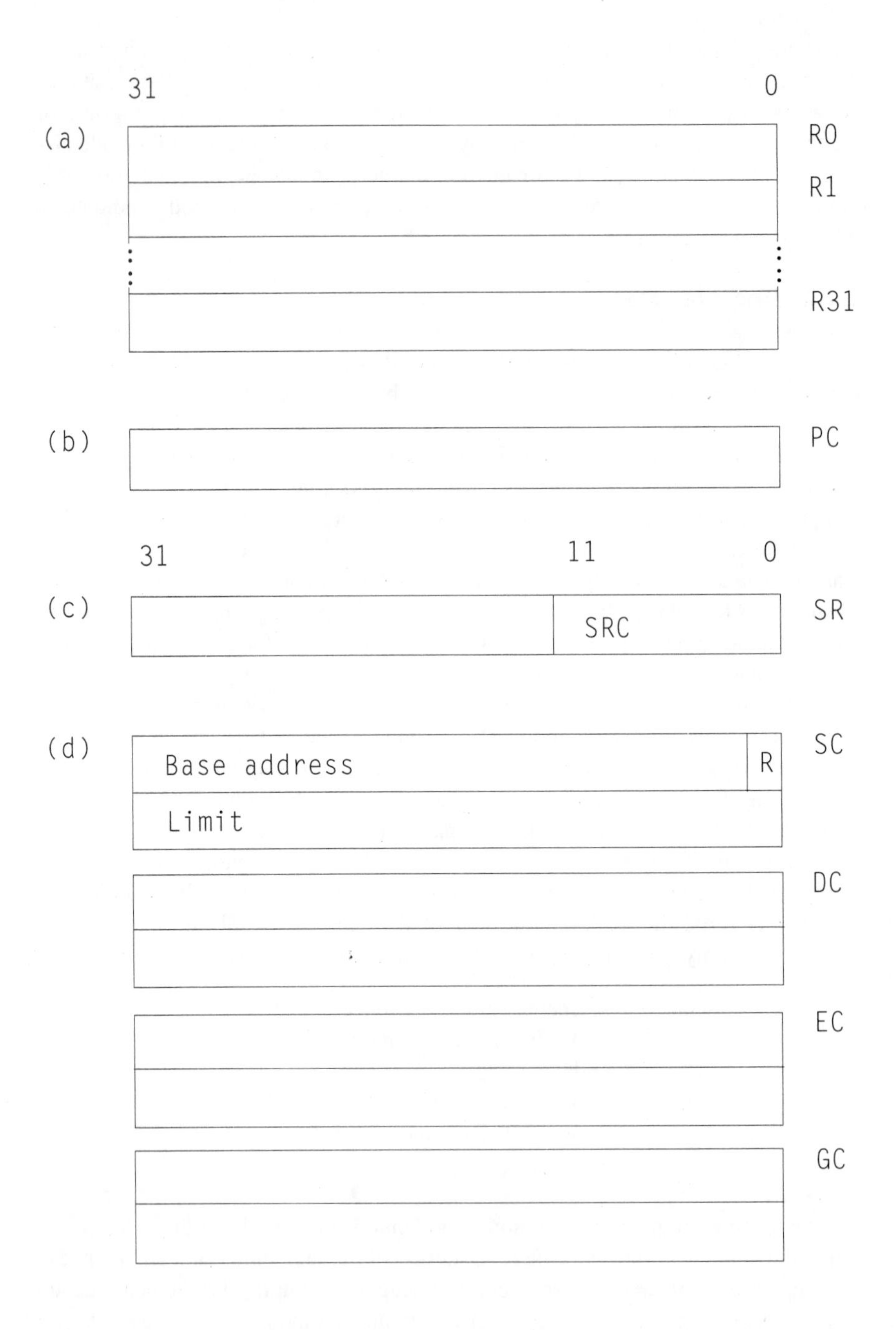

Figure 9.5 RssC register set: (a) 32 GPRs, (b) PC, (c) SR, and (c) four RA registers, SC, DC, EC, and GC.

9.4.3 Instruction Set Architecture

Because the instruction can be 8, 16, or 32 bits, the code size is reduced by one-third. An eight-bit instruction contains opcode only, as shown in Figure 9.6a. There are two types of 16-bit instruction. First, an instruction may have an 11-bit op and one operand, as shown in Figure 9.6b. Such an instruction may perform a unary operation on a register. Second, the instruction has a six-bit op followed by a five-bit R1 and R2 for binary operations, as shown in Figure 9.6c. The R1 is the first source operand as well as the destination, and the R2 is the second source operand.

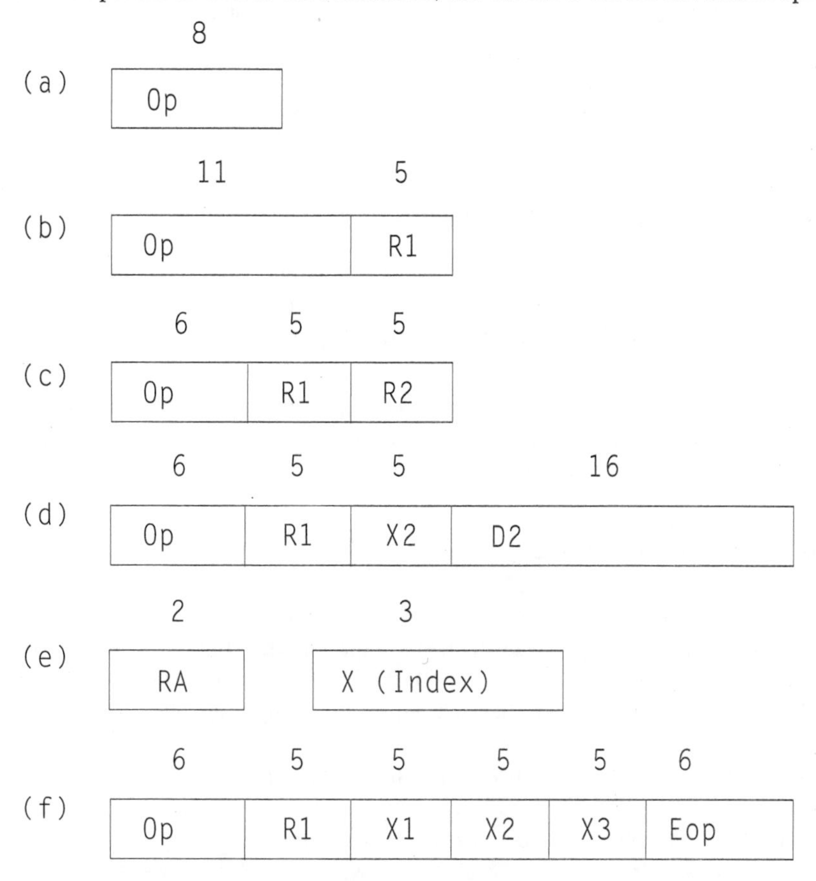

Figure 9.6 RssC instructions: (a) 8-bit op, (b) 11-bit op and 5-bit R1, (c) 6-bit op and 5-bit R1 and R2, (d) 6-bit op, 5-bit R1, 5-bit X2, and 16-bit D2, (e) the 5-bit X2 field, and (f) 6-bit op followed by R1, X1, X2, X3, and eop.

As shown in Figure 9.6d, a 32-bit instruction contains two addresses. The R1 field specifies a register, and the five-bit X2 and the 16-bit D2 represent a memory address. For example, the instruction may load an operand from memory into a register or store an operand from a register to memory. A scalar instruction operates on a single register, while a limited vector instruction operates on a vector of eight registers. Each X2 field represents a 2-bit RA and a 4-bit index as shown in 9.6e. This is because the second bit in X2 is shared by both the RA and the index. That is, the trailing bit of the RA is the leading bit of the index register field, so the index register address ranges from zero to 15. Thus, a total of 16 index registers can be specified. The D2 field contains an offset that is added to the index register to form the effective address in a segment. The proposed vector instruction format is shown in Figure 9.6g where eop means extended op and R1 specifies a register containing the vector length. Each of the X1, X2, and X3 fields represents a memory address, and the RA field selects one of four segment base registers as shown below:

00 : SC (static code)
01 : DC (dynamic code)
10 : EC (extra code)
11 : GC (global code)

The index registers are based on the following rules:

1. Registers 1 – 7 are index registers if SC or EC are specified.
2. Registers 9 – 15 are index registers if DC or GC are specified.
3. For string or vector operations, the index register may contain an operand or a pointer to an operand.
4. R0 or R8 implies zero which also means that no indexing is specified.
5. The SP (stack pointer) means R31, which is specially designed for stack operations in the dynamic code segment.

9.4.4 Addressing Space

The VAS is 4 GB per task, and the physical addressing space is 64 GB. That is to say, each task may take up to four segments and each segment may be up to 1 GB. In the case that the program is larger than 4 GB, the user task can issue a system call to dynamically load the segment. In other words, the composite code segment with data can be fetched from disk into memory and overlaid. The PAS is determined by the block address in the RA. Because the physical block address has 32 bits and each block has 16 bytes, the PAS is 64 GB (2^{36}). That is, a segment must reside in 16 B block boundary in memory. If the accessed unit of memory is an 8 B aligned block, the physical block address on the bus has 33 bits, which is also the number of address pins on the chip. If the memory addresses are two-way interleaved, the bandwidth is effectively doubled.

9.5 OTHER DISTINCTIONS

The CPU is designed as a decoupled pipe that consists of an instruction queue, instruction cache, data cache, fetch/decode unit, microcode pool, dispatch/execute unit, and a retire unit.

9.5.1 Decoupled Pipe

The bus interface unit issues memory cycles to access the Icache (instruction cache) and Dcache (data cache). The Ifetch/Idecode unit is decomposed into three units: the Ifetch (instruction fetch) unit and two Idecode (instruction decode) units. The Ifetch unit retrieves a 64-bit instruction block from the Icache into the instruction queue. Conceptually, the instruction block flows into the 16 B queue from right to left. After an instruction is justified to the left, the queue passes the instruction to an Idecode unit. At the same time, the queue is shifted to the left by the instruction length.

As the hole in the instruction queue becomes bigger or equal to 64 bits, the next instruction block is loaded to fill up the hole. The 512 B Icache uses the direct-mapping method to search for its entries. After a branch condition is detected, search the Icache to see if it has the target instruction. If a miss is encountered, the Icache issues a memory read cycle to fetch two consecutive 64-bit instruction blocks from the current one. The instruction retrieval time is shortened if the Icache can supply the instruction stream on a continuous basis. The two Idecode units generate μops (micro-operations) in parallel and store them in the microcode pool that is content addressable. A 16 KB data cache on-chip uses two-way set associative search, and each set contains two LRU (least recently used) entries.

The dispatch/execute unit has two units, dispatch and execution. The EU (execution unit) consists of six functional units, namely two fixed point units, load/store, branch, multiply, and float. The dispatch unit dispatches μops from the microcode pool to the execution unit provided that both the data and the functional unit are ready. The pipe can execute an operation out of order so another μop is returned to the microcode pool after execution. Based on its programming order, the RU (retire unit) retires the executed instructions and honors the interrupt.

9.5.2 Compact Load Module

The load module is compact because it contains pure code if the segment attributes are placed in the file descriptor. The code is memory position independent and fully protected in that a user task can not crash another user task or the OS. Without the page frame table, page tables, and external page tables, the BIOS and nucleus can be embedded in ROM less than 1 MB. Consequently, the nucleus of the OS immediately executes after the power is turned on. In addition, because of its smaller size, a load module can be down loaded from a server computer in the network in less time. If the so-called sliding window protocol is implemented to transfer files in the network, bits remain intact after down loading. Moreover, the network maintains the latest version of a program on the server computer, which alleviates the maintenance, piracy, and virus problems.

9.5.3 Special Features

As the RssC supports four explicit segments, it takes one add cycle to compute the linear program address. However, this penalty is alleviated by two design features. First, the bus interface unit keeps two physical address counters: one for instruction and one for data. The instruction fetch overhead is reduced because the next instruction block address is increased in the address counter. That is to say, after an 8 B instruction block is fetched into the Icache, the instruction address counter is increased by eight, so it points to the next instruction block without going through an add cycle again. The data fetch overhead is reduced by supporting vector load and store instructions. A vector is limited to eight registers, and the type can be byte, word, double word, or quad word. Each time an operand is fetched, the operand address counter is increased by 2^N where N varies from 0 to 3 depending on the operand size. Second, when the OS operates in supervisor mode, no base register is used, so the extra add cycle is bypassed meaning V=R.

The machine supports no speculative branch prediction for the following reasons. The leading two bits of the opcode tell the instruction length, so the instruction boundary is determined quickly. In addition, the branch execution unit is hardwired, and a conditional branch, condition can be detected just like an unconditional branch as explained below.

9.5.3.1 Branch Instruction

The machine supports register indirect branches: unconditional and conditional. The machine supports relational compares, and the execution of such an instruction sets a system branch flag in the SR to either true or false. A conditional branch register on true (BRT) or false (BRF) tests the flag so as to pass the branch address from a register to the bus interface unit. A programming loop is constructed below:

```
        LD   R1, =A'L0010'  ;load doubleword literal address
                            ;constant L0010 in R1 as index
L0010:  ...
        ...
        ...
        EQ   R24, R25       ;if R24 .EQ. R25, flag <— 1
        BRT  [R1]           ;branch register indirect on true
```

If the branch unit decoder is hardwired at the instruction queue level, a conditional branch executes just like an unconditional branch because the opcode and the system branch flag are decoded at the same time. That is, if the test condition is true, the instruction branches and the instruction queue is flushed. Otherwise, no branch occurs and the instruction executes like a nop. In either case, current operations in the pipe remain valid.

9.5.3.2 Privileged Instructions

A set of privileged instructions are designed to perform system control functions, such as load/store RA registers, load SR, return from interrupt, and I/O instructions, in order to make the machine fully protected. Any I/O operation request must be issued via a system call. Therefore, the OS interprets and initiates the I/O operation. That is to say, the I/O supervisor leaves messages in low memory for an I/O processor to retrieve, interpret, and execute.

9.5.3.3 Vector Instructions

The machine supports a set of limited vector instructions to load or store a fixed number (4 or 8) of registers wrapped around. The generic operand can be byte, word, double word, or quad word. For example, the L8D instruction can load an eight-double word memory block into eight registers, and the ST8D instruction can store eight registers into an eight double word block. The machine also supports string processing instructions such as move block and search block on equal or not equal. In addition, if a floating point unit is implemented on-chip, a single float instruction can operate on vectors of large sizes.

Because the load modules are short, the required memory size is reduced. In addition, the loading time is reduced and so is the transmission time. For PC applications, the memory and I/O are the bottlenecks, not the CPU. Thus, a predicate-based CPU would increase the code size but would not increase its throughput by much.

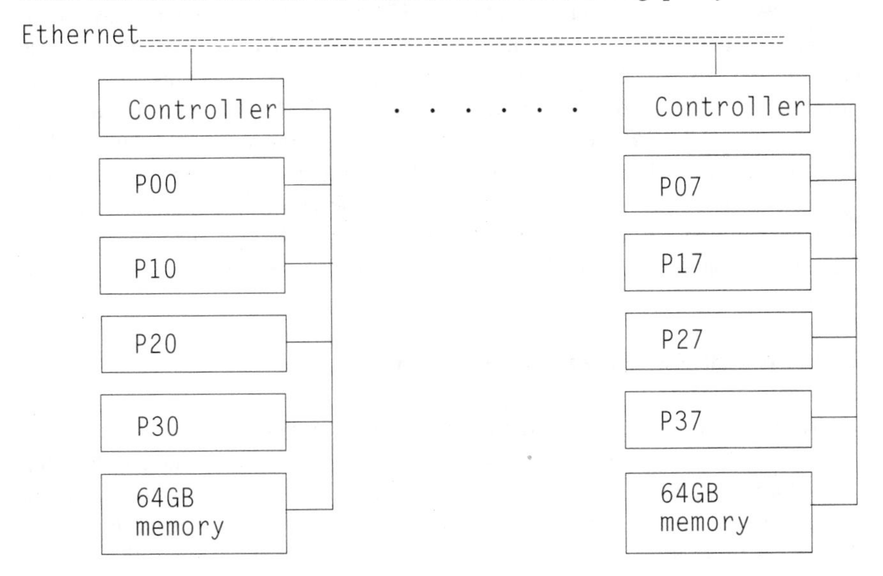

Figure 9.7 Eight clusters of RssC computers.

Note that a large register set also hurts speed if context switching is frequently performed. In contrast, the RssC machine runs the OS more efficiently in a multitasking environment. For more computing power, a system may contain a vector unit for floating point operations. Of course, a machine may have many clusters and each cluster may have many RssCs.

For example, a SMP (symmetric multiple processor) design of 4 x 8 nodes is shown in Figure 9.7. The I/O structures are omitted in the block diagram. Horizontally, there are eight clusters connected by an Ethernet bus. In each cluster, there are four vertical RssC machines sharing one memory of 64 GB and one controller driving the Ethernet bus. Each processor has a node address XY, where X is the row number and Y is the column number. For example, P37 is the processor in the fourth row and eighth column. The number of clusters and the number of nodes in a cluster may be increased for a particular application.

It is our hope that future desktop PCs should work like vacuum cleaners or cars. Turn them on; use them; and turn them off. The PC should consume less power and cost less, yet run an efficient OS in small memory. There should be no waiting time, no virus, and no system crash.

9.6 VIRTUAL MACHINES

The term virtual machine (VM) was coined by IBM. Suppose there are three physical machines in a room. Each machine shares the same architecture but runs under its own operating system, e.g., OS1, OS2, and OS3. The memory layouts of three physical machines are shown in Figure 9.8a. The VM tends to replace all three old machines with one new machine and make the user believe that his old machine is still in the machine room. The memory layout of a VM is shown in Figure 9.8b. That is to say, all the old programs running on the new machine require no modifications whatsoever. The new machine has a VM monitor which is the native OS running in supervisor mode. Any old OS and its user programs are running under the VM in user mode. Because the architecture of the old machines and the new machine is the same, it is possible for a new machine to execute more than 95% of the instructions in an old program and interpret less than 5%.

9.6.1 Virtual Program Status Word

Any old OS, while running under a VM, has its own PSW (program status word) in memory. Therefore, the PSW in an old machine is virtual because it does not occupy hardware. In the virtual PSW, the problem state bit (b14) is zero denoting supervisor mode, but again it is virtual. The VM has a real PSW in the hardware register that has b14 equal to zero, so the VM runs in real supervisor mode. The idea is clear. Whenever an interrupt occurs, the VM monitor interprets the event. If a system call is issued by the user program, interrupt occurs so control is passed to the VM monitor. As a result, after the VM interprets the SVC instruction, it passes control to

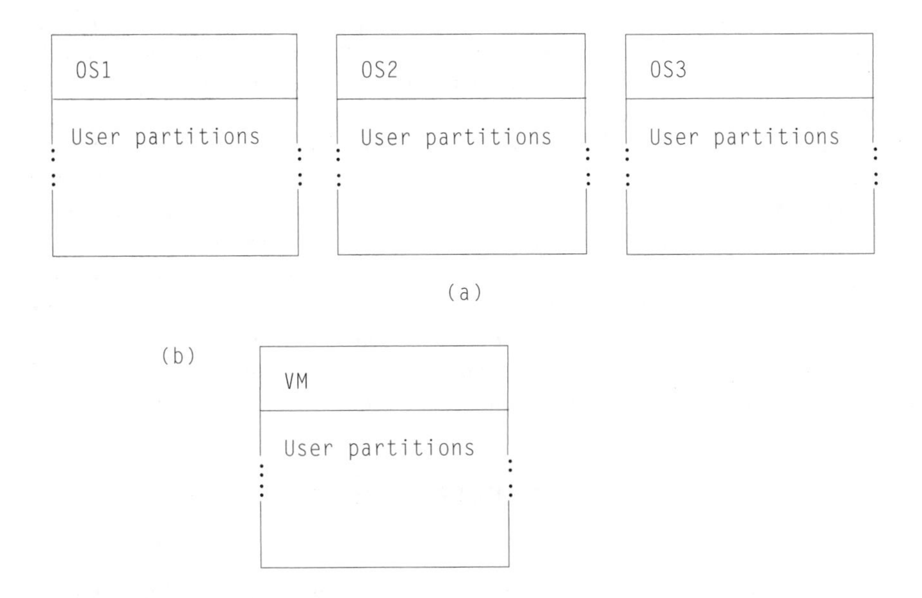

Figure 9.8 Memory layouts: (a) the memories of three real machines and (b) the memory of a virtual machine.

the old OS and execution continues. Because a privileged instruction may be issued by the user task or the old OS, its execution results in an interrupt because the PSW in the hardware register specifies the user mode. After context switching, the VM monitor examines the virtual PSW in memory. There are two possible scenarios. If the virtual PSW indicates supervisor mode, the privileged instruction is legal because it is issued by the old OS. Thus, the VM carries out the privileged operation for the old OS. If the virtual PSW indicates the user mode, the privileged instruction is a privileged exception, i.e., illegal, because it is issued by the user task. Accordingly, the VM monitor aborts the user task. How are program interrupts handled? Since such an interrupt is caused by a programming error, the user task should be terminated.

All three old OSs can run concurrently in the memory of the new machine and they are all page swappable. This implies a lot of disk swapping if the old OSs are moved in and out of memory all the time. Therefore, for the sake of performance, all the user programs under the same OS should run as a batch under the VM because the old OS can stay in memory without being swapped.

9.6.2 Java Virtual Machine

A computer simulator or interpreter is a virtual machine in that it generates computation results. A user may not even know that the target machine does not exist. That is to say, a VM can be 100% software based. More often than not, the target machine is chosen as a stack machine whose architecture is described in the last chapter.

The JAVA virtual machine is divided into two parts, and each part is a job step. The first part is the JAVAC (JAVA compiler), which translates the JAVA language statements into bytecode.[9,33] The second part is the JAVA (JAVA interpreter), which interprets bytecode in order to generate computation results.

The bytecode is so named because its length varies from one to five bytes. As a matter of fact, the bytecode is stack machine code of variable length. The JAVA VM or engine has gained momentum for two reasons. First, it can run on any host that is a superscalar machine. Second, it is extremely flexible. Because of the fast raw speed of a CPU, the flexibility of an interpreter outweighs its slower speed in interpretation. That is to say, the application platform is OS independent for the programmer because the interpreter handles all communication services provided by NOS (network operating system). In other words, a programmer can write JAVA programs without knowing much about system calls or network protocols.

9.7 SUMMARY POINTS

1. A CISC machine has instructions of different lengths and a smaller register set.
2. An IBM mainframe is a two-address machine with variable length instructions.
3. A RISC machine has fixed length instructions and a large set of general registers.
4. The IA-64 processor uses a 128-bit instruction bundle that contains three 41-bit instructions and one 8-bit template. A typical RR instruction has opcode, predicate, and three register addresses.
5. The IA-64 processor has a 64-bit data bus and it is backward-compatible with IA-32.
6. The RssC machine supports instructions of 1, 2, or 4 bytes in order to reduce code size. Each task may take up to four explicit segments, and each segment can go up to 1 GB.
7. A VM can replace many physical machines in a room without software changes.

PROBLEMS

1. Use one sentence to describe why the IBM mainframe is popular in corporate computing.
2. Briefly describe the execution units of a Power PC.
3. What is the instruction length of a MIPS, Power PC, Alpha, and IA-64?

4. Briefly describe the 128-bit instruction bundle of the IA-64.
5. Use one sentence to explain why each register address in an instruction above R32 is virtual, as demonstrated by the routine being called.
6. Briefly describe why the RssC machine supports a VAS of 4 GB per task and a PAS of 64 GB?
7. Why can the RssC machine alleviate the conditional branch in the pipe?
8. What is a virtual machine?
9. On an IBM VM, why is the percentage of execution much larger than the percentage of interpretation?
10. Explain why the user OS runs in user mode instead of supervisor mode on an IBM VM?

CHAPTER 10

Stack Machine Principles

It is a pleasure to teach the bright minds on earth.
Mencious (390 - 305 B.C.)

10.1 STACK MACHINE BASICS

A stack machine means that no address is coded in an arithmetic or logical instruction, unary or binary. That is to say, any operand, source or destination, is implied on the stack. Usually, a software based virtual machine is divided into two parts: translator and interpreter. First, the translator software routine translates the source statements into intermediate code. Next, the interpreter software routine interprets the intermediate code to obtain computation results. The translator can be written in a high-level language, but the interpreter is often written in assembly code.

The stack machine code is an ideal intermediate language for two reasons. First, the compiling or translation job step is fast because the generated code has little room for optimization. Second, the interpretation process is sequential in nature. That is to say, fetch an instruction, interpret the opcode, and execute. The next instruction is not fetched until the execution of the current instruction is completed. Therefore, a target machine with two or three addresses in its instruction would not improve the speed. As far as hardware design is concerned, a stack machine can not match the performance of a register machine due to the sequential nature of stack operations. Efforts were made, to no avail, to execute stack instructions in parallel.

A stack machine uses a working stack in lieu of working registers. This working or operand stack is designed for both unary and binary operations. Typically, in an arithmetic or logical instruction, the opcode field has an op and an eop (extended op). If the eop specifies a unary operation, the CPU pops the source operand off the stack, performs an operation on the operand, and pushes the result back on the stack. If the eop specifies a binary operation, the CPU pops the stack once to fetch sopd 2, pops it again to fetch sopd 1, performs the operation, and pushes the result back on the stack. However, some of the stack machine instructions do have addresses. As an example, a push instruction has a memory address to tell the CPU where the operand is stored. Our main goal is to write a simulator for a stack machine. As programming is concerned, on a three-address, two-address, or one-address machine, we think in infix notation, but on a zero-address machine we think in postfix Polish notation as introduced below.

10.2 POSTFIX POLISH NOTATION

In programming language C, the infix notation is used to represent an assign statement, as shown below:

<var> = <arith exp>;

The LHS is a single variable, the RHS is an arithmetic expression, and the equal sign (=) is an assign operator. An arithmetic expression contains both unary and binary operators. For a unary operation, the operator is placed before the operand. For a binary operation, the operator resides between two operandi as shown below:

```
X = A + B * C;
        --------
      -------------
  -------------------
```

In the statement, X, A, B, and C are symbolic names for addresses and the semicolon is a delimiter to signal the end of a statement. Based on operator precedence, the underscore line describes the sequence of operations: multiply first, add next, and assign last.

A postfix notation has different rules for positioning the operators. For unary operation, we place the operator after the operand. For binary operation, we place the first operand first, the second operand next, and the operator last. Note that in an infix notation, parentheses are needed to resolve the operator precedence problem. In a postfix notation, no parentheses are needed because the operator precedence is built in stack operations. Let us translate the above infix notation into a postfix notation, as shown below:

#X A B C * + =;

Note that the operand X is preceded by a pound (#) sign, but not the operandi A, B, and C. The # sign has special meaning in that the address of X is of interest. The symbol #X means to push the address of X, and the symbol X means to push the content or value of X. There is no confusion in parsing because of the equal sign (=) right after the symbol X. The parser knows to take special action to output #X instead of X. The semicolon is left in the notation to signal the end of a statement.

This ingenious notation was invented by the Polish mathematician Lukaciewiz; each symbol can be mapped into a stack machine instruction, as shown below:

```
PH#    X    ;push address of X
PH     A    ;push content of A
PH     B    ;push content of B
PH     C    ;push content of C
MUL         ;multiply
ADD         ;add
STW         ;store word
```

Starting from the left, these instructions show the opcode, the address field, and the comment. Except for PH# and PH, there is no address coded in the instructions. The PH# instruction means to push memory immediate, so the address field in this instruction is pushed on the stack. In contrast, the PH instruction means to push memory direct, so the content at this address is pushed on the stack. Each of the instructions MUL (multiply), ADD (add), and STW (store word) has opcode only. The execution sequence is described as follows:

1. Push the address of X.
2. Push the content of A.
3. Push the content of B.
4. Push the content of C.
5. Pop the second source operand, pop the first source operand, multiply, and push result.
6. Pop the second source operand, pop the first source operand, add, and push result.
7. Pop the source operand, pop the destination address, and store the source operand at this address.

The stack frame changes dynamically, as shown in Figure 10.1. A symbol with # means an address and a symbol alone means content or value. The stack frame after four pushes is shown in Figure 10.1a and SP points to operand C. After MUL, the stack looks like Figure 10.1b and SP points to the product of B and C. After ADD, the stack looks like Figure 10.1c. After executing the STW instruction, the stack goes back to its original state before the assign statement, as shown in Figure 10.1d. Suppose that we wish to perform A + B before multiply in the above statement. The infix notation below shows that the add operation in enclosed in a pair of parentheses that has the highest precedence.

```
X = (A + B) * C;
         ---------
      --------------
----------------------
```

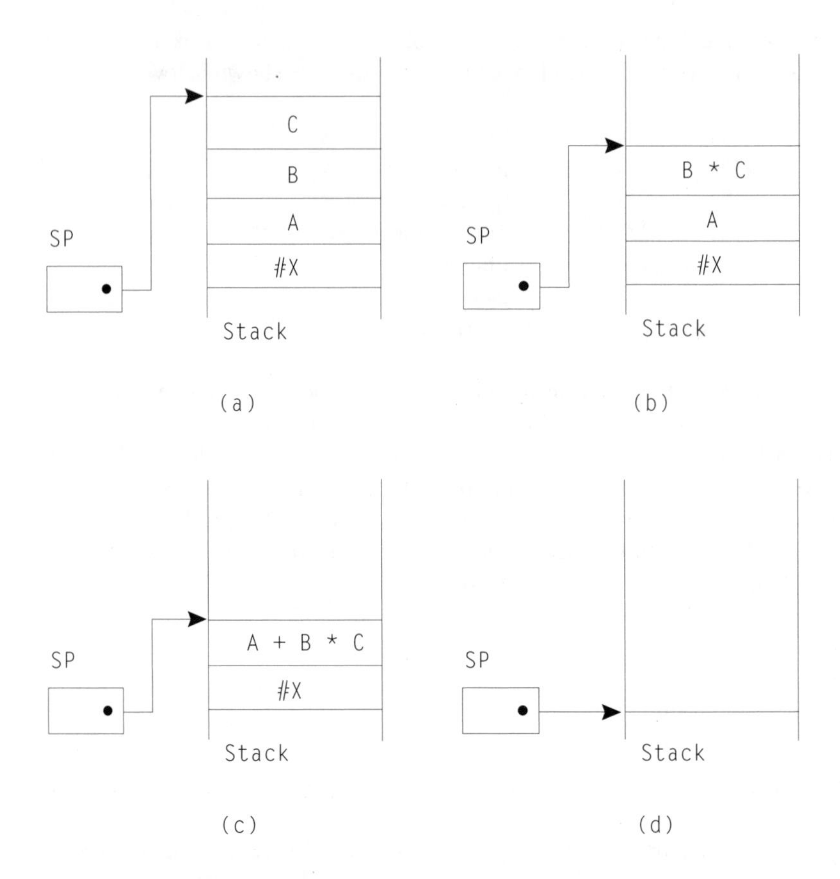

Figure 10.1 Stack frames: (a) after four pushes, (b) after mul, (c) after add, and (d) after store.

After parsing, we obtain a different postfix notation for this assignment statement as follows:

#X A B + C * =;

Two important notions must be observed. First, the relative positions of all operandi should be the same. Second, there is no parentheses in the notation, and the operator positions may be changed as determined by precedence. In this case, the add operation is performed before multiply. Interestingly, we can parse a statement from infix notation into postfix notation by inspection or by programming. Either way, we need to understand operator precedence as introduced in the following section.

10.3 OPERATOR PRECEDENCE

In compiler design, we have had two generations. A parser is a software routine that translates a language statement into some kind of intermediate form. The first generation compiler uses operator precedence parsing that is syntax controlled. The second generation compiler uses syntax directed parsing based on grammar. That is to say, the syntax of a language statement is defined by a set of grammatical rules. The syntax directed parser follows the flow of grammar and is more viable because of its correctness. After generating the intermediate code, a compiler takes another step to translate it into target machine code that is stored in an object module. The interpreter, on the other hand, stores the intermediate code in an object module on disk. Later, in another job step, the interpreter interprets the intermediate code to generate computation results.

Two famous syntax directed parsing techniques are top-down and bottom-up.[83] As far as implementation is concerned, we need only to choose one. Top-down and bottom-up, which is better and more intuitive? The answer may be subjective. However, my former student, an IBM associate, took a long time to convince me that top-down is better for two reasons. First, there is one-to-one mapping between the grammar and the language statement. Second, since the programming language C provides interactive output, it is easy to examine the printout after each parsing step. In consequence, all bugs can be detected easily, and the code is easy to maintain.

The interpreter approach has gained popularity for two reasons. First, the intermediate code and the operating environment are host machine independent. Second, the intermediate code can also be saved on disk in an object module or a data file. Later, the interpreter can interpret the code without going through the translation phase again. Our focus is to write an interpreter for the stack machine. For the sake of simplicity, we discuss operator precedence parsing to generate stack machine code. The operator precedence concept is described in Table 10.1. The priority of each operator is listed in descending order. The pair of parentheses on top has the highest precedence and the assign operator at bottom has the lowest.

Table 10.1 Operator Precedence

Symbol	Description
()	Parentheses for grouping
.NEG.	Negate
.EXP.	Exponential
*, /	Multiply, divide
+, -	Add, subtract
^	Concatenate
{.EQ., .NE., .GT., .GE., .LT., .LE.}	All relational ops have equal precedence.
.NOT.	NOT
.AND.	AND
.OR.	OR
.EOR.	Exclusive OR
=	Assign

10.3.1 Parsing By Inspection

Using operator precedence, we can generate the postfix notation by inspection. That is, by scanning an assignment statement from left to right, we write down the operandi, one by one. Next, based on the operator precedence, we can insert operators one by one after the operand pair. Thus, the following parsing algorithm is based on operator precedence by inspection:

```
REPEAT
Scan all the operators from left to right;
  IF the operator has a higher precedence than the subsequent
  operator;
  THEN
  Move the operator after the operand pair to form a triple
  treated as a single result operand; ENDIF;
UNTIL all the operators are moved to the proper places;
```

Example 1: The assignment statement is given below.

```
X4 = (X0 - X1) / (X2 - X3);
     ------------  ------------
     ---------------------------
----------------------------------
```

The underscore lines indicate the operation sequence based on operator precedence. Obviously, the two subtract operations between X0 and X1 and between X2 and X3 should be performed first in the form of triples. A triple means an operand pair followed by an operator. Next, a divide operator is placed after the two results. Finally, the assign operator is placed at the end in a triple. Therefore, the postfix notation is as follows:

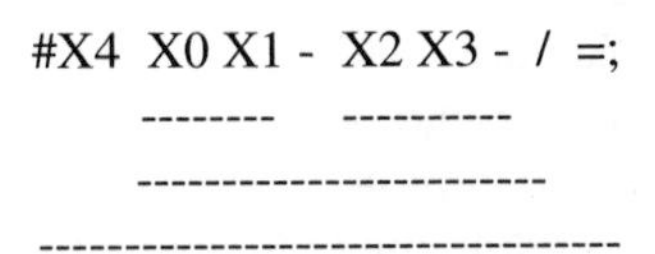

It should be stressed that no parentheses are needed in a postfix notation. However, in order to grasp the concept, we may use parentheses to enclose each triple, as shown below:

(#X4 ((X0 X1 -) (X2 X3 -) /) =);

Example 2: The assignment statement is given below.

```
I5 = I1 + I2 - I3 + I4 / 2;
    ----------      --------
    ---------------
    -------------------------
------------------------------
```

The postfix notation is derived as follows:

```
#I5 I1 I2 + I3 - I4 2 / + =;
    ---------  --------
    --------------
    --------------------
------------------------------
```

The + operator between I1 and I2 and the / operator between I4 and 2 should be performed first. Therefore, we move + after I2 and / after 2. Note that the numeric 2 is treated like a literal, a memory address where the integer two is stored. Subsequently, the - operator should be moved after I3, and the + operator after I3 is moved after the triple (I4 2 /). Finally the assign operator = is moved to the end.

Example 3: The quadratic equation below has A, B, and C as real coefficients.

```
A*X2 + B*X + C = 0;
```

In FORTRAN, we may use one assignment statement to compute one of the two roots, as shown below:

```
X1 = (-B + (B * B - 4.0 * A * C) ** .5) / (2.0 * A);
```

We use two meta-symbols in the postfix notation below. First, the symbol .neg. means negate. Next, the symbol .exp. means exponential, and the exponent of .5 means the square root.

```
#X1 B .neg. B B *  4.0 A * C *  - .5 .exp. +  2.0  A *  / = ;
    -------- ------ ---------                --------
                    -------------
             ------------------------
             -----------------------------
    --------------------------------------
    ------------------------------------------------------
    --------------------------------------------------------
------------------------------------------------------------------
```

The spaces inserted between symbols are for clarification. The float constants are in literal form, and each represents a memory address where the float constant is stored. Next, we will write a software parser based on operator precedence.

10.3.2 Parsing by Software Algorithm

Operator precedence parsing by inspection can be done by humans. Parsing by software, though intricate, is also possible based on the same principles. Thus, by scanning the statement from left to right, we take appropriate actions. Intuitively, if the symbol is an operand, we output. If the symbol is an operator, we can not output if its precedence is lower than the next operator to follow. Thus, we use an operator stack to contain all operators whose actions are yet to be taken.

10.3.2.1 Operator Stack

The operator stack is a data structure in memory for the parser to generate the postfix notation. The operator stack is different from the operand stack. The latter is a working stack for execution. In practice, the operand stack on a stack machine can be simulated in memory or designed as hardware. The operand stack is used to store temporary results during the execution of a postfix notation, while the operator stack is a software tool for parsing. That is, the infix notation is examined from left to right, which is known as scanning. Actions taken are based on the current symbol. If the current symbol is an operand, we always output. If the symbol is an operator, we either output or push it on the operator stack depending on the status of the operator stack, i.e., the precedence of its TOS. Suppose we are parsing an assign statement with arithmetic operators. The operator stack precedence is specified in Table 10.2.

Table 10.2 Precedence of Operator Stack

TOS	Priority	Description
(	6,0	Left paren for grouping
.neg.	5	Negate, a unary op
.exp.	4	Exponential
*, /	3	Multiply, Divide
+, -	2	Add, Subtract
=	1	Assign operator
<nul>	0	Stack empty

The left parentheses has the highest precedence. After being pushed on stack, its priority drops too, so any actions must be deferred until more operators in the parentheses are examined. The meta-symbol <nul> has the lowest precedence indicating an empty stack, so during parsing the first operator symbol is always pushed. For other operators, actions are taken as determined by the current operator and the status of stack. Note that the right parentheses is not in the stack because it is never pushed. In fact, a popping action must follow until a left parentheses is popped out as a match.

10.3.2.2 Operator Precedence Parsing Algorithm

The following algorithm has two assumptions. First, the assign statement is syntactically correct. Second, its postfix notation is generated as output after parsing.

1. If the current symbol is an operand, we have two cases. If the operand symbol is followed by an equal sign (=), we output the # sign and the symbol so its address is of interest; otherwise we output the symbol alone.
2. If the current symbol is a left parentheses, push it onto the stack.
3. If the current symbol is an operator, not a right parentheses and it has strictly higher precedence than the TOS, push it onto the stack; otherwise do the following: First, pop the operator stack and output until the TOS has strictly lower precedence. Then, push the current operator onto stack. Interestingly, if the operator has equal precedence, we should pop the stack and output because the left to right rule applies.
4. If the current symbol is a right parentheses operator, pop the operator stack one by one, and output until a left parentheses is popped out. Then, discard the pair so that the right parentheses is matched with a left parentheses.
5. If the current symbol is a semicolon, pop the operator stack and output one by one until the stack is empty. Then, output the semicolon as the final delimiter.

Therefore, the algorithm is as follows.

```
Program;
REPEAT;
Scan the infix notation statement from left to right;
   CASE current symbol of,
   Operand followed by an equal sign:
      Output a pound (#) sign first and the operand;
   Operand with no equal sign:
      Output the operand;
   Operator:
      CASE operator of,
      Left paren:  Push on the operator stack;
      Right paren: Pop the operator stack and output until the
                   matching a left paren is popped out;
                   Discard the pair of parentheses;
```

```
            Semicolon: REPEAT
                        Pop the operator stack and output;
                        UNTIL the stack is empty;
                        Output the semicolon;
                        Exit;
            ELSE:
               IF its precedence is higher than TOS,
               THEN
               Push the operator on the operator stack;
               ELSE
               REPEAT
               Pop the operator stack and output
               UNTIL TOS has strictly a lower precedence;
            Push the operator on the operator stack; ENDIF;
         ENDCASE;
       ENDCASE;
     UNTIL forever;
     END.
```

The parsing steps of an assign statement and its operator stack are shown in Table 10.3.

Table 10.3 Assignment Statement Parsing

Current Symbol	Action	Operator Stack
X	Output #X	
=	Push	=
(	Push	= (
(	Push	= ((
I1	Output I1	= ((
+	Push	= ((+
I2	Output I2	= ((+
)	Pop + and output	= ((
	Pop (and discard	= (
/	Push	= (/
(	Push	= (/ (
I3	Output I3	= (/ (
-	Push	= (/ (-
I4	Output I4	= (/ (-
)	Pop - and output	= (/ (
	Pop (and discard	= (/
)	Pop / and output	= (
	Pop (and discard	=

Current Symbol	Action	Operator Stack
/	Push	= /
2	Output 2	= /
;	Pop / and output	=
	Pop = and output	empty
	Output semicolon ;	

Note: Input: X = ((I1 + I2) / (I3 - I4)) / 2;
Output: #X I1 I2 + I3 I4 - / 2 / =;

10.4 SIMPLE STACK MACHINE

For pedagogical reasons, this section defines a simple stack machine code named SSM315. Each word is 16 bits long denoted by <15:0>, and bit zero is the LSB. This is a little endian computer, so the low order byte is stored in memory before the high order byte. Twos complement notation is used to represent negative integers. The execution unit can handle both arithmetic and logical operations via a stack simulated in memory.

10.4.1 Register Set

As described in Table 10.4, the SSM315 has seven registers of 16 bits each primarily for address control.

Table 10.4 SSM315 Register Set

Register	Description
FL	Field length contains the length of the entire program in bytes.
IR	Instruction register contains the current instruction in execution.
PC	Program counter points to the next instruction in memory with respect to the segment base.
RA	Relocation address contains the physical base address of the entire program segment.
SF	Stack frame points to the lowest address of the stack frame.
SP	Stack pointer points to the TOS.
SR	Status register contains the status of the current running environment.

Before executing a program, the OS sets the RA register that contains the base address of the program segment. Note that RA is an implicit base register because it is not specified in the opcode. Any address in an instruction or other register is a program address, also called virtual address. The program segment includes code, data, and stack, and the sole reason is to simplify interpreter design. The total length of the composite segment is specified in the FL register. To fetch an operand instruction or data in memory, the relative address is added to RA to form its absolute memory address. That is, the program address is relative, and a relative address greater than the FL triggers an interrupt, i.e., address exception. Therefore, a program can only cause damage within its own partition.

The SF (stack frame) is a relative index pointing to the lowest address of the stack segment. The SP (stack pointer) is also a relative index pointing to the top of the stack frame.

10.4.2 Stack Machine Ops

Stack machine opcodes, s-ops, and stack machine instructions are all synonymous. All the SSM315 instructions have a uniform length of 16 bits. The push direct and push immediate instructions are shown in Figure 10.2a. The four-bit opcode is followed by a 12-bit disp (displacement) relative to the segment base in the RA. There are many operator call instructions and each one has a four- bit op followed by a 12-bit eop as shown in Figure 10.2b. In Figure 10.2c, the 12- bit SVC code in an SVC instruction is defined by the OS.

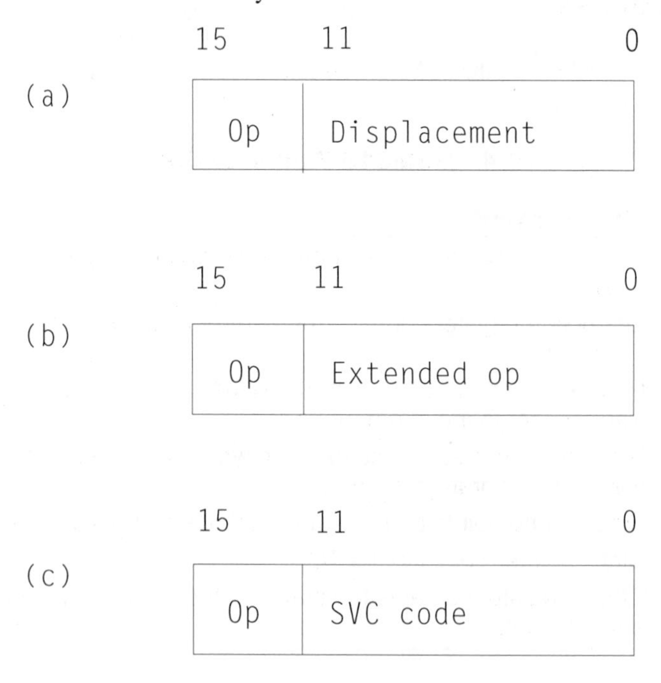

Figure 10.2 The SSM 315 instruction format: (a) push instructions, (b) operator call, and (c) supervisor call.

10.4.2.1 Stack Operations

The symbol S[0] denotes the TOS, S being a mnemonic for stack. The stack is simulated in memory and it grows from high address to low address as depicted in Figure 10.3. The entry S[0] represents M[RA + SP + 0], and 0 is an index with respect to the stack top. Thus, S[2] represents the next word below the TOS with the integer 2 as an index. At any time during execution, the SP can not be greater than the FL or less than the SF. The four-bit main opcodes are defined in Table 10.5. Except for SP+, SP-, PH (Push) and PH# (Push Immediate), no other instructions contain addresses.

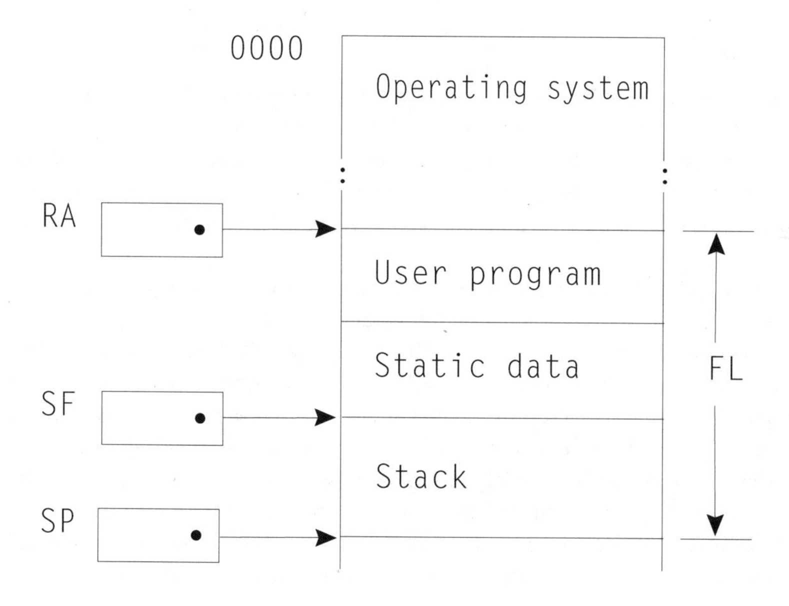

Figure 10.3 Memory layout: operating system and user partition.

Table 10.5 SSM315 Instruction Set

Opcode	Mnemonic	Descriptions
0000	ILL	Illegal opcode
0001	SP+	Increase SP SP <-- SP + Disp;
0010	SP-	Decrease SP SP <-- SP - Disp;
0011	SVC	Supervisor call where the 12-bit disp is an SVC code.

0100	-	Operator call and the 12-bit disp is an eop and each one has a different mnemonic.
1000	PH	Push word, memory direct SP <-- SP - 2; S[0] <-- M[RA + Disp] <15:0>;
1100	PH#	Push address, memory immediate SP <-- SP - 2; S[0] <-- disp <11:0>; S[0] <-- 0 <15:12>;

If an instruction has four leading zeroes, it is an illegal op. The SP+ instruction increases the SP by a length that is its 12-bit displacement. Similarly, the SP- instruction decreases the SP by a length that is its 12-bit displacement. For both SP+ and SP-, the second source operand is memory immediate. The push and push immediate instructions contain a 12-bit displacement relative to the segment base as specified in the RA.

The SVC instruction is a supervisor call, and the SVC code design is determined by software. Each SVC code tells the OS what to do. In order to write a test program, we merely study two SVC codes, 10 and 0. The SVC 10 instruction is issued to request termination, and SVC 0 is issued to display a string on the screen. If there are any passing arguments, they are pushed on the stack before the call. Take the output driver call as an example; first we push an address constant pointing to the buffer. The address constant is symbolically represented by =a 'buffer' denoting the address of buffer. The string in the buffer has a property such that the first character is not printed because it is the length or byte count of the string. The first instruction pushes the argument, and the second is a supervisor call, as shown below:

```
ph      =a `buffer′  ;push 2B address constant
svc     0            ;Output driver call
```

It must be emphasized that the interpreter does not simulate the interrupt mechanism at the machine level. Instead, it performs the service and returns control to the user program. The interpreter pops the passing argument off the stack before returning to the calling task.

An opcode of four or 0100 in binary, means an operator call. That is, the lower 12-bit displacement is an eop as shown in Table 10.6. The three-digit hex code is followed by its mnemonic and description. The BAL (branch and link) instruction replaces a call instruction, the B (branch) instruction is the same as a return instruction, and the BF (branch on false) instruction handles the IF – THEN – ELSE construct.

At the end of the table, we show the privileged instructions; each one is denoted by an asterisk. They are IRT (interrupt return), LRA (load RA), LFL (load FL), InB (input byte), OutB (output Byte), InW (input word), and OutW (output word). These instructions can be executed only if the supervisor bit in PSWR is set. The IRT instruction restores the register set from low memory. The input byte instruction places the I/O address (i.e., TOS) on the I/O bus, and the input byte replaces the old TOS. Before executing any input instruction, the I/O address must first be pushed. The execution of an output instruction is a little different. Before its execution, we push not only an I/O address but also an operand. Both the address and the operand are placed on the I/O bus and, consequently, the stack is popped off four bytes.

Table 10.6 SSM315 Extended Opcode

Eop	Mnemonic	Descriptions
000	ADD	Integer add The operation is as follows: Temp2 <-- S[0]; SP <-- SP + 2; Temp1 <-- S[0]; S[0] <-- Temp1 + Temp2; Statically, we have S[2] <-- S[2] + S[0]; {SP <-- SP + 2 implied}
001	SUB	Integer subtract S[2] <-- S[2] - S[0];
002	MUL	Signed multiply S[2] <-- S[2] * S[0] <15:0>;
003	DIV	Signed divide S[2] <-- Quo(S[2] / S[0]);
004	NOT	Not S[0] <-- .NOT. S[0];
005	AND	And S[2] <-- S[2] .AND. S[0];
006	OR	Or S[2] <-- S[2] .OR. S[0];
007	EOR	Exclusive or S[2] <-- S[2] .EOR. S[0];
008	EQ	Equal IF S[2] .EQ. S[0], THEN S[2] <-- 1; ELSE S[2] <-- 0; ENDIF;
009	NE	Not equal

00A	LT	Less than
00B	LE	Less equal
00C	GT	Greater than
00D	GE	Greater equal
00E	BT	Branch on true
		IF (S[2] .AND. 1) .EQ. 1,
		THEN PC <-- S[0]; ENDIF;
		{SP <-- SP + 4 implied}
00F	BF	Branch on false
		IF (S[2] .AND. 1) .EQ. 0,
		THEN PC <-- S[0]; ENDIF;
010	B	Branch {same as return}
		PC <-- S[0];
		{SP <-- SP + 2 implied}
011	BAL	Branch and link {same as call}
		PC <--> S [0]; {interchange}
012	NEG	Negate
		S[0] <-- -S[0];
013	ABS	Absolute
		S[0] <-- |S[0]|; {absolute value}
014	LW	Load word
		S[0] <-- M[RA + S[0]];
015	STW	Store word
		M[RA + S[2]] <-- S[0];
016	LC	Load character
		S[0] <-- M[RA + S[0]] <7:0>;
017	STC	Store character
		M[RA + S[2]] <-- S[0] <7:0>;
018	MOD	Modulus
		S[2] <-- S[2] .MOD. S[0];
		{ S[2] <-- Rem(S[2] / S[0]) }
01A	INC	Increment the TOS by one.
		S[0] <-- S[0] + 1;
01B	DEC	Decrement the TOS by one.
		S[0] <-- S[0] - 1;
030	MVC	Move characters from memory to memory
		{The length is specified in S[0],

		the source address is RA + S[2], and the destination address is RA + S[4]. SP <-- SP + 6 implied}
031	CVB	Convert from five decimal digits in ASCII to binary. M[RA+S[2]]<15:0> <-- Convert M[RA+S[0]]<39:0>; {ASCII to binary} {SP <-- SP + 4 implied}
032	CVC	Convert from binary to five decimal digits in ASCII. M[RA+S[2]]<39:0><--Convert M[RA+S[0]]<15:0>; {binary to ASCII} {SP <-- SP + 4 implied}
033	SW	Switch the TOS S[0] <--> S[2];
04X	SHL	Shift logical left the TOS X bits
05X	SHR	Shift logical right the TOS X bits
06X	SAR	Shift arithmetic right the TOS X bits
800	*IRT	Interrupt return to restore registers SP,SF,RA,FL,SR,PC <-- S[0] <95:0>; {SP <-- SP + 12 implied}
801	*LRA	Load relocation address register RA <-- S[0]; {SP <-- SP + 2 implied}
802	*LFL	Load Field Length register FL <-- S[0]; {SP <-- SP + 2 implied}
803	*InB	Input byte S[0] <-- IOR[S[0]] <7:0>;
804	*OutB	Output byte IOR[S[2]] <-- S[0] <7:0>; {SP <-- SP + 4 implied}
805	*InW	Input word S[0] <-- IOR[S[0]];
806	*OutW	Output word IOR[S[2]] <-- S[0]; {SP <-- SP + 4 implied}

10.4.3 Interrupt Mechanisms

Upon interrupt, an interrupt code is set in the SR. The CPU then performs context switching by saving the entire register set on the stack and loading the new set from a memory block at address zeroes. The sequence operations are specified as follows,

```
SR <15:8> <-- interrupt code;
S[0] <95:0> <-- SP, SF, RA, FL, SR, PC;
SP <-- SP - 12;SP, SF, RA, FL, SR, PC <-- M[0000] <95:0>;
```

10.4.4 Status Register

The SR has 16 bits as defined below:

Bit 1 – 0:	Interrupt enable: b0 for channel 0, b1 for channel 1
2:	Machine check enable
3:	External interrupt enable
4:	Wait
5:	Privileged
6 – 7:	Reserved
8 – 15:	Interrupt code

A hardware timer is implemented in memory at address 0080 in hex whose content is decreased by one every 10 ms. When the timer changes from positive to negative, an external interrupt is triggered if bit three in the SR is set. For debugging purposes, a programmer may store an ID onto the stack before executing a branch instruction. If a branch address is outside the limit (i.e., greater than the FL), interrupt is triggered, and the ID on the stack indicates the branch instruction that caused the error.

10.5 STACK MACHINE ASSEMBLY LANGUAGE

All the SSM315 assembly statements are case insensitive, and they are divided into machine ops and pseudo ops. The following sections discuss the pseudo ops, the load and store ops, the instructions to translate an IF statement, and the instructions to access arrays.

10.5.1 Pseudo Ops

The pseudo ops are start, end, org, DC, DS, stack, and equ. In the define constant statement, the operand field has the following syntax:

[<dup>] <data type> [L<length>] `<value>′

The first optional integer tells the number of times that the data structure will be duplicated. There are two kinds of data types: generic and explicit. The generic types are B (byte), W (word), and D (double word) whose length is implied. The explicit types are X (hex), O (octal), C (character), A (address), I (integer), etc. If no data type is specified, the default is integer in decimal. The data type may have a explicit length specifier, that is, the keyword L followed by a decimal integer indicating the length in bytes. Any specific value or bit pattern must be enclosed in single quotes. For example, a decimal integer 315 of two bytes can be specified as iL2‘315’, W‘315’, or xL2‘13B’. An address constant has its symbol enclosed with the understanding that it will be translated into an address by the assembler. Note that if a define constant has no value specified, it is equivalent to a define storage. Some coding examples are listed in Table 10.7.

Table 10.7 Data Operand Types

Example	Description
iL2`1021′	16-bit integer whose value is 1021 in decimal
xL2`3FD′	16-bit integer whose hex value is 3FD, i.e. 1021 in decimal
w`100’	16-bit word with decimal 100
100w`0′	100-word block of zeroes
cL12` is a prime.′	ASCII string of 12 bytes
iL1`80′, 80c″	A byte count of decimal 80 is followed by 80 ASCII space characters so the total storage area is 81 bytes
wL2	Define two bytes in word boundary
cL81	Define an 81 byte storage block
2048w	Reserve a 2084 word storage block with any bit pattern
a`buffer′	Address constant of two bytes, e.g., buffer has a value 010C in hex
0d	Zero double word so align the next address to the double word boundary

10.5.2 Push Direct vs. Push Immediate

The opcode of a push (direct) instruction is different from the opcode of a push immediate instruction. To push a 16-bit zero on the stack, we can use push immediate or push, as shown below:

```
ph#     0        ;push immediate 0
ph      ='0'     ;push literal 0
```

The first instruction pushes a 12-bit immediate operand zero with zero extended on the stack. If an unsigned operand has a value between 4 K and (64 K-1), the push immediate instruction can not do the work. Therefore, the second instruction is used instead to push a literal. The literal symbol = '0' means =iL2'0' where i stands for integer, L2 means length of two bytes, and zero is the value. Note that the literal operand must reside in the first 4 KB of the segment to be accessible. As far as the source operand is concerned, the first instruction uses the addressing mode of memory immediate. The second instruction uses the addressing mode of memory direct. After executing either instruction, the TOS (top of stack) is zero. The reason that we call the ='0' symbol a literal is because the assembler recognizes the symbol, prepares the bit pattern in the code, and inserts its memory address in the push instruction. That is, ='0' represents an address where the 16-bit zero is stored. A literal symbol is no different from a symbolic name except it is not explicitly defined. Some assemblers may allow an equal sign to be placed before the literal value, such as =0, =300, etc. By scanning the literal symbol, the assembler knows to construct the bit pattern for the integer and insert its address in the instruction. A literal is not mandatory if the constant is explicitly defined.

10.5.3 Store vs. Pop

The following block of code is executed to push immediate X, evaluate an expression, and store the result.

```
ph#   X
<expression>
stw
```

The two instructions, first and last, have caused some debate concerning whether a pop instruction can replace the pair, as shown below:

```
<expression>
pop     X   ;M[RA + X] <-- S[0]
```

The pop instruction has an opcode followed by a 12-bit displacement. After its execution, the TOS is popped into a memory location named X, so the SP is increased by two. The pop instruction is more intuitive, but it changes the relative operand positions in the postfix notation, so it is not pure stack machine code. In addition, its address field is not a full 16 bits, so it can not access the entire 64 KB. In contrast, if we push a full literal address on the stack under the assumption that it is within 4 KB, we can use the STW instruction to access the entire 64 KB addressing space, as shown below:

```
ph      =a`X'   ;S[0] <-- M[RA + =a`X']
<expression>
stw
```

Recall that the literal symbol =a`X′ represents an address where the address X is stored. In other words, the literal symbol =a`X′ is a pointer variable that contains the address of X.

10.5.4 Translation of IF Statements

In a short or long IF statement, a logical expression is evaluated first. Control may be passed to one location or another based on the outcome. A short IF statement has only the THEN clause, but a long IF statement also has an ELSE clause, as shown below:

```
IF (n .mod. i) .eq. 0,
THEN  <statement 1>;
ELSE  <statement 2>; ENDIF;
```

If the condition is true, the statement 1 block is executed, otherwise the statement 2 block is executed. The two keywords THEN and ELSE are treated as operators, and after parsing, the postfix notation becomes:

```
n i .mod. 0 .eq. THEN <statement 1> ELSE <statement 2>;
```

In this notation, there is a one to one relationship between each symbol and the stack machine code. Note that both the statement 1 block and the statement 2 block are expressed in postfix notations. Inside the block, it is possible to have more nested IF statements. When the statement is translated into assembly code, an internally generated label F00010 is placed before <statement 2>, where F means forward jump and the suffix 00010 is a randomly selected number to indicate some sort of ordering. The THEN operator is translated into BF (branch on false), so control is passed to <statement 2> in the case that the condition is false. The ELSE operator after <statement 1> is translated into B (unconditional branch) to pass control to the next statement after IF. This is designed to skip the statement 2 block after executing the statement 1 block. Thus, the stack machine code is as follows.

```
ph      n       ;push n
ph      i       ;push i
mod             ;push remainder of (n/i)
ph      ='0'    ;push literal integer 0
eq              ;IF S[2] .eq. S[0],
                ;THEN push TRUE flag,
```

```
                                        ;ELSE push FALSE flag, ENDIF.
            ph#         F00010          ;push address F00010
            bf                          ;IF S[2] .eq. FALSE,
                                        ;THEN PC <-- S[0], ENDIF.
            <statement 1>
            ph#         F00020          ;push address F00020
            b                           ;PC <-- S[0]
F00010:
            <statement 2>
F00020:
```

Interestingly, the THEN operator rule is applied to a short IF statement. That is, control is passed to the next statement if the condition is false. In the comment, the stack pointer (SP) is updated accordingly after each push or pop.

10.5.5 Translation of Subscripted Variables

The stack machine does not use index registers, so any subscripted variable must be taken into account with special attention. Study the postfix notation for the assignment statement given below:

```
array[i] = array[i] + 1;
```

Assume that zero indexing is used and each entry in the array is n bytes long. Array[i] on the LHS represents an effective address which is the sum of the starting address of the array and a relative displacement computed as (i * n). Hence, the pair of square brackets is treated like a subscript operator. After parsing, we obtain the following postfix notation,

```
#array i sub #array i sub 1 + =;
```

where the sub operator requires special action in order to generate the correct stack machine code as shown below:

```
ph#     array       ;push starting address of array
ph      i           ;push index
ph      n           ;push element length

mul                 ;compute displacement
add                 ;add to starting address
ph#     array
ph      i
ph      n
mul
add
ex                  ;exchange--pop address, fetch the
```

```
                                    ;operand at this address and push it
                                    ;on the stack
        push        ='1'            ;push literal 1
        add
        stw
```

Our goal is to implement a software based VSM (virtual stack machine). First, we write a test program. Next, we hand translate the test program into assembly code, and then into binary s-ops.

10.6 TEST PROGRAM FOR SSM315

The common programming language (CPL) is a direct translation from the chinese programming language.[23] It is a tool to describe the logical flow at the programming level. Some of the syntactic and semantic rules are highlighted below:

- A declaration block is placed in the front enclosed by the delimiters, DCL (declare) and ENDDCL (end declare).
- The declaration block consists of single variables, each of which is followed by a colon (:) and attributes. Grouping variables is also allowed.
- Structured constructs are supported just like PDL or any other high-level languages.
- The assign operator is an equal sign, and each statement ended with a semicolon. Some of the operators may be enclosed by two periods, like RTL.
- Each I/O statement is associated with a file along with a format to describe the ASCII layout in a record. A read file statement is shown below.

```
Readfile x into (PC, codelen, datalen, stacklen)
   format(x4, 3(skip 1 char, x4));
```

The first line reads the record in the sequential file represented by x. The record has fours fields, and the separator between two fields can be a space character or a comma. The record is read into four variables, namely PC (program counter), codelen (code length), datalen (data length), and stacklen (stack length). The second line for format control, tells how to read the record and convert the data. The first control field x4 means that four hex characters will be converted into an integer and stored in the PC. The next control field has a three followed by a parentheses to mean grouping. The integer three is the duplicate factor so whatever enclosed in the parentheses is repeated three times. Each time, as one character is skipped, the next four hex characters are converted into an integer and stored in codelen, datalen, and stacklen respectively. All four variables are declared as 16-bit short integers. The print statement has a built in default conversion format, as shown below:

```
Print n, `is not a prime.';
```

The integer n is converted into a decimal ASCII string in a buffer, and the rest of characters are padded with whatever string enclosed in single quotes. For example, if n is 1023, the message '1023 is not a prime.' will be displayed.

10.6.1 Test Program Source

Given an integer, the program tests whether that integer is a prime. The test program is hand translated into assembly code and then into s-ops. Finally, an interpreter interprets the binary s-ops to generate computation results. A compiler, if available, can translate the source into s-ops. First, the test program is translated to assembly code. Second, the assembly code is translated into s-ops. The high-level source is listed below:

```
PROGRAM Test;
                                    {Declare variables.
                                    n: integer to be tested,
                                    i: integer as the divisor,
                                    done: logical flag,
                                    TRUE: logical constant 1,
                                    FALSE: local constant 0.}
DCL     (n, i): integer;
        done:   logical; ENDDCL;
done = FALSE;
n = 1021;
i = 2;
DO WHILE (.not. done)               {In this loop, divide n by i. If
                                    remainder is 0, it is not a prime.
                                    Otherwise,find out how big i is. If i is
                                    greater than the quotient, then n is a
                                    prime. Otherwise, increase i by 1, loop
                                    back, and divide again.}
IF (n .mod. i) .eq. 0,
THEN
   DO;
   Print n, `is not a prime.';
   done = TRUE;

   ENDDO;
ELSE
   IF i .gt. n/i,
   THEN
```

```
        DO;
        Print n, `is a prime.';
        done = TRUE;
        ENDDO;
      ELSE
        i = i + 1;
      ENDIF; ENDIF; ENDDO;
    END.
```

10.6.2 S-ops

The test source program is in the file test.asm. Each statement is either a machine op or a pseudo op. If no assembler is available, each mop is hand translated into an s-op and the executable file is test.exe. The test.asm file contains ASCII characters. Normally, the test.exe file contains pure binary code without <cr> and <lf> as delimiters. However, for the sake of easy preparations, we make two deviations. First, each 16-bit s-op is represented by four hex digits in ASCII in the test.exe file. Second, we must type four hex digits and press the enter key on the line via an editor. Thus, the end of each line has <cr> and <lf> as delimiters. However, each line or group represents 16 bits after being loaded into memory. The C compiler can generate code in the interpreter that reads four hex digits in ASCII and converts them into a 16-bit integer in memory. If the interpreter is written in MASM, the conversion job must be handled by the interpreter routine. Note that by using ASCII digits, the code size on the disk is doubled.

The assembly source and the executable code of the test program are compared in Table 10.8. The left-hand side shows the assembly source, and the right-hand side shows the code in hex. The remaining translation work is left as an exercise. The load module has a header, which is the first record. The line has four initial addresses, PC, FL, SF, and SP, followed by <cr> and <lf>.

Table 10.8 Test.asm vs. Test.exe

test.asm			test.exe
pgm1	start		0000 0106 005a 1000
	ph#	done	c106
	ph#	0	c000
	stw		4015
	ph#	n	c108
	ph	=`1021'	808e
	stw		4015
	ph#	i	c10a
	ph	='2'	8090

```
        stw                     4015
L01     ph      done            8106
        not                     4004
        ph#     L05             c08c
        bf
        ph      n
        ph      i
        mod
        ph      ='0'
        eq
        ph#     L02
        bf
        ph#     buffer
        ph#     space
        ph#     81
        mvc
        ph#     buffer + 1
        ph#     n
        cvc
        ph#     buffer + 11
        ph#     str1
        ph#     16
        mvc
        ph      =a'buffer'
        svc     0               ;Output driver call
        ph#     done
        ph#     1
        stw
        ph#     L04
        b
L02     ph      i
        ph      n
        ph      i
        div
        gt
        ph#     L03
        bf
        ph#     buffer
        ph#     space
```

```
           ph#    81
           mvc
           ph#    buffer + 1
           ph#    n
           cvc
           ph#    buffer + 11
           ph#    str2
           ph#    12
           mvc
           ph     =a`buffer'
           svc    0
           ph#    done
           ph#    1
           stw
           ph#    L04
           b
L03        ph#    i
           ph     i
           ph     ='1'
           add
           stw
L04        ph#    L01
           b
L05        svc    10         ;exit to OS     300a
='1021'    dc     iL2`1021'                  03fd
='2'       dc     iL2`2'                     0002
='0'       dc     iL2`0'                     0000
=a 'buffer' dc    a`buffer'                  010c
='1'       dc     iL2`1'                     0001
space      dc     iL1`80', 80c' '            2050 2020 2020 ...
str1       dc     cL16' is not a prime.'
str2       dc     cL12' is a prime.'
done       ds     wL2        ;word address
n          ds     wL2
i          ds     wL2
buffer     ds     cL81
           ds     0d  ;align double word address
           stack  2048w
           end    pgm1       ;entry point
```

During the assembly process, a symbol table is constructed to contain all the symbols and their addresses. Any symbol reference is translated into a binary address in the code by the assembler via a symbol table look-up. All the symbols in the test program and their hexadecimal addresses are listed in Table 10.9.

Table 10.9 Symbol Table for the Test Program

Symbol	12-bit address in hex
pgm1	000
L01	012
L02	04c
L03	07e
L04	088
L05	08c
=`1021′	08e
=`2′	090
=`0′	092
=a`buffer′	094
=`1′	096
space	098
str1	0e9
str2	0f9
done	106 {codelen = 0106}
n	108
i	10a
buffer	10c {datalen = 005a; stacklen = 1000}

The address for str2 is computed as 0f9 in hex, and the length of the string is 12 in decimal. Thus, the address for the next variable should be 105 (0f9 + 0c) in hex. However, the done flag is declared as a word, so the address for that variable is aligned as an even number. Consequently, one extra byte is padded at the end of the block, so its address becomes 106 in hex (i.e., the code length). The data length is the length of the defined storage that does not need to be loaded. Nevertheless, this length is passed to the OS. If the stack is aligned in double words, the data length is computed as:

$$\begin{aligned} \text{data length} &= 2 + 2 + 2 + 81 + 3 \text{ (padded bytes)} \\ &= 90 \text{ (decimal)} \\ &= 5a \text{ (hex)} \end{aligned}$$

Since a stack is defined as 2048 words or 4 KB at the end of a program, the stack length is 1000 in hex. The header in the load module has four fields. The first field is the entry point specified as 0000 in hex. The codelen (code length) is specified as 0106, the datalen (data length) as 005a, and the stacklen (stack length) as 1000. After loading, the OS computes the initial values for the PC, FL, SF, and SP as follows:

PC = 0000
FL = codelen + datalen + stacklen = 1160
SF = codelen + datalen = 0160
SP = 1160

10.7 VIRTUAL STACK MACHINE

The software based VSM (virtual stack machine) means a computer simulator or interpreter, as shown in Figure 10.4. The inner circle represents the host machine hardware that can be a PC running with its OS. The VSM executes on top of the OS as an application program. All the registers of the target machine are declared as memory variables and all the target machine instructions are treated as data by the VSM. There are many software components in the VSM. The main program is LGO (load and go), i.e., after loading, go ahead and interpret. There are two procedures SVC (supervisor call) and OPC (operator call). The former passes control to the OS requesting a system service. The latter just interprets s-ops. There is no OS in SSM315 memory, so the interrupt mechanism is not interpreted at the machine level.

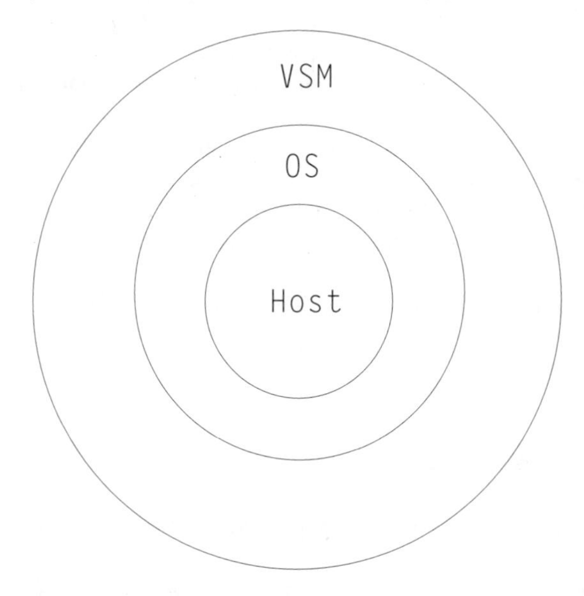

Figure 10.4 Virtual stack machine layers. Figure A.1 The 2-variable Karnaugh maps for: (a) \A B + A \B and (b) A + B.

That is to say, the interpreter carries out the service for the test program by means of the real OS services on the host. Using PDL and RTL jointly, we describe the design of the VSM as follows.

```
; Author:  John Y. Hsu
; Date:    Nov. 12, 1998
; Program design:
;  Program Lgo;
;  Declare global variables and local procedures;
;    Proc Load;
;    Proc Go;
;    Proc Svc;
;    Proc Opc;
;    Function StackOF;
;    Function StackUF;
;  Initialize RA;
;  Call Load;       {Load and initialize PC, FL, SF, SP.}
;  Initialize SR;
;  Call Go;
; End.
```

The main program is Lgo (load and go) which also simulates the OS functions. All the registers are declared as global variables in memory. At the second level, there are four procedure subroutines and two function subroutines. Each function returns a value, TRUE or FALSE. The first function checks stack overflow, and the second checks stack underflow. We merely simulate the condition in a few instructions as a token in design. After initializing the RA register, the main program calls Load (loader) to read the executable file into an array named M. The executable file has a header containing the running information about the program, such as the program counter, code length, data length, and stack length. All the addresses in the code segment are relative and relocatable. After loading, all the address registers are set. After initializing the status register, the main program calls Go to interpret and execute. After compiling and linking, we run the program by typing:

```
Lgo test.exe
```

10.7.1 Interpreter Source

The program listing in CPL is self explanatory, as shown below.

```
PROGRAM Lgo( argc, argv);  {This program loads an executable
                            file in memory, interprets and
                            executes. Global variables are:
```

```
                                   endpgm - end program flag;
                                   FL     - Field Length;
                                   IR     - Instruction Register;
                                   M      - Memory array;
                                   PC     - Program Counter;
                                   RA     - Relocation Address;
                                   SF     - Stack Frame;
                                   SP     - Stack Pointer;
                                   SR     - Status Register;
                                   temp   - temporary; }
DCL;
  (argc, FL, IR, PC, RA, SF, SP, SR, temp): short integer;
  M[0 .. (64 * 1024 - 1)]: short integer;
                 {Declare memory array as short integer
                 with index ranging from 0 to (64K-1).  }
  argv[], x      : pointer;
  endpgm         : logical; ENDDCL;

PROC Load;                         {count   - counter,
                                   codelen  - code length,
                                   datalen  - data length,
                                   stacklen - stack length.}
DCL; (count, codelen, datalen, stacklen): integer; ENDDCL;
x = argv[1];
Openfile x;                        {x points to the file name
                                   'test.exe' in our design.    }
Readfile x into (PC, codelen, datalen, stacklen)
     format(x4, 3(skip 1 char, x4));
                                   {In the format, we have 4 char in
                                   hex for PC, and 3 is the duplicate
                                   factor. Each time, skip 1 char and
                                   the 4-char in hex is for codelen,
                                   datalen, and stacklen.        }
count = 0;
REPEAT;
Readfile x into M[RA + count]
     format(x4); {Read 4 hex char into a 16-bit memory word.}
count = count + 2;
UNTIL (count .GE. codelen);
FL = codelen + datalen + stacklen;
SF = codelen + datalen
```

```
SP = FL;
return;
ENDPROC;

PROC Go;
DCL; (mop, disp): short integer; ENDDCL;{mop - machine op,
                                         disp - displacement}
REPEAT;
IR = M[RA + PC];              {Fetch instruction.}
PC = PC + 2;
mop = IR .SHR. 12;            {Shift logical right 12 bits.}
disp = IR .AND. x'0fff';
  CASE mop of
  x'0': DO;                   {Illegal op}
      Writefile display from ('Illegal opcode at PC = ',
       PC-2) format(c, skip 4 char, x4);
      {The string uses the variable char format, followed
      by 4 spaces and 4-char address in hex.}
    endpgm = TRUE;
    ENDDO;
  x`1′: SP = SP + disp;        {SP+}
    IF StackUF, THEN return; ENDIF;
  x`2′: SP = SP - disp;  {SP-}
    IF StackOF, THEN return; ENDIF;
  x`3′: Call Svc;              {Supervisor call}
  x`4′: Call Opc;              {Operator call}
  x`8′: DO;                    {Push memory direct.}
    SP = SP - 2;
      IF StackOF, THEN return;
      ELSE M[RA + SP] = M[RA + disp]; ENDIF;
    ENDDO;
  x`c′: DO;                    {Push immediate.}
    SP = SP - 2;
      IF StackOF, THEN return;
      ELSE M[RA + SP] = disp; ENDIF;
    ENDDO;
  ELSE: DO;                   {Undefined op}
    Writefile display from ('Undefined opcode at PC = ',
      PC - 2) format(c, skip char, x4);
    endpgm = TRUE;
    ENDDO;
```

```
    ENDCASE;
    IF endpgm, THEN return; ENDIF;
UNTIL forever;
ENDPROC;

PROC Svc;                          {Supervisor Call}
CASE disp of
x`000': DO;                        {Write buffer on screen.}
      Writefile display from M[RA + SP] -> format( c);
                                   {The symbol -> means the object
                                   pointed by.}
      SP = SP + 2;                 {Pop argument off stack.}
      ENDDO;
x`00A': endpgm = TRUE;
ELSE : interpret Svc;              {Not written.}
ENDCASE;
ENDPROC;

PROC Opc;            {Operator Call.}
DCL; shiftCount: short integer; ENDDCL;
CASE disp of
x`000': IF StackUF,
      THEN return;
      ELSE
         DO;          {Add}
         M[RA + SP + 2] = M[RA + SP + 2] + M[RA + SP];
         SP = SP + 2; ENDDO; ENDIF;
x`001': IF StackUF,
         THEN return;
         ELSE
         DO;          {Subtract}
         M[RA + SP + 2] = M[RA + SP + 2] - M[RA + SP];
      SP = SP + 2; ENDDO; ENDIF;
x `002': DO;   {Multiply--stack underflow check implied.}
    M[RA + SP + 2] = M[RA + SP + 2] * M[RA + SP] <15:0>;
    SP = SP + 2; ENDDO;
x`003': DO;       {Divide}
    M[RA + SP + 2] = M[RA + SP + 2] / M[RA + SP];
      SP = SP + 2; ENDDO;
x`004': M[RA + SP] = .NOT. M[RA + SP];
x`005': DO;
```

```
      M[RA + SP + 2] = M[RA + SP + 2] .AND. M[RA + SP];
      SP = SP + 2; ENDDO;
x`006′: DO;
      M[RA + SP + 2] = M[RA + SP + 2] .OR. M[RA + SP];
      SP = SP + 2; ENDDO;
x`007′: DO;
      M[RA + SP + 2] = M[RA + SP + 2] .EOR. M[RA + SP];
      SP = SP + 2; ENDDO;
x`008′: DO;          {Equal}
         IF M[RA + SP +2] .EQ. M[RA + SP],
         THEN M[RA + SP + 2] = x'0001';    {TRUE flag}
         ELSE M[RA + SP + 2] = x'0000'; ENDIF;
      SP = SP + 2; ENDDO;
x`009′: DO;          {Not Equal}
         IF M[RA + SP +2] .NE. M[RA + SP],
         THEN M[RA + SP + 2] = x'0001';
         ELSE M[RA + SP + 2] = x'0000'; ENDIF;
      SP = SP + 2; ENDDO;
x`00A′: DO;          {Less Than}
         IF M[RA + SP +2] .LT. M[RA + SP],
         THEN M[RA + SP + 2] = x'0001';
         ELSE M[RA + SP + 2] = x'0000'; ENDIF;
      SP = SP + 2; ENDDO;
x`00B′: DO;          {Less Equal}
         IF M[RA + SP +2] .LE. M[RA + SP],
         THEN M[RA + SP + 2] = x'0001';
         ELSE M[RA + SP + 2] = x'0000'; ENDIF;
      SP = SP + 2; ENDDO;
x`00C′: DO;             {Greater Than}
         IF M[RA + SP +2] .GT. M[RA + SP],
         THEN M[RA + SP + 2] = x'0001';
         ELSE M[RA + SP + 2] = x'0000'; ENDIF;
      SP = SP + 2; ENDDO;
x`00D′: DO;             {Greater Equal}
         IF M[RA + SP +2] .GE. M[RA + SP],
         THEN M[RA + SP + 2] = x‘0001’;
         ELSE M[RA + SP + 2] = x‘0000’; ENDIF;
      SP = SP + 2; ENDDO;
x`00E′: DO;          {Branch on True}
         IF (M[RA + SP + 2] .AND. x‘0001’) .EQ. x‘0001’,
```

```
            THEN PC = M[RA + SP]; ENDIF;
        SP = SP + 4; ENDDO;
x`00F': DO;             {Branch on False}
            IF (M[RA + SP + 2] .AND. x'0001') .EQ. x'0000',
            THEN PC = M[RA + SP]; ENDIF;
        SP = SP + 4; ENDDO;
x`010': DO;             {Branch}
        PC = M[RA + SP];
        SP = SP + 2; ENDDO;
x`011': DO;             {Branch and link.}
        temp = M[RA + SP];
        M[RA + SP] = PC;
        PC = temp; ENDDO;
x`012': M[RA + SP] = - M[RA + SP];     {Negate}
x`013': M[RA + SP] = | M[RA + SP] |;    {Absolute}
x`014': M[RA + SP] = M[RA + M[RA + SP]];{Load word.}
x`015': DO;             {Store word.}
        M[RA + M[RA + SP + 2]] = M[RA + SP];
        SP = SP + 4; ENDDO;
x`016': M[RA + SP] = M[RA + M[RA + SP]] <7:0>;    {Load char.}
x`017': DO;             {Store char.}
        M[RA + M[RA + SP + 2]] = M[RA + SP] <7:0>;
        SP = SP + 4; ENDDO;
x`018': DO;             {Modulus}
       M[RA + SP + 2] = M[RA + SP + 2] .MOD. M[RA + SP]);
       SP = SP + 2; ENDDO;

x`01A': M[RA + SP] = M[RA + SP] + 1;  {Increase}
x`01B': M[RA + SP] = M[RA + SP] - 1;  {Decrease}
x`030': DO;             {Move characters.}
        Call Mvc( RA + M[RA + SP + 4]), RA + M[RA + SP + 2],
                        M[RA + SP]);
        SP = SP + 6; ENDDO;
x`031': DO;             {Convert to binary.}
        Call Cvb( RA + M[RA + SP + 2], RA + M[RA + SP]);
        SP = SP + 4; ENDDO;
x`032': DO;                {Convert to char.}
        Call Cvc( RA + M[RA + SP + 2], RA + M[RA + SP]);
        SP = SP + 4; ENDDO;
x`033': DO;             {Switch TOS.}
```

```
        temp = M[RA + SP];
        M[RA + SP] = M[RA + SP + 2];
        M[RA + SP + 2] =  temp; ENDDO;
x`040′ to x`04f′:       {Shift Logical Left.}
        DO;  shiftCount = disp .AND. x`000f′;
        M[RA + SP] = M[RA + SP] .SHL. shiftCount; ENDDO;
x`050′ to x`05f′:  DO;       {Shift Logical Right.}
        shiftCount = disp .AND. x'000f';
        M[RA + SP] = M[RA + SP] .SHR. shiftCount; ENDDO;
x`060′ to x`06f′:  DO;       {Shift Arithmetic Right.}
        shiftCount = disp .AND. x`000f′;
        M[RA + SP] = M[RA + SP] .SAR. shiftCount; ENDDO;
ELSE:   interpretOpc;        {Not written.}
ENDCASE;
return;
ENDPROC;

FUNCTION StackOF: logical;
   IF (SP .LT. SF),
   THEN
   Writefile display from (`Stack overflow at PC = `, PC - 2)
      format(c, skip 4 char, x4);
   endpgm = TRUE; ENDIF;
return;
ENDFUNC;

FUNCTION StackUF: logical;
   IF (SP .GT. FL),
   THEN
   Writefile display from (`Stack underflow at PC = `, PC - 2)
      format(c, skip 4 char, x4);
   endpgm = TRUE; ENDIF;
return;

ENDFUNC;
; -------------Main program starts here.
RA = x`1000′;
Call Load;
SR = x`000f′;  {Interrupts are enabled, user mode.}
endpgm = FALSE;
Call Go;
END.
```

10.8 CONCLUSIONS

In the interpreter design, the entry in an array is referenced by an index or relative offset in bytes. It presents no problems in assembly code. However, if the interpreter is written in C, every other entry in the array is wasted because the index is measured in 2 B short integers. Nonetheless, the routine works fine as the code mimics the machine specifications in RTL.

We can make several improvements. First, the main opcode field can be reduced to two bits for push direct and push immediate so the displacement is increased to 14 bits. Second, the operand size can be increased to 32 bits, or 48 bits. Third, the top two words in the stack can be data cache.[2] Fourth, if the RA register contains a 16 B paragraph number, the PAS is 1 MB.

10.9 SUMMARY POINTS

1. In a stack machine, no addresses is coded in a unary or binary instruction.
2. To program a stack machine, think in postfix notation.
3. Each symbol in the postfix notation is translated into an S-op.
4. For a unary operation, pop the stack to fetch sopd, perform an operation, and push the result back on the stack.
5. For a binary operation, pop the stack to fetch sopd 2, pop the stack again to fetch sopd 1, perform an operation, and push the result back on the stack.
6. A software-based VM has a translator and an interpreter.
7. Stack machine code is often used as an immediate language.
8. In a postfix notation, no parentheses are required.
9. The VSM has three layers: host machine hardware, OS, and interpreter.

PROBLEMS

1. Find the postfix notations for the following infix notations:

   a. X = A + B * C;
   b. X = (A + B) * C;
   c. X = (A - B) / C + 2;
   d. X = (A - 4095) / C - 2;
   e. X = A - 4095 / (C - 2);

2. Find the postfix notations for the following infix notations:

a. X4 = (X0 - X1) / (X2 - X3);
b. X4 = X0 - X1 / X2 - X3;
c. X4 = X0 - (X1 / X2 - X3);
d. X4 = X0 * (X1 / (X2 - X3));
e. X4 = X0 * X1 / (X2 - X3);

3. Find the postfix notations for the infix IF statement as shown below:

   a. IF i .le. 100, THEN sum = sum + i; ENDIF;
   b. IF (n .mod. i) .eq. 0,
      THEN i = i + 1; ELSE i = i - 1; ENDIF;

4. State the software parsing algorithm based on operator precedence.
5. A uniform syntax notation is used to define a constant in SSM315, as shown below.

   [<dup>] <data type> [L <length>] '<value>'

   a. Use the o (octal) data type and define a block of 16 16-bit integers (each one is 177770 in octal).
   b. Use the x (hexadecimal) data type and define a block of 16 16-bit integers (each one is FFF8 in hex).
   c. Does either notation prepare the same block in memory?

6. What is the difference between the instructions, STW (store word) and pop?
7. Write down the register transfer language for the s-ops as shown below.
   a. BAL (Branch and link). b. BF (Branch on false).
   c. BT (Branch on true). d. B (Branch)

8. If the mnemonic Ret is used as a return instruction, what should be its opcode, BAL, BF, BT, or B?
9. The stack machine instruction is short. If we place the address constants in the lower 4 KB and the operand size is 32 bits, what is the total addressing space using byte address?
10. Complete the translation of all the hex code in the test.exe file shown in Table 10.8. Explain why the code is relocatable after being loaded in memory.
11. If we allow the composite segment to be written in, the code is not reentrant, e.g., test.exe. However, let us change the design to have two segment base registers and obey the rule as follows. One segment base is used to store static code, instruction and data, which is read-only. The other base is used to store dynamic code, data and stack, that is writable but private. We say the code is reentrant if each user has his own private addressing space. Why?

12. Translate the following statements into s-ops.

    a. sum = sum + i;

b. i = j / (k - 1) + 2;

c. IF (i .LE. 100),
   THEN GOTO L0001; ENDIF;
   {Hint: After optimization, only five s-ops are required for this statement provided that you use bt (branch on true) after the test.}

13. In practice, the target machine in a JAVA engine or Pascal engine is stack based. Why?
14. Form a project team of up to three students and write an interpreter for SSM315 in C, assembly, or any other language. Select your own leader who calls design meetings and coordinates the effort. You can write machine spec of whatever kind and design a creative test program. The project report should include design specifications and list the source as an appendix.

References

1. Amdahl, G., Adventures in the mainframe trade, *IEEE Design Test Comput.*, 5, 1997.

2. Burroughs Corp., *B5000 System Reference Manual*, Blue Bell, PA.

3. Blauuw, G.A. and Brooks, F., *Computer Architecture: Concepts and Evolution,* Addison-Wesley, Reading, MA, 1997.

4. Burks, A. et al., Preliminary discussion of the logical design of an electronic computing instrument, Institute for Advanced Study, Princeton, NJ, 1946.

5. Control Data Corporation, *CDC6600/7600 System Reference Manual,* St. Paul, MN.

6. A VLIW architecture for a trace scheduling compiler, Proc. 2nd Int. Conf. Architectural Support for Programming Lang. and Operating Syst., IEEE CS Press, Los Alamitos, CA, 180, 1987.

7. Corbato, F.J. and Vyssotsky, V.A., Introduction and overview of the MULTICS system, 185, 1995.

8. Dacey, G.C. and Ross, I.M., The field effect transistor, *Bell Syst. Tech. J.*, 37, 1149, 1955.

9. Deitel, H.M. and Deitel, P.J., *Java: How to Program,* Prentice-Hall, Englewood Cliffs, NJ, 1997.

10. Dexter, A., *Microcomputer Bus Structures and Bus Interface Design,* Marcel Dekker, New York, 1986.

11. Dulong, C., The IA-64 architecture at work, *IEEE Comput.*, 24, 1988.

12. Fabricius, E.D., *Modern Digital Design and Switching Theory*, CRC Press, Boca Raton, FL, 1992.

13. Flynn, M.J., Some computer organizations and their effectiveness, *IEEE Trans. Comput.*, C-21, 948, 1972.

14. Flynn, M.J., *Computer Architecture: Pipelined and Parallel Processor Design,* Jones and Bartlett, Sudbury, MA, 1995.

15. Forrester, J.W., Digital information in three dimensions using magnetic cores, *J. Appl. Phys.*, 22, 44, 1951.

16. Goodman, J. and Miller, K., *A Programmer's View of Computer Architecture: MIPS,* W.B. Saunders, Philadelphia 1993.

17. Grehan, R., Embedded PCs: presents from above, *Comput. Design,* 1997, 73.

18. Hennessy, J.L., VLSI processor architecture, *IEEE Trans. Comput.,* C-33, 1221, 1984.

19. Henessy, J.L. and Patterson, D.A., *Computer Architecture: A Quantitative Approach,* 2nd ed., Morgan Kaufmann, 1996.

20. Hsu, J.Y., A versatile core memory with content-addressability, Ph.D. thesis, University of California, Berkeley, 1969.

21. Hsu, J.Y., A driver-receiver design for data communication, Proc. Int. Comput. Symposium, Taiwan, 821 1973.

22. Hsu, J.Y., On the design of an integrated banking network, Proc. Int. Comput. Symposium, vol. 1, 501, Taiwan, 1980.

23. Hsu, J.Y. et al., CPL—Chinese programming language, Proc. Int. Comput. Symposium, vol. 2, 1208, Taiwan, 1980.

24. Hsu, J.Y., *Computer Networks: Architecture, Protocols, and Software,* Artech House, Norwood, MA, 1996.

25. Hwang, K., *Advanced Computer Architecture: Parallelism, Scalability, Programmability,* McGraw-Hill, New York, 1993.

26. IBM Corporation, *System/370 Principles of Operation,* Armonk, NY.

27. IBM Corporation, *Personal Computer Hardware Technical Reference Manual,* Armonk, NY.

28. Intel Corporation, *Pentium Processor – Architecture and Programming Manual,* No. 241428, vols. 1, 2, and 3, Santa Clara, CA.

29. Intel Corporation, *iAPX 8088/8086 User's Manual,* Santa Clara, CA.

30. Intel Corporation, *IA-64 Application Developer's Architectural Guide,* No. 245188-001, available from http://developer.intel.com/design/IA64, Santa Clara, CA.

31. Irvine, K.R., *Assembly Language for Intel-Based Computers,* 3rd ed., Prentice-Hall, Englewood Cliffs, NJ, 1999.

32. Jacob, B. and Mudge, T., Virtual memory issues of implementation, *IEEE Comput.,* 33,1998.

33. Sun Microsystems, Inc., *The Java Virtual Machine Specification*, 1995.

34. Kilburn, T. et al., One-level storage system, *IRE Trans. Electron. Comput.,* 223, 1962.

35. Koerner, M. et al., *PowerPC: An Inside View,* Prentice-Hall, Englewood Cliffs, NJ, 1996.

36. Kozyrakis, C.E. and Patterson, D.A., A new direction for computer architecture research, *IEEE Comput.,* 24, 1998.

37. Leiberman, H., The debugging scandal and what to do about it, *CACM,* 40(4), 27, 1997.

38. Lim, R., personal communications, Ames Research Center, Moffet Field, CA, 1973.

39. Liptay, J.S., Structural aspects of the System/360 Model 85, part II: the cache, *IBM Syst. J.,* 7(1) 15.

40. Motorola, Inc., *M68000 Microprocessor Reference Manual,* Phoenix.

41. Motorola, Inc., *M6850 Asynchronous Communications Interface Adaptor (ACIA)*, Phoenix.

42. Mallach, E. and Sondak, N., *Advances in Microprogramming,* Artech House, Norwood, MA, 1983.

43. Mano, M., *Computer System Architecture,* 3rd ed., Prentice-Hall, Englewood Cliffs, NJ, 1993.

44. Moto-Oka, T., Ed., *Fifth Generation Computer Systems,* Springer-Verlag, Amsterdam, 1988.

45. National Semiconductor Corp., *8250 Universal Asynchronous Receiver/Transmitter (UART) Specifications*, Sunnyvale, CA.

46. Needham, J., *Science and Civilization in China, Vol. III,* Cambridge University Press, New York, 1959.

47. Patterson, D.A., Reduced instruction set computers, *Comm. ACM,* 28, 348, 1985.

48. Patterson, D.A. and Hennessy, J.L., *Computer Organization and Design: The Hardware/Software Interface,* 2nd ed., Morgan Kaufmann, 1998.

49. Digital Equipment Corp., *PDP-10 Architecture Reference Manual,* Bedford, MA.

50. Peleg, A. et al., Intel MMX for multimedia PCs, *CACM,* 40(1), 25, 1997.

51. Pressman, M.R., *Assembly Language Programming for VAX-11,* Mayfield Publishing, Palo Alto, CA, 1985.

52. Ryan, B., RISC drives Power PC, *Byte,* 79, 1993.

53. Schaller, R.R., Moore's law: past, present, and future, *IEEE Spectrum,* 53,1997.

54. Schlett, M., Trends in embedded-microprocessor design, *Computer,* 44, 1998.

55. Schmid, H., *Decimal Computation,* John Wiley & Sons, New York, 1975.

56. Sebern, M.J., A minicomputer-compatible microcomputer system: the DEC SLI-11, *Proc. IEEE,* 881, 1976.

57. Segee, B. and Field, J., *Microprogramming and Computer Architecture,* John Wiley & Sons, New York, 1991.

58. Shanley, T., *Power PC System Architecture,* Addison-Wesley, Reading, MA, 1995.

59. Siewiorek, D.P. et al., *Computer Structures: Principles and Examples,* McGraw-Hill, New York, 1982.

60. Site, R.L., Ed., *Alpha Architecture Reference Manual*, Digital Press, Bedford, MA.

61. Solari, E. and Willse, G., *PCI: Hardware and Design*, 4th ed., Annabooks, San Diego, 1998.

62. Stallings, W., *Computer Organization and Architecture: Designing for Performance,* 4th ed., Prentice-Hall, Englewood Cliffs, NJ, 1996.

63. Stenstrom, P., A survey of cache coherence schemes for multiprocessors, *IEEE Comput.,* 23(6), 12, 1990.

64. Tanenbaum, A.S., *Structured Computer Organization,* 4th ed., Prentice-Hall, Englewood Cliffs, NJ, 1999.

65. Thornton, J.E., *Design of a Computer: The Control Data 6600,* Scott, Foresman and Co., Glenview, IL, 1970.

66. Varhol, P., Overhaul the Pentium, *Comput. Design,* 38, 1997.

67. Wallace, C.E., A suggestion for a fast multiplier, *IEEE Trans. Comput.,* 13, 14, 1964.

68. Wilkes, M.V., The best way to design an automatic calculating machine, Manchester University Computer Inaugural Conf., 16, 1951.

69. Wilkes, M.V., *Computer Perspectives,* Morgan Kaufmann, 1995.

70. World Wide Web, http://www.netlib.org/performance/html/dhrystone.data.co10.html

71. World Wide Web, http://www.scl.ameslab.gov/Projects/HINT/user.html

72. Zaks, R. *Programming the 6502,* Sybex Inc., 1980.

73. *Computer, 50 Years of Computing Issue,* IEEE Press, Los Alamitos, CA, 1996.

74. Kernighan, B.W. and D.M. Ritchie, *The C Programming Language,* 2nd ed., Prentice-Hall, Englewood Cliffs, NJ, 1988.

75. Struble, G., *Assembler Language Programming: The IBM System/370 Family,* 3rd ed., Addison-Wesley, Reading, MA, 1984.

76. Case, R.P. and Padegs, A., Architecture of the IBM system/370, CACM, 21(1), 1978.

77. Holt, C.A., *Microcomputer Systems: Hardware, Assembly Language, and Pascal,* Macmillan, New York, 1986.

78. Hsu, J.Y., The reduced software computer, Proc. 5th World Conf. Integrated Design and Process Technology, Texas, 2000.

79. Microsoft, *DOS Technical Reference Manual,* Redmond, WA, 1986.

80. DATA COMMUNICATION NETWORKS: Services and Facilities, Interfaces, Fascicle VIII.2, ITU-T.

81. Microsoft, *Macro Assembler Programmer's Guide,* Redmond, WA, 1992.

Index

D

E

F

G

H

I

J

K

L

M

N

O

P

R

S

T

U

V

W

Z